D1206682

Graduate Texts in Mathematics 154

Graduate Texts in Mathematics

1 TAKEUTI/ZARING. Introduction to Axiomatic Set Theory. 2nd ed.
2 OXTOBY. Measure and Category. 2nd ed.
3 SCHAEFFER. Topological Vector Spaces.
4 HILTON/STAMMBACH. A Course in Homological Algebra.
5 MAC LANE. Categories for the Working Mathematician.
6 HUGHES/PIPER. Projective Planes.
7 SERRE. A Course in Arithmetic.
8 TAKEUTI/ZARING. Axiomatic Set Theory.
9 HUMPHREYS. Introduction to Lie Algebras and Representation Theory.
10 COHEN. A Course in Simple Homotopy Theory.
11 CONWAY. Functions of One Complex Variable. 2nd ed.
12 BEALS. Advanced Mathematical Analysis.
13 ANDERSON/FULLER. Rings and Categories of Modules. 2nd ed.
14 GOLUBITSKY/GUILEMIN. Stable Mappings and Their Singularities.
15 BERBERIAN. Lectures in Functional Analysis and Operator Theory.
16 WINTER. The Structure of Fields.
17 ROSENBLATT. Random Processes. 2nd ed.
18 HALMOS. Measure Theory.
19 HALMOS. A Hilbert Space Problem Book. 2nd ed.
20 HUSEMOLLER. Fibre Bundles. 3rd ed.
21 HUMPHREYS. Linear Algebraic Groups.
22 BARNES/MACK. An Algebraic Introduction to Mathematical Logic.
23 GREUB. Linear Algebra. 4th ed.
24 HOLMES. Geometric Functional Analysis and Its Applications.
25 HEWITT/STROMBERG. Real and Abstract Analysis.
26 MANES. Algebraic Theories.
27 KELLEY. General Topology.
28 ZARISKI/SAMUEL. Commutative Algebra. Vol.I.
29 ZARISKI/SAMUEL. Commutative Algebra. Vol.II.
30 JACOBSON. Lectures in Abstract Algebra I. Basic Concepts.
31 JACOBSON. Lectures in Abstract Algebra II. Linear Algebra.
32 JACOBSON. Lectures in Abstract Algebra III. Theory of Fields and Galois Theory.
33 HIRSCH. Differential Topology.
34 SPITZER. Principles of Random Walk. 2nd ed.
35 WERMER. Banach Algebras and Several Complex Variables. 2nd ed.
36 KELLEY/NAMIOKA et al. Linear Topological Spaces.
37 MONK. Mathematical Logic.
38 GRAUERT/FRITZSCHE. Several Complex Variables.
39 ARVESON. An Invitation to C^*-Algebras.
40 KEMENY/SNELL/KNAPP. Denumerable Markov Chains. 2nd ed.
41 APOSTOL. Modular Functions and Dirichlet Series in Number Theory. 2nd ed.
42 SERRE. Linear Representations of Finite Groups.
43 GILLMAN/JERISON. Rings of Continuous Functions.
44 KENDIG. Elementary Algebraic Geometry.
45 LOÈVE. Probability Theory I. 4th ed.
46 LOÈVE. Probability Theory II. 4th ed.
47 MOISE. Geometric Topology in Dimensions 2 and 3.
48 SACHS/WU. General Relativity for Mathematicians.
49 GRUENBERG/WEIR. Linear Geometry. 2nd ed.
50 EDWARDS. Fermat's Last Theorem.
51 KLINGENBERG. A Course in Differential Geometry.
52 HARTSHORNE. Algebraic Geometry.
53 MANIN. A Course in Mathematical Logic.
54 GRAVER/WATKINS. Combinatorics with Emphasis on the Theory of Graphs.
55 BROWN/PEARCY. Introduction to Operator Theory I: Elements of Functional Analysis.
56 MASSEY. Algebraic Topology: An Introduction.
57 CROWELL/FOX. Introduction to Knot Theory.
58 KOBLITZ. p-adic Numbers, p-adic Analysis, and Zeta-Functions. 2nd ed.
59 LANG. Cyclotomic Fields.
60 ARNOLD. Mathematical Methods in Classical Mechanics. 2nd ed.
61 WHITEHEAD. Elements of Homotopy Theory.
62 KARGAPOLOV/MERLZJAKOV. Fundamentals of the Theory of Groups.
63 BOLLOBAS. Graph Theory.
64 EDWARDS. Fourier Series. Vol. I. 2nd ed.

continued after index

Arlen Brown Carl Pearcy

An Introduction to Analysis

Springer-Verlag
New York Berlin Heidelberg London Paris
Tokyo Hong Kong Barcelona Budapest

Arlen Brown
460 Kenwood Place
Bloomington, IN 47401
USA

Carl Pearcy
Department of Mathematics
Texas A&M University
College Station, TX 77843-3368
USA

With 7 Illustrations

Mathematics Subject Classifications (1991): 46-01, 11Axx

Library of Congress Cataloging-in-Publication Data
Brown, Arlen, 1926–
An introduction to analysis / Arlen Brown, Carl Pearcy.
p. cm. – (Graduate texts in mathematics ; 154)
Includes bibliographical references and index.
ISBN 0-387-94369-2
1. Mathematical analysis. I. Pearcy, Carl M., 1935–
II. Title. III. Series.
QA300.B73 1994
515 – dc20 94-22509

Printed on acid-free paper.

Production managed by Bill Imbornoni; manufacturing supervised by Genieve Shaw.
Photocomposed pages prepared from the authors' TeX file.
Printed and bound by R.R. Donnelley & Sons, Harrisonburg, VA.
Printed in the United States of America.

9 8 7 6 5 4 3 2 1

ISBN 0-387-94369-2 Springer-Verlag New York Berlin Heidelberg
ISBN 3-540-94369-2 Springer-Verlag Berlin Heidelberg New York

Preface

As its title indicates, this book is intended to serve as a textbook for an introductory course in mathematical analysis. In preliminary form the book has been used in this way at the University of Michigan, Indiana University, and Texas A&M University, and has proved serviceable. In addition to its primary purpose as a textbook for a formal course, however, it is the authors' hope that this book will also prove of value to readers interested in studying mathematical analysis on their own. Indeed, we believe the wealth and variety of examples and exercises will be especially conducive to this end.

A word on prerequisites. With what mathematical background might a prospective reader hope to profit from the study of this book? Our conscious intent in writing it was to address the needs of a beginning graduate student in mathematics, or, to put matters slightly differently, a student who has completed an undergraduate program with a mathematics major. On the other hand, the book is very largely self-contained and should therefore be accessible to a lower classman whose interest in mathematical analysis has already been awakened.

The contents of the book may be briefly summarized. Chapters 1 through 3 constitute an overview of the preliminary material on which the rest of the book is built, viz., set theory, the number systems, and linear algebra. In no case do we imagine that this brief summary of material can serve as the reader's initial encounter with these ideas. Rather we have gathered together here the basic terminology and facts to be employed in all that follows. In particular, in Chapters 2 and 3 we introduce only material that is assumed to be already familiar to the reader, though perhaps in different form, and these two chapters may in most cases be treated quite lightly. Chapter 1, on the other hand, dealing with the rudiments of set theory, acquaints the reader with inductive proofs based on the *maximum principle* in its various forms, and is deserving of more careful attention.

In Chapters 4 and 5 we present the essentials from the theory of transfinite numbers. This treatment, while concise, presents all of the ideas and results that will actually be employed in the sequel, and is, in any case, fuller than is to be found in most other texts. In this connection we note that the various number systems, formally introduced in Chapter 2, actually make a few brief cameo appearances in Chapter 1 as well. This minor logical embarrassment could easily be averted, of course, but only at the cost of unwelcome circumlocutions.

Chapters 6 through 8 constitute the heart of the book. In them we explore in thoroughgoing fashion the structure of various metric spaces and the mappings defined on or taking values in such spaces. The topics and facts adduced are largely standard, though our choice of examples, problems, and manner of presentation may make some modest claim to freshness if not to novelty, but many of these lines of inquiry are pursued in greater detail than will be found in most other recent texts.

The final chapter (Chapter 9) consists of a treatment of general topology. In this chapter we equip the reader with the full panoply of topological equipment needed for the transition from the world of classical analysis, set in metric spaces, to "modern" or "abstract" analysis, the realm of maximal ideal spaces, kernel-hull topologies, etc.

In formulating the sets of problems that follow each chapter we have followed current practice. Each problem, or part of a problem, is, in effect, a theorem to be proved, and it is our intention that the solutions should be written out with that in mind. Thus a problem posed as a simple yes–or–no question has for its proper solution not a simple yes–or–no answer, but rather an argument showing which is, in fact, correct. Similarly, a problem posed as a statement of fact is really a disguised invitation to the reader to establish the validity of that fact. No conscious attempt was made to grade the problems according to difficulty, but they are arranged in loosely chronological order, so that the first problems in each chapter relate to the earlier parts of that chapter and subsequent problems to later parts. Thus the earlier problems in any one chapter do turn out, in general, to be somewhat easier than the later ones. (The problem sets are an integral part of the text; an independent reader is advised to begin to look into the problem set at the end of a chapter as soon as he begins the perusal of the chapter itself, just as he would do if assigned homework problems in a formal classroom setting.)

Finally, the authors take this opportunity to express their appreciation to the Mathematics Department of Texas A&M University for its support during the preparation of the manuscript. In particular, the existence of the associated TEX file is due almost entirely to the efforts of Professor N. W. Naugle, a leading expert in this area, and Ms. Jan Want, who cheerfully and conscientiously produced the entire file.

Arlen Brown
Carl Pearcy
June 1994

Contents

Preface	v
1 The rudiments of set theory	3
2 Number systems	25
3 Linear analysis	46
4 Cardinal numbers	65
5 Ordinal numbers	80
6 Metric spaces	96
7 Continuity and limits	135
8 Completeness and compactness	174
9 General topology	224
Bibliography	277
Index	279

An Introduction to Analysis

1 The rudiments of set theory

Sets and relations

We assume the reader to be familiar with the basic concepts of *set* and *element* (or *member* or *point*) of a set, as well as with the idea of a *subset* of a set, and the notions of *union* and *intersection* of a collection of sets. We write $x \in A$ to mean that x is an element of a set A, $x \notin A$ to mean that x is not an element of A, and $B \subset A$ (or $A \supset B$) to mean that B is a subset of A. We also use the standard notation $\cup$ and $\cap$ for unions and intersections, respectively.

If $p(\ )$ is some *predicate* that is either true or false for every element of some set X, then the notation $\{x \in X : p(x)\}$ will be used to denote the subset of X consisting of all those elements of X for which $p(x)$ is true. If A and B are sets, we write $A \backslash B$ for the *difference*

$$A \backslash B = \{x \in A : x \notin B\},$$

$A \nabla B$ for the *symmetric difference* $A \nabla B = (A \backslash B) \cup (B \backslash A)$, and $A \times B$ for the (*Cartesian*) *product* consisting of the set of all ordered pairs (a, b) where $a \in A$, $b \in B$. It will also be convenient to reserve certain symbols throughout the book for certain sets. Thus the *empty set* will consistently be denoted by $\varnothing$, the *singleton* on an element x, i.e., the set whose sole element is x, by $\{x\}$, the *doubleton* having x and y as its only elements by $\{x, y\}$, etc. The set of all positive integers will be denoted by $\mathbb{N}$, the set of all nonnegative integers by $\mathbb{N}_0$, and the set of all integers by $\mathbb{Z}$. Similarly, we consistently use the symbols $\mathbb{Q}$, $\mathbb{R}$, and $\mathbb{C}$ to denote the systems of *rational*, *real*, and *complex numbers*, respectively. (Some explanatory remarks concerning these basic *number systems* will be given in the next chapter.)

Suppose now that X and Y are sets. By a *relation* between X and Y is meant any rule R with the property that if x is an arbitrary element of X and y an arbitrary element of Y, then it is possible to say with certainty that x and y either *satisfy* the rule R (so that x and y are *related by* R), or that they do not. (A relation between a set X and itself is customarily called a relation *on* X.) We write $x\,R\,y$ to indicate that x and y are related by R.

Example A. Let X be a set and let 2^X denote the set consisting of all the subsets of X. (The set 2^X will be called the *power class* on X; this notation will be justified in due course.) Then $\in$ (is an element of) is a relation between X and 2^X—a relation of fundamental importance in all set-theoretic considerations.

Example B. The familiar relation $<$ (less than) is a relation on the system $\mathbb{Z}$ of all integers, as well as on the system $\mathbb{N}$ of all positive integers.

Let X and Y be sets and let R be a relation between X and Y. The set $G_R = \{(x,y) \in X \times Y : x\,R\,y\}$ is called the *graph* of the relation R. Thus for any relation R between X and Y, x and y are related by R if and only if $(x,y) \in G_R$. It is apparent that each relation R between X and Y determines uniquely the subset G_R of $X \times Y$, and that, conversely, each subset G of $X \times Y$ determines uniquely a relation $R = R_G$ between X and Y having G for its graph (simply declare $x\,R\,y$ to be true when and only when $(x,y) \in G$). Thus there is no logical necessity for distinguishing between a *relation* R between X and Y and the *graph* of R (a subset of $X \times Y$). Nevertheless, it is frequently psychologically desirable to maintain this distinction. (For example, it seems psychologically, if not logically, supportable to distinguish between the relation $<$ on the set of positive integers and the subset $G_< = \{(m,n) : m < n\}$ of $\mathbb{N} \times \mathbb{N}$ that is its graph.)

Example C. On any set X *equality* $(=)$ is a relation having for its graph the *diagonal* $\Delta = \{(x,x) : x \in X\}$.

Notation and terminology. Let X and Y be sets, let R be a relation between X and Y, and let A be a subset of X. Then the subset of Y consisting of all those elements y for which there exists some element x of A such that $x\,R\,y$ is called the *image* of A *under* R (or *with respect to* R) and is denoted by $R(A)$. Dually, if B is a subset of Y, then the subset of X consisting of all those elements x such that xRy for some y in B is called the *inverse image* of B *under* R (or *with respect to* R) and is denoted by

$R^{-1}(B)$.

Definition. A relation ϕ between a set X and a set Y is called a *mapping* of X into Y (or a *function* from X to Y) if for each element x of X there exists *precisely one* element y of Y such that $x\,\phi\,y$. Thus ϕ is a mapping if for each element x of X the product $\{x\} \times Y$ *meets* (that is, intersects) the graph G_ϕ in a singleton, or, equivalently, if for each element x of X the image $\phi(\{x\})$ of the singleton $\{x\}$ under ϕ is a singleton in Y.

Notation and terminology. Other terms that are sometimes used as synonyms for *mapping* or *function* are *map* and *transformation*. We shall usually write $\phi : X \to Y$ to indicate that ϕ is a mapping of X into Y. If X and Y are sets and if $\phi : X \to Y$, it is customary to write $y = \phi(x)$ instead of $x\,\phi\,y$ (so that $x\phi(\phi(x))$ for every element x of X). If $y = \phi(x)$ we say that ϕ *maps* x to y, or that y is the *value* of the function ϕ at x, or again that y is the *image* of x under ϕ.

While a mapping is really a special kind of relation, it is clear from the foregoing discussion of the terminology habitually employed in connection with mappings that they are not ordinarily thought of in that light. In particular, if $\phi : X \to Y$ and if $y = \phi(x)$ for some element x of X, one does not say that "x and y are related by ϕ."

Definition. Let ϕ be a mapping of a set X into a set Y. Then X is the *domain* (*of definition*) of ϕ, and Y is the *codomain* of ϕ, while the *range* of ϕ is the image $\phi(X)$ of X under ϕ. When the range of ϕ coincides with the entire codomain Y, the mapping ϕ is said to map X *onto* Y, or to be *onto*. If the mapping ϕ has the property that $\phi(x) = \phi(x')$ implies that $x = x'$ for all elements x and x' of the domain X, then ϕ is a *one-to-one* mapping. A one-to-one mapping of a set X onto a set Y is frequently called a *one-to-one correspondence* between X and Y. An element x of the domain X of ϕ is a *fixed point* of ϕ if $\phi(x) = x$ (such a point must also be an element of Y of course).

Example D. For each element x of an arbitrary set X the product $\{x\} \times X$ meets the diagonal Δ in the singleton $\{(x, x)\}$. Thus the relation of equality is a one-to-one mapping of X onto itself. When the relation of equality on a set X is regarded as a mapping, it is reierred to as the *identity mapping* on X and will ordinarily be denoted by ι or, when necessary, by ι_X.

Example E. Let X be a set and let A be a subset of X. The mapping of A into X that leaves each point of A fixed, that is, that has each point of A as a fixed point, is the *inclusion mapping* of A into X. In this case it is best to maintain the distinction between the inclusion mapping of A into

X and the identity mapping on A, but in the future we will not always be entirely scrupulous about distinguishing between two mappings that differ only in that one has a larger codomain than the other.

Notation. We shall sometimes write $x \to \phi(x)$ to indicate the action of a mapping ϕ. It is also sometimes highly convenient, if a trifle illogical, to use the compound symbol $\phi(x)$ to denote the mapping ϕ itself (rather than an element of the codomain of ϕ). It should be noted that these notational conventions are principally of use when the domain and codomain of the mapping are agreed upon in advance.

Example F. If X is a set, then a mapping $\Box$ of $X \times X$ into X is sometimes called a *binary operation* on X. If this point of view is adopted, and if x and y are elements of X, then one thinks of $\Box\ (x, y)$ as the result of combining x and y according to the rule $\Box$, and one writes

$$\Box(x, y) = x \Box y.$$

If a binary operation $\Box$ on X has the property that

$$x \Box y = y \Box x$$

for all x and y in X, then $\Box$ is said to be *commutative*. If $\Box$ has the property that

$$x \Box (y \Box z) = (x \Box y) \Box z$$

for all x, y and z in X, then $\Box$ is said to be *associative*. (When $\Box$ is associative, we may and shall write simply $x \Box y \Box z$ for this threefold composition.) If $\Box$ is a binary operation on X and if e is an element of X such that

$$x \Box e = e \Box x = x$$

for every element x of X, then e is said to be a *neutral element* with respect to $\Box$. It is clear that if e and e' are both neutral elements with respect to the same binary operation on X, then $e = e'$, so that the neutral element with respect to a binary operation is always unique if it exists.

It should be noted that the symbol $\Box$, here used to stand for an arbitrary abstract binary operation, was deliberately chosen for its artificiality, and will not, in fact, be used to designate any binary operation in the sequel. In actual practice it is usual to think of an *associative* binary operation on a set X as a kind of *multiplication* on X, and when this is done the *product* of two elements x and y of X is regularly denoted by xy, and the neutral element with respect to this multiplication (if there is one) by 1. Likewise, an associative *and commutative* binary operation on a set X is frequently thought of as an *addition* on X, and when this is done, the *sum* of two elements x and y of X is denoted by $x+y$, and the additive neutral element (if there is one) by 0.

Another type of relation that is of considerable importance in mathematics is the *equivalence relation.*

Definition. Let R be a relation on a set X. Then R is said to be *reflexive* if $x\,R\,x$ for every element x of X. Likewise, R is *symmetric* if for all x and y in X it is the case that $y\,R\,x$ whenever $x\,R\,y$, and R is *transitive* if $x\,R\,y$ and $y\,R\,z$ imply $x\,R\,z$ for all x, y, z in X. A relation $\sim$ on a set X that is reflexive, symmetric and transitive is an *equivalence relation* on X.

Example G. The *identity relation* (equality) of Examples C and D is an equivalence relation on an arbitrary set X. Likewise, the entire product $X \times X$ is the graph of an equivalence relation $\sim$ on X. (In this relation it is the case that $x \sim y$ for all elements x and y of X.) Every equivalence relation on X lies between these two extremes in the sense that its graph contains the diagonal Δ and is contained in the entire product $X \times X$. (In this connection see Problem L.)

Definition. If $\sim$ is an equivalence relation on a set X, then for each element x of X the *equivalence class* of x (with respect to $\sim$) is the set $[x] = \{y \in X : y \sim x\}$.

Example H. Let ϕ be a mapping of a set X into a set Y, and define $x \sim x'$ to mean that $\phi(x) = \phi(x')$. Then $\sim$ is an equivalence relation on X. The equivalence classes of the various elements of X with respect to this equivalence relation are called the *level sets* of ϕ.

Definition. A collection $\mathcal{C}$ of sets *covers* a set X, or is a *covering* of X, if $\bigcup \mathcal{C} \supset X$, i.e., if every element of X belongs to some set belonging to $\mathcal{C}$. Likewise, $\mathcal{C}$ is said to be *disjoint* if the sets belonging to $\mathcal{C}$ are *pairwise disjoint*, i.e., if $E \cap F = \varnothing$ for any two distinct sets E and F belonging to $\mathcal{C}$. A collection $\mathcal{P}$ of nonempty subsets of X is a *partition* of X if it is a disjoint covering of X.

It is an immediate consequence of these definitions that if $\sim$ is an equivalence relation on a set X, and if for each x in X we write $[x]$ for the equivalence class of x, then $\mathcal{P}_\sim = \{[x] : x \in X\}$ is a partition of X. (This collection $\mathcal{P}_\sim$ of equivalence classes *modulo* the equivalence relation $\sim$ is called the *quotient space* of X modulo $\sim$, and is sometimes denoted by $X/\sim$.) Conversely, if $\mathcal{P}$ is an arbitrary partition of X, then the relation $\sim_\mathcal{P}$ defined by setting $x \sim_\mathcal{P} y$ when and only when x and y belong to the same set in $\mathcal{P}$ is an equivalence relation on X called the equivalence relation *determined* by the partition $\mathcal{P}$. Clearly the quotient space of X modulo the equivalence relation determined by $\mathcal{P}$ coincides with $\mathcal{P}$ itself, just as the equivalence relation determined by the quotient space modulo some given equivalence relation $\sim$ on X coincides with $\sim$. Thus the collection of all equivalence relations on an arbitrary set X is in this way in one-to-one correspondence with the collection of all partitions of X.

Example I. If $\sim$ is an equivalence relation on a set X and if $X/\sim$ is the corresponding quotient space, then the rule π that assigns to each element x of X its equivalence class $[x]$ modulo $\sim$ is a mapping of X onto $X/\sim$. This mapping π is called the *natural projection* of X onto $X/\sim$. (The level sets of π are the equivalence classes modulo $\sim$.)

Example J. Let $\mathcal{C}$ be a collection of nonempty sets. If C and D are sets belonging to $\mathcal{C}$, we say that C and D are *chained* in $\mathcal{C}$ if there exists a positive integer p and sets $C_0, \ldots, C_p$ in $\mathcal{C}$ such that $C_0 = C$, $C_p = D$, and $C_{i-1} \cap C_i \neq \varnothing, i = 1, \ldots, p$. (The collection $\mathcal{C}$ itself is said to be *chained* if every pair of elements of $\mathcal{C}$ is chained in $\mathcal{C}$.) If for each pair C, D of elements of $\mathcal{C}$ we write $C \sim D$ to indicate that C and D are chained in $\mathcal{C}$, then $\sim$ is an equivalence relation on $\mathcal{C}$. Moreover, if $[C]$ and $[D]$ are any two distinct equivalence classes in $\mathcal{C}$ with respect to this equivalence relation, and if $V = \bigcup[C]$ and $W = \bigcup[D]$, then V and W are disjoint subsets of the union $U = \bigcup\mathcal{C}$. Indeed, if $a \in V \cap W$, then there exists a set C_0 in $[C]$ and a set D_0 in $[D]$ such that $a \in C_0 \cap D_0$, whence it follows that $C \sim D$, and hence that $[C] = [D]$, contrary to hypothesis. Thus the collection $\{\bigcup[C] : C \in \mathcal{C}\}$ of all unions of equivalence classes with respect to the equivalence relation $\sim$ is a partition of U. This partition will be called the partition of U *corresponding to* the relation $\sim$ of being chained in the covering $\mathcal{C}$.

Indexed families

There are many situations in mathematics in which it is convenient to think of a function ϕ not as a mapping carrying one set X into another set Y, but rather as a scheme for labeling certain elements of Y by means of the elements of X. In such cases it is customary to vary both the terminology and the notation, calling the set X an *index set*, the function ϕ itself an *indexing*, and writing $\{y_x\}_{x \in X}$ in place of ϕ, where, of course, $y_x = \phi(x)$ for each x in X. In this book we shall use the terminology and notation of indexed families without further explanation or apology whenever it is convenient to do so.

The most important instance of this usage is provided by the familiar infinite sequence. Recall that an *infinite sequence* (or, more simply, *sequence*, as we shall usually prefer to say) of elements of a set Y is just a mapping of the set $\mathbb{N}$ (or perhaps $\mathbb{N}_0$) into Y, but that a sequence is ordinarily denoted by $\{y_n\}_{n\in\mathbb{N}}$, or, more usually still, by $\{y_n\}_{n=1}^{\infty}$.

Example K. For an absolutely arbitrary set X the identity mapping on X may be viewed as an indexing of X, the *self-indexing* of X. This observation is trivial but important. It shows that we may, at any time, assume a given set of objects to be (the range of) an indexed family without any loss of generality. In the sequel we shall use this fact whenever it is convenient to do so.

Suppose given a family of sets $\{X_\gamma\}_{\gamma\in\Gamma}$ indexed by a set Γ. Then the union $\bigcup_{\gamma\in\Gamma} X_\gamma$ (or $\bigcup_\gamma X_\gamma$) and (if $\Gamma \neq \varnothing$) the intersection $\bigcap_{\gamma\in\Gamma} X_\gamma$ (or $\bigcap_\gamma X_\gamma$) are defined in the usual way to consist of the union and intersection, respectively, of the collection of all sets X_γ appearing in the indexed family. (The union of the empty collection of sets is easily seen to be empty; the intersection of the empty collection of sets is undefined.) Somewhat less familiar perhaps is the *indexed product*.

Definition. Let $\{X_\gamma\}_{\gamma\in\Gamma}$ be an indexed family of sets. Then the *Cartesian product*

$$\prod_{\gamma\in\Gamma} X_\gamma$$

is the set consisting of all those indexed families $\{x_\gamma\}_{\gamma\in\Gamma}$ with the property that $x_\gamma \in X_\gamma$ for every index γ in Γ. (If Γ happens to be the set $\{1, 2, \ldots, n\}$, then $\prod_{\gamma\in\Gamma} X_\gamma$ coincides with the set $X_1 \times X_2 \times \ldots \times X_n$ of all n-tuples $(x_1, \ldots, x_n)$ where $x_i \in X_i, i = 1, \ldots, n$. Technically speaking, the indexed product $X_1 \times X_2$ of two sets defined in this manner does not coincide with the Cartesian product defined earlier, but it is obvious how these two concepts of product are to be identified, and we shall ignore this minor distinction.)

Example L. Let Γ be an index set, and suppose $X_\gamma = X$ for all indices γ in Γ, where X is some fixed set. Then the product $\prod_{\gamma\in\Gamma} X_\gamma$ simply coincides with the collection of all mappings of the set Γ into X. In the sequel it will sometimes be convenient to denote this set by X^Γ.

Example M. Let $\{X_\gamma\}_{\gamma\in\Gamma}$ be an indexed family of sets and let γ' be any one fixed index. The mapping that assigns to each element $\{x_\gamma\}$ of $\Pi = \prod_{\gamma\in\Gamma} X_\gamma$ its *term* or *coordinate* $x_{\gamma'}$ having index γ' is called the *projection* of Π onto $X_{\gamma'}$, and is denoted by $\pi_{\gamma'}$. The indexed family $\{\pi_\gamma\}_{\gamma\in\Gamma}$ of all such projections is *separating* on Π in the sense that if $\{x_\gamma\}$

and $\{y_\gamma\}$ denote two distinct elements of Π, then $\pi_{\gamma'}(\{x_\gamma\}) \neq \pi_{\gamma'}(\{y_\gamma\})$ for at least one index γ'. (If X and Y are sets, and if Y^X is defined as in the preceding example, then for each element x of X the projection π_x is simply the *evaluation* of the elements of Y^X at the point x.)

In connection with indexed unions, intersections and products, we shall assume the reader to be familiar with the standard associative, commutative, and distributive laws (see Problems D and G). There is, however, one particular observation in this context that must not be omitted, since it is an axiom, and, in fact, an axiom of great importance.

Axiom of Products. *If $\{X_\gamma\}_{\gamma\in\Gamma}$ is an indexed family of nonempty sets, where $\Gamma \neq \varnothing$, then $\prod_{\gamma\in\Gamma} X_\gamma$ is also nonempty.*

In the sequel we shall employ this axiom without further explanation, even though, as we shall see, some of its consequences are remarkable. Note that the axiom of products says, in the language of algebra, that there are no divisors of zero in the formation of indexed products. It is also not hard to see that the axiom of products is equivalent to the following somewhat more familiar axiom (Prob. H).

Axiom of Choice. *Let $\mathcal{C}$ be a collection of nonempty sets, and let U denote the union of the collection $\mathcal{C}$. Then there exists a mapping z of $\mathcal{C}$ into U (called* a selection function *or* choice function *on $\mathcal{C}$) such that $z(E) \in E$ for each set E in $\mathcal{C}$.*

Example N. For any set X there exists a mapping s that assigns to each subset E of X such that $E \neq X$ an element $s(E)$ of $X\backslash E$. Indeed, if $\mathcal{C}$ denotes the collection of all nonempty subsets of X, and if z is a choice function on $\mathcal{C}$, then $s(E) = z(X\backslash E)$ is a mapping having the required property.

Ordered sets

A relation $\leq$ on a set X is a *partial ordering* of X if it is reflexive and transitive, and if, for all x and y in X,

$$x \leq y \quad \text{and} \quad y \leq x \quad \text{imply} \quad x = y.$$

We adopt the standard practice of writing $y \geq x$ to mean $x \leq y$, and we also write $x < y$ to mean $x \leq y$ and $x \neq y$, and $y > x$ to mean $x < y$. A set X equipped with a partial ordering is a *partially ordered set*. (Thus a partially ordered set is a pair $(X, \leq)$ where X is a set and $\leq$ is a partial ordering of X.) Observe that every subset of a given partially ordered set is also a partially ordered set with respect to (the restriction to that subset of) the same ordering. Whenever a subset of a partially ordered set is regarded as a partially ordered set in its own right, it is this restricted ordering that is understood unless some other ordering is expressly stipulated.

Example O. Let X be an arbitrary set, and let 2^X denote the power class on X. The set 2^X is partially ordered by the *inclusion ordering* $\subset$. Thus an arbitrary collection $\mathcal{C}$ of subsets of a set X is also a partially ordered set in the inclusion ordering, and it is this partial ordering that is intended whenever such a collection $\mathcal{C}$ is viewed as a partially ordered set unless some other ordering is expressly indicated.

The most fundamental concepts in the theory of partially ordered sets are those of upper and lower bounds. If A is a subset of a partially ordered set X, then an element x of X is an *upper bound* for A in X if $y \leq x$ for every element y of A. Dually, x is a *lower bound* for A in X if $x \leq y$ for every y in A. A subset A of a partially ordered set X may possess upper bounds in X or it may not. If it does, we say that A is *bounded above* in X; likewise, if A possesses a lower bound in X, we say that A is *bounded below* in X. Finally, if A is bounded both above and below in X, then we say that A is *bounded* in X.

Let X be a partially ordered set and let A be a subset of X. If A is bounded above in X and if some upper bound of A belongs to A, then this element (which is obviously unique) is the *greatest* or *maximum* element of A. There is also the different and weaker notion of a *maximal* element of A. An element x of A is *maximal* in A if there exists no element y of A such that $x < y$. (The difference between the maximum element of A and a maximal element of A is subtle but important. The maximum element of A is unique if it exists, and is an upper bound for A in X; maximal elements of A, on the other hand, are not necessarily upper bounds of A and may exist in abundance.) Dually, one defines the concepts of *least*, or *minimum* element and *minimal* elements of a subset A of X.

Another important concept arising from the relation between a subset A of a partially ordered set X and the set A_u of all upper bounds of A in X is that of *least* upper bound. If A_u is nonempty and possesses a least element z, then z is called the *least upper bound*, or *supremum*, of A in X (notation: $z = \sup A$). Dually, one defines the concept of *greatest lower bound*, or *infimum*, of A in X (notation: $\inf A$). For finite subsets $\{x_1, \ldots, x_n\}$ of X we shall also write $x_1 \vee \ldots \vee x_n$ for $\sup\{x_1, \ldots, x_n\}$ and $x_1 \wedge \ldots \wedge x_n$ for $\inf\{x_1, \ldots, x_n\}$.

Example P. Let X be a partially ordered set with partial ordering $\leq$ and let us write $x \leq^* y$ to mean $y \leq x$. Then it is easily verified that $\leq^*$ is also a partial ordering of X. This partial ordering is called the ordering of X *inverse* to the given ordering (cf. Problem B). Note that the notions of bounds and of suprema and infima are exactly reversed in passing from $\leq$ to $\leq^*$. Thus x_0 is an upper bound in X for a subset A of X with respect to $\leq$ when and only when x_0 is a lower bound in X for A with respect to $\leq^*$, etc. The partially ordered set $(X, \leq^*)$ will be denoted by X^*.

Example Q. Let X be a set and let $\mathcal{C}$ be an arbitrary collection of subsets of X. Then $\mathcal{C}$ is a partially ordered set in the inclusion ordering (Ex. O), and is also a partially ordered set in the *inverse inclusion ordering* $\leq^*$, in which we set $A \leq^* B$ when and only when $B \subset A$.

If Y is a partially ordered set and $\phi : X \to Y$ is a mapping of a set X into Y, then an element y of Y is an *upper bound* of ϕ if it is an upper bound for the range $\phi(X)$, and if such an upper bound exists, ϕ is said to be *bounded above*. Dually, y is a *lower bound* of ϕ if it is a lower bound for $\phi(X)$, and if such a lower bound exists, ϕ is said to be *bounded below*. If ϕ is bounded both above and below, it is said to be *bounded*. Similarly, we define the *supremum* of ϕ (notation: $\sup_{x \in X} \phi(x)$) to be the supremum of the range of ϕ if it exists, and the *infimum* of ϕ (notation: $\inf_{x \in X} \phi(x)$) to be the infimum of the range of ϕ if it exists.

A mapping $\phi : X \to Y$ of one partially ordered set X into another partially ordered set Y is *monotone increasing* if $x_1 \leq x_2$ in X implies $\phi(x_1) \leq \phi(x_2)$ in Y for all x_1 and x_2 in X. If ϕ is monotone increasing and also one-to-one, then ϕ is said to be *strictly increasing*. Dually, one defines *monotone decreasing* and *strictly decreasing* mappings between partially ordered sets. A one-to-one mapping ϕ of a partially ordered set X onto a partially ordered set Y is an *order isomorphism* if ϕ has the property that $x_1 \leq x_2$ in X when and only when $\phi(x_1) \leq \phi(x_2)$ in Y (equivalently, if both ϕ and ϕ^{-1} are monotone increasing; see Problem B). If there exists an order isomorphism of X onto Y, then X and Y are said to be *order isomorphic*.

If a partially ordered set X has the property that $x \vee y$ and $x \wedge y$ both exist for every pair x, y of elements of X, then X is said to be a *lattice*. Clearly (mathematical induction) every nonempty finite subset of a lattice has a supremum and an infimum. If X is a partially ordered set with the property that *every* subset of X has a supremum and an infimum, then X is a *complete lattice*. If X is a partially ordered set with the property that every *bounded nonempty* subset of X has a supremum and an infimum, then X is a *boundedly complete lattice*.

Example R. The power class 2^X on an arbitrary set X is a complete lattice in the inclusion ordering. (The supremum of a subcollection of 2^X is its union; the infimum of a nonempty subcollection is its intersection.)

If A_0 is a fixed subset of X, then the collection $\mathcal{C}$ of all those subsets of X that contain A_0 is also a complete lattice, as is the collection $\mathcal{C}'$ of all those subsets of X contained in A_0. A subset M of a lattice L with the property that $x \vee y$ and $x \wedge y$ belong to M whenever x and y do is called a *sublattice* of L; thus $\mathcal{C}$ and $\mathcal{C}'$ are sublattices of 2^X. An example of a sublattice of 2^X that is not a complete lattice (unless X itself is a finite set) is the collection $\mathcal{C}_f$ of all finite subsets of X. (The sublattice $\mathcal{C}_f$ is boundedly complete, however.)

Proposition 1.1. *If X is a complete lattice, then X is nonempty and possesses a greatest element* 1 *and a least element* 0. *Indeed, the supremum of the empty subset of X is* 0; *dually,* $\inf \varnothing = 1$.

PROOF. Since, by definition, the empty subset of X has both supremum and infimum, X itself cannot be empty. Since every element of X is both an upper and a lower bound for $\varnothing$, the result follows. □

Proposition 1.2 (Banach-Knaster-Tarski Lemma). *Let X be a complete lattice, and let ϕ be a monotone increasing mapping of X into itself. Then ϕ has a fixed point.*

PROOF. Set $A = \{x \in X : \phi(x) \leq x\}$, let $x_0 = \inf A$, and suppose $x \in A$. Then $x_0 \leq x$ and therefore $\phi(x_0) \leq \phi(x) \leq x$. Thus $\phi(x_0)$ is also a lower bound of A, and therefore $\phi(x_0) \leq x_0$, so that $x_0 \in A$. But it follows at once from the monotonicity of ϕ that $\phi(A) \subset A$, so that, in particular, $\phi(x_0) \in A$, and therefore $x_0 \leq \phi(x_0)$. Thus we see that $\phi(x_0) = x_0$. □

A weaker notion than that of a lattice, also of great importance, is that of a directed set. A nonempty partially ordered set Λ is said to be *directed upward* if every doubleton in Λ is bounded above in Λ; likewise, Λ is said to be *directed downward* if every doubleton in Λ is bounded below in Λ. If Λ is directed either upward or downward, then Λ is a *directed set*. (If nothing is said to the contrary, a directed set will be understood to be directed upward; a directed set Λ is directed upward [downward] if and only if Λ^* is directed downward [upward] (Ex. P).)

Example S. Every nonempty lattice is directed both upward and downward.

Definition. If Λ is a directed set and X is an arbitrary set, then an indexed family $\{x_\lambda\}_{\lambda \in \Lambda}$ of elements of X indexed by Λ is called a *net* in X *indexed* (or *directed*) by Λ.

Example T. The sets $\mathbb{N}$ and $\mathbb{N}_0$, consisting, respectively, of all positive integers and all nonnegative integers, are directed sets in their natural orderings. Thus any sequence $\{x_n\}_{n=1}^{\infty}$ or $\{x_n\}_{n=0}^{\infty}$ in a set X is a net in X. Indeed, these nets are the very prototypes out of which the general concept of a net arose. In the sequel we shall have occasion to deal with nets having a considerably more complex structure.

In many discussions dealing with sequences the following terminology is very useful.

Definition. Let $p(\ )$ be a predicate that pertains to the terms of an infinite sequence $\{x_n\}$. Then we shall say that $p(\ )$ is true of $\{x_n\}$ *eventually* if there exists an index n_0 such that $p(x_n)$ is true for all $n \geq n_0$, that is, if there exists an entire *tail* $T_{n_0} = \{x_n : n \geq n_0\}$ of $\{x_n\}$ for every term of which $p(x_n)$ is true. Likewise, we shall say that $p(\ )$ is true of $\{x_n\}$ *infinitely often* if there are infinitely many indices n for which $p(x_n)$ is true or, equivalently, if for any index n_0, no matter how large, there is an index $n > n_0$ such that $p(x_n)$ is true.

If $\{x_n\}_{n=1}^{\infty}$ is a sequence and $\{n_k\}$ is a strictly increasing sequence of positive integers, the sequence $\{x_{n_k}\}$ is a *subsequence* of $\{x_n\}$. Clearly a predicate $p(\ )$ is true of a sequence $\{x_n\}$ infinitely often if and only if $\{x_n\}$ has a subsequence $\{x_{n_k}\}$ such that $p(x_{n_k})$ is true for every k. Moreover, it is a matter of pure logic that a predicate $p(\ )$ fails to be true of a sequence $\{x_n\}$ eventually if and only if the contrary predicate (not p)( ) is true of $\{x_n\}$ infinitely often.

Example U. Given an arbitrary sequence $\{E_n\}_{n=1}^{\infty}$ of sets there is a natural way to construct from it a monotone increasing sequence. Indeed, if we define

$$D_n = \bigcap_{k=n}^{\infty} E_k, \quad n \in \mathbb{N},$$

then the sequence $\{D_n\}_{n=1}^{\infty}$ is monotone increasing. It is customary to call the union $\underline{L} = \bigcup_{n=1}^{\infty} D_n$ the *limit inferior* of the given sequence $\{E_n\}$ (notation: $\underline{L} = \liminf_n E_n$). Dually, if we set

$$S_n = \bigcup_{k=n}^{\infty} E_k, \quad n \in \mathbb{N},$$

then the sequence $\{S_n\}_{n=1}^{\infty}$ is monotone decreasing and the intersection $\overline{L} = \bigcap_{n=1}^{\infty} S_n$ is called the *limit superior* of $\{E_n\}$ (notation: $\overline{L} = \limsup_n E_n$). It is easily seen that $\underline{L}$ is the set of all elements that belong eventually to the sequence $\{E_n\}$, while $\overline{L}$ is the set of all those elements that belong to infinitely many sets E_n. Hence, in particular,

$$\underline{L} \subset \overline{L}.$$

In the event that $\underline{L} = \overline{L}$ we say that the sequence $\{E_n\}$ *converges* to the *limit* $L = \underline{L} = \overline{L}$ (notation: $L = \lim_n E_n$).

Suppose now that the originally given sequence $\{E_n\}$ is itself monotone increasing. Then it is apparent that $D_n = E_n$ for each index n, while $S_n = \bigcup_n E_n = \sup_n E_n$ for all n. Thus $\underline{L} = \overline{L} = \bigcup_n E_n$—a monotone increasing sequence of sets converges to its union. Dually, a monotone decreasing sequence of sets $\{E_n\}$ converges to $\bigcap_n E_n = \inf_n E_n$. Note that this shows that the terms "limit inferior" and "limit superior" are fully justified; it is literally true that for an arbitrary sequence $\{E_n\}$ of sets,

$$\liminf_n \ E_n = \lim_n \left(\inf_{k \geq n} E_k \right)$$

and

$$\limsup_n E_n = \lim_n \left(\sup_{k \geq n} E_k \right).$$

It follows easily from these definitions that for any sequence $\{E_n\}$ of subsets of a fixed set X we have

$$X \backslash \limsup_n E_n = \liminf_n (X \backslash E_n) \quad \text{and} \quad X \backslash \liminf_n \ E_n = \limsup_n (X \backslash E_n).$$

Hence the sequence $\{E_n\}$ is convergent when and only when the sequence $\{X \backslash E_n\}$ of *complements* is, and when this is the case,

$$\lim_n (X \backslash E_n) = X \backslash \lim_n E_n.$$

Moreover, as with any useful notion of limit, it is the case that deleting, changing, or adjoining any finite number of terms to a sequence $\{E_n\}$ of sets does not change either $\liminf_n E_n$ or $\limsup_n E_n$, and hence has no effect on either the convergence of $\{E_n\}$ or on $\lim_n E_n$ when it exists.

Example V. For each positive integer m let us write T_m for the tail $T_m = \{n \in \mathbb{N} : n \geq m\}$. Then the system $\{T_m\}_{m=1}^{\infty}$ of all tails in $\mathbb{N}$ is a directed set in the inverse inclusion ordering (see Example Q), and the mapping $m \to T_m$ is an order isomorphism of $\mathbb{N}$ onto this directed set.

If X is a partially ordered set with the property that for every pair x, y of elements of X either $x \leq y$ or $y \leq x$, then X is a *simply ordered* (or *linearly* or *totally ordered*) set. The most familiar examples of simply ordered sets are the set $\mathbb{R}$ of real numbers and its various subsets. (Clearly any subset of a simply ordered set is itself simply ordered.)

Example W. The simply ordered sets $\mathbb{N}_0$ and $\mathbb{N}$ are order isomorphic, the unique order isomorphism of $\mathbb{N}_0$ onto $\mathbb{N}$ being the mapping $n \to n+1$. (The uniqueness of this order isomorphism will be established later, in Lemma 5.2; no use will be made of the stated uniqueness in the interim.)

Example X. A finite set X containing n elements can be simply ordered in $n!$ different ways (there being $n!$ permutations of X), but in each of these orderings X is order isomorphic to the set $\{1, 2, \ldots, n\}$ of the first n positive integers.

If W is a partially ordered set with the property that every nonempty subset of W possesses a least element, then W is a *well-ordered* set. In a well-ordered set every doubleton, in particular, possesses a least element, whence it follows that every well-ordered set is simply ordered. Conversely, as was seen in the preceding example, every finite simply ordered set is well-ordered. The best known examples of infinite well-ordered sets are the set $\mathbb{N}_0$ of all nonnegative integers, and various of its subsets. (It is, once again, obvious that every subset of a well-ordered set is well-ordered.)

Example Y. If X is either the empty set or a singleton, then there is but one partial ordering of X, and this ordering is automatically a well-ordering. If X is a doubleton, then X admits exactly three partial orderings, namely, the identity relation and two well-orderings. Any set X containing three or more elements admits various partial orderings that are not simple orderings.

We conclude this introduction to the theory of ordered sets by stating two more axioms. (These axioms are, in fact, equivalent to one another (Prob. V), and each is equivalent to the axiom of choice (see Problems 5R–5V).)

The Maximum Principle. *Every partially ordered set X contains a maximal simply ordered subset.*

It may be observed that the name *maximum* principle is slightly inappropriate, since all that is asserted is the existence of a *maximal* linearly ordered subset. Note also that the ordering of the subsets of X referred to here is, as usual, the inclusion ordering (Ex. O). A frequently encountered variant of the maximum principle is the following proposition, known in the literature, inappropriately, as "Zorn's lemma".

Zorn's Lemma. *Let X be a partially ordered set, and suppose that every* nested (*i.e., simply ordered*) *subset of X is bounded above in X. Then X possesses a maximal element.*

PROBLEMS

A. If S is a relation between a set X and a set Y, and if R is another relation between Y and a third set Z, then the *composition* of R and S is the relation $R \circ S$ between X and Z defined by setting

$$x(R \circ S)z$$

when and only when there exists an element y of Y such that

$$xSy \quad \text{and} \quad yRz.$$

(i) Prove that for every subset A of X we have $(R \circ S)(A) = R(S(A))$. Similarly, for every subset B of Z, $(R \circ S)^{-1}(B) = S^{-1}(R^{-1}(B))$.

(ii) Show that if either $R \circ (S \circ T)$ or $(R \circ S) \circ T$ is defined, then the other is also defined, and $R \circ (S \circ T) = (R \circ S) \circ T$. Thus in this situation we may and shall write simply $R \circ S \circ T$ for the triple composition. Likewise, if R is a relation on a set X and $n \in \mathbb{N}$, we shall write R^n for the composition

$$\overbrace{R \circ \ldots \circ R}^{n}.$$

Show also that if R is an arbitrary relation between a set X and a set Y, and if ι_X and ι_Y denote the identity mappings on X and Y, respectively (Ex. D), then $R \circ \iota_X = \iota_Y \circ R = R$.

(iii) Verify that if $\phi : X \to Y$ and $\psi : Y \to Z$ are both mappings, then $\psi \circ \phi$ is a mapping of X into Z.

B. Let R be a relation between nonempty sets X and Y. Then the *inverse* relation R^{-1} between Y and X is defined by setting $yR^{-1}x$ when and only when xRy.

(i) Show that $R = R^{-1}$ if and only if $X = Y$ and R is a symmetric relation on X.

(ii) If $\phi : X \to Y$ is a mapping of X into Y, then ϕ^{-1} is a mapping of the range B of ϕ onto X when and only when for each y in B there exists *exactly one* x in X such that $\phi(x) = y$. (Recall that such a mapping is said to be one-to-one.) Verify that if ϕ is a one-to-one mapping of X onto Y, then $\phi^{-1} \circ \phi$ and $\phi \circ \phi^{-1}$ are the identity mappings on X and Y, respectively. Show, too, that the composition of two one-to-one mappings is one-to-one.

(iii) Let ϕ be a mapping of X into Y. Prove that there exists a mapping ψ of Y onto X such that $\psi \circ \phi = \iota_X$ if and only if ϕ is a one-to-one mapping of X into Y. Prove also that there exists a mapping ω of

Y into X such that $\phi \circ \omega = \iota_Y$ if and only if ϕ maps X onto Y, and conclude that if there exist mappings ψ and ω of Y into X such that

$$\psi \circ \phi = \iota_X \quad \text{and} \quad \phi \circ \omega = \iota_Y,$$

then ϕ is a one-to-one mapping of X onto Y, and $\psi = \omega = \phi^{-1}$. (Hint: Recall the axiom of choice.)

C. Let X be a nonempty set and let $\mathcal{G}$ denote the collection of all one-to-one mappings of X onto itself. Verify that composition $\circ$ is an associative binary operation on $\mathcal{G}$ having the identity mapping ι_X for neutral element (Ex. F), and that for each element ϕ of $\mathcal{G}$ the mapping ϕ^{-1} has the property that

$$\phi \circ \phi^{-1} = \phi^{-1} \circ \phi = \iota_X.$$

(These facts are customarily summarized by saying that $\mathcal{G}$ is a *group* with respect to the operation $\circ$. The elements of the group $\mathcal{G}$ are called *permutations* of the set X; the group $\mathcal{G}$ itself is sometimes referred to as the *symmetric group* on X.)

D. Let Γ be a nonempty index set and let $\{A_\gamma\}_{\gamma\in\Gamma}$ be a family of sets indexed by Γ. Show that if $\{\Gamma_\delta\}_{\delta\in\Delta}$ is an arbitrary indexed partition of Γ, then

$$\bigcup_{\gamma\in\Gamma} A_\gamma = \bigcup_{\delta\in\Delta}\ \bigcup_{\gamma\in\Gamma_\delta} A_\gamma \quad \text{and} \quad \bigcap_{\gamma\in\Gamma} A_\gamma = \bigcap_{\delta\in\Delta}\ \bigcap_{\gamma\in\Gamma_\delta} A_\gamma,$$

and likewise that if π is an arbitrary permutation of Γ, then

$$\bigcup_{\gamma\in\Gamma} A_{\pi(\gamma)} = \bigcup_{\gamma\in\Gamma} A_\gamma \quad \text{and} \quad \bigcap_{\gamma\in\Gamma} A_{\pi(\gamma)} = \bigcap_{\gamma\in\Gamma} A_\gamma.$$

Verify also that if B is an arbitrary set then

$$B \cup \bigcup_\gamma A_\gamma = \bigcup_\gamma (B \cup A_\gamma) \quad \text{and} \quad B \cup \bigcap_\gamma A_\gamma = \bigcap_\gamma (B \cup A_\gamma).$$

Similarly, verify that

$$B \cap \bigcup_\gamma A_\gamma = \bigcup_\gamma (B \cap A_\gamma) \quad \text{and} \quad B \cap \bigcap_\gamma A_\gamma = \bigcap_\gamma (B \cap A_\gamma)$$

and that

$$B \times \bigcup_\gamma A_\gamma = \bigcup_\gamma (B \times A_\gamma) \quad \text{and} \quad B \times \bigcap_\gamma A_\gamma = \bigcap_\gamma (B \times A_\gamma).$$

E. Let $\{A_\gamma\}$ and $\{B_\gamma\}$ be two similarly indexed families of subsets of a set X. Verify that both

$$\left(\bigcup_\gamma A_\gamma\right) \setminus \left(\bigcup_\gamma B_\gamma\right) \quad \text{and} \quad \left(\bigcap_\gamma A_\gamma\right) \setminus \left(\bigcap_\gamma B_\gamma\right)$$

are subsets of $\bigcup_\gamma (A_\gamma \setminus B_\gamma)$. Verify, likewise, that

$$\left(\bigcup_\gamma A_\gamma\right) \nabla \left(\bigcup_\gamma B_\gamma\right) \quad \text{and} \quad \left(\bigcap_\gamma A_\gamma\right) \nabla \left(\bigcap_\gamma B_\gamma\right)$$

are subsets of $\bigcup_\gamma (A_\gamma \nabla B_\gamma)$.

F. Let X and Y be sets and let ϕ be a mapping of X into Y.

(i) The mapping $A \to \phi(A)$ of 2^X into 2^Y is said to be *induced* by ϕ. Show that this induced mapping preserves unions, i.e., $\phi(\bigcup \mathcal{C}) = \bigcup_{A\in\mathcal{C}} \phi(A)$ for any collection $\mathcal{C}$ of subsets of X. Does this assertion remain valid if ϕ is replaced by an arbitrary relation R between X and Y? What if unions are replaced by intersections?

(ii) The mapping Φ of the power class on Y into the power class on X obtained by setting $\Phi(B) = \phi^{-1}(B), B \subset Y$ is said to be *inversely induced* by ϕ. Show that Φ preserves all set operations on 2^Y, i.e., arbitrary unions, intersections, differences and symmetric differences. Does this assertion remain valid if ϕ is replaced by an arbitrary relation R between X and Y?

(iii) Show that $\phi^{-1} \circ \phi$ is an equivalence relation on X, and find the equivalence classes of this relation.

G. The indexed product of sets is neither commutative nor associative. Verify, however, that if $\{X_\gamma\}_{\gamma\in\Gamma}$ is an arbitrary nonempty indexed family of sets, and if π is a permutation of the index set Γ, then there is a simple and natural one-to-one correspondence between

$$\prod_{\gamma\in\Gamma} X_\gamma \quad \text{and} \quad \prod_{\gamma\in\Gamma} X_{\pi(\gamma)}.$$

Show similarly that if $\{\Gamma_\delta\}_{\delta\in\Delta}$ is an arbitrary indexed partition of Γ, then there is also a simple and natural one-to-one correspondence between

$$\prod_{\gamma\in\Gamma} X_\gamma \quad \text{and} \quad \prod_{\delta\in\Delta}\prod_{\gamma\in\Gamma_\delta} X_\gamma.$$

H. Verify that the axiom of products and the axiom of choice are equivalent. (Hint: It suffices to treat the case of an indexed collection $\{X_\gamma\}_{\gamma\in\Gamma}$ of nonempty sets; recall Example K.)

I. A subset M of a partially ordered set X is said to be *cofinal* in X if for every element x of X there is an element x_0 of M such that $x \leq x_0$. Show that if M is a cofinal subset of a partially ordered set X, and if M is bounded above in X, then X has a greatest element x_1 and $x_1 \in M$. If a partially ordered set X has a greatest element x_1, then the singleton $\{x_1\}$ is cofinal in X. If x_1 is merely a maximal element in X, then $\{x_1\}$ need not be cofinal in X.

J. Suppose given a transitive and reflexive relation $\prec$ on a set X. Show that if, for all x and y in X, $x \sim y$ is defined to mean that $x \prec y$ and $y \prec x$, then $\sim$ is an equivalence relation on X with the property that if $x \sim x'$ and $y \sim y'$, then $x \prec y$ if and only if $x' \prec y'$. Hence if $[x]$ and $[y]$ denote the equivalence classes of x and y, respectively, with respect to the relation $\sim$, then either $x' \prec y'$ for *every* pair of elements selected from $[x]$ and $[y]$, respectively, or $x' \prec y'$ is true of no such pair. In the former of these two situations we write $[x] \leq [y]$. Show that this relation is a partial ordering of the quotient space $X/\sim$. (The set $X/\sim$ equipped with this partial ordering is the partially ordered set *associated* with the given pair $(X, \prec)$; a set X equipped with a relation such as $\prec$ is called a *weakly partially ordered set*.)

K. Let X be a partially ordered set with the property that every subset of X has a supremum in X. Show that every subset of X also has an infimum in X, and hence that X is a complete lattice. Dually, if every subset of a partially ordered set X has an infimum in X, then X is a complete lattice. Find and prove appropriate versions of these results for boundedly complete lattices.

L. If R and S are two relations between a set X and a set Y, then it is customary to define $R \subset S$ to mean that for every pair x and y of elements of X and Y, respectively, xRy implies xSy. (If $R \subset S$ then R is a *restriction* of S and S is an *extension* of R.)

(i) Verify that this is a partial ordering of the set $\mathcal{R}$ of all relations between X and Y, and that the mapping $R \to G_R$ assigning to each relation R in $\mathcal{R}$ its graph is an order isomorphism of $\mathcal{R}$ (in this *extension ordering*) onto the power class on $X \times Y$ (in the inclusion ordering). Conclude that $\mathcal{R}$ is a complete lattice in the extension ordering. Show too that the collection of all equivalence relations on an arbitrary set X is also a complete lattice in the extension ordering, though not, in general, a sublattice of the lattice $\mathcal{R}$.

(ii) If R is a relation on a set X, then $R^2 = R \circ R \subset R$ if and only if R is transitive. Similarly, if R is reflexive, then $R \subset R^2$. Is the converse of this latter assertion true?

M. Let X and Y be sets. The collection $\mathcal{M}$ consisting of all mappings ϕ such that the domain of ϕ is a subset of X and the range of ϕ is a subset of Y is a subset of the lattice $\mathcal{R}$ of all relations between X and Y and is therefore a partially ordered set (in the extension ordering introduced in the preceding problem). If $\phi \subset \psi$ in $\mathcal{M}$, and if the domain of ϕ is A, then ϕ will be denoted in the sequel by $\psi|A$.

(i) Show that every pair of mappings ϕ and ψ in $\mathcal{M}$ possesses an infimum $\phi \cap \psi$ in $\mathcal{M}$. (In particular, if the domains of ϕ and ψ are disjoint, then $\phi \cap \psi$ is the *empty mapping*, that is, the unique mapping having for its domain of definition the empty set $\varnothing$.)

(ii) If ϕ and ψ are elements of $\mathcal{M}$ with domains A and B, respectively, then ϕ and ψ are *coherent* if the domain of $\phi \cap \psi$ is $A \cap B$. Likewise, an indexed family $\{\phi_\gamma\}_{\gamma \in \Gamma}$ of elements of $\mathcal{M}$ is *coherent* if each pair of mappings belonging to the family is coherent. Show that an indexed family $\{\phi_\gamma\}$ in $\mathcal{M}$ is bounded above in $\mathcal{M}$ if and only if it is coherent, and that, in this situation, the family possesses a supremum. Show also that if the family $\{\phi_\gamma\}$ is coherent and if, for each index γ, D_γ and R_γ denote the domain and range, respectively, of the mappings ϕ_γ, then the supremum $\bigcup_\gamma \phi_\gamma$ has for domain the union $\bigcup_\gamma D_\gamma$, and for range the union $\bigcup_\gamma R_\gamma$.

(iii) Let $\{\phi_\gamma\}$ be an indexed family in $\mathcal{M}$, and let D_γ and R_γ denote, respectively, the domain and range of the mapping ϕ_γ. If the family $\{\phi_\gamma\}$ is nested, then it is coherent. Show that if the family is nested, and if each of the mappings ϕ_γ is one-to-one, then $\bigcup_\gamma \phi_\gamma$ is a one-to-one mapping of $\bigcup_\gamma D_\gamma$ onto $\bigcup_\gamma R_\gamma$. Likewise, if the domains D_γ are pairwise disjoint, then the family $\{\phi_\gamma\}$ is coherent. Show that if the domains D_γ and the ranges R_γ are both pairwise disjoint, and if each mapping ϕ_γ is one-to-one, then, once again, $\bigcup_\gamma \phi_\gamma$ is a one-to-one mapping.

N. Let $(X, \leq)$ be a simply ordered set.

(i) Show that $<$ is a transitive relation on X that satisfies the condition that if x and y are any two elements of X, then exactly one of the following three statements is true:

$$x < y, \quad x = y, \quad x > y.$$

(Recall that we write $x < y$ to mean that $x \leq y$ and $x \neq y$, and that $x > y$ means $y < x$.) Show, conversely, that if $<$ is a given transitive relation on X satisfying the last stated condition (known as the *trichotomy law*), and if we define a relation $\leq$ on X by setting $x \leq y$ whenever either $x < y$ or $x = y$, then $\leq$ is a simple ordering on X.

(ii) Show that if ϕ is a mapping of X into another simply ordered set $(Y, \leq)$, then the following conditions are equivalent:

(1) ϕ is strictly increasing,
(2) $x_1 < x_2$ implies $\phi(x_1) < \phi(x_2)$, $\quad x_1, x_2 \in X$,
(3) ϕ is an order isomorphism of X onto $\phi(X)$.

(iii) Show that X is a lattice and that if $F = \{x_1, \ldots, x_n\}$ is an arbitrary nonempty finite subset of X, then $x_1 \vee \ldots \vee x_n$ and $x_1 \wedge \ldots \wedge x_n$ belong to the set F.

O. Show that the simple orderings of a set X are distinguished as the maximal elements in the partially ordered set consisting of all the partial orderings of X (in the extension ordering; cf. Problem L).

P. Let W be a well-ordered set, and let ψ be an arbitrary strictly increasing mapping of W into itself. Show that $\psi(w) \geq w$ for all w in W, and conclude that $\psi(W)$ is necessarily cofinal in W (Prob. I).

Q. If Γ is an index set, X is a partially ordered set, and ϕ and ψ are mappings of Γ into X, we write $\phi \leq \psi$ to mean that $\phi(\gamma) \leq \psi(\gamma)$ for every γ in Γ. Show that this relation is a partial ordering of the set X^Γ of all mappings of Γ into X. More generally, the same definition introduces a partial ordering on every indexed product $\Pi = \prod_{\gamma \in \Gamma} X_\gamma$ of partially ordered sets. Show that if each partially ordered set X_γ is a (complete) lattice, then Π is a (complete) lattice.

R. Let X be a fixed set. For each subset A of X the *characteristic function* of A is that function χ_A that takes the value one at every point of A and the value zero at every point of $X \backslash A$. If A and B are subsets of X, then $\chi_{A \cap B} = \chi_A \chi_B$ and $\chi_{A \cup B} = \chi_A \vee \chi_B$. Furthermore, $A \subset B$ if and only if $\chi_A \leq \chi_B$, and $\chi_{A \cup B} = \chi_A + \chi_B$ if and only if A and B are disjoint.

S. If $\mathcal{P}$ and $\mathcal{P}'$ are two partitions of the same set X, then $\mathcal{P}'$ is said to be *finer than* $\mathcal{P}$, or to *refine* $\mathcal{P}$ (and $\mathcal{P}$ is said to be *coarser* than $\mathcal{P}'$; notation: $\mathcal{P} \leq \mathcal{P}'$) if every set in $\mathcal{P}'$ is a subset of some set in $\mathcal{P}$.

(i) Show that if $\mathcal{P}$ and $\mathcal{P}'$ are partitions of a set X, then $\mathcal{P} \leq \mathcal{P}'$ if and only if every set E in $\mathcal{P}$ is partitioned by the subcollection of $\mathcal{P}'$ consisting of the sets in $\mathcal{P}'$ contained in E. Show too that the relation $\leq$ is a partial ordering on the collection of all partitions of X, and that the mapping $\mathcal{P} \to \sim_P$ assigning to each partition of X the equivalence relation on X that it determines is an *order anti-isomorphism* of that collection onto the complete lattice of equivalence relations on X (in the extension ordering; see Problem L). Show, that is, that $\mathcal{P} \leq \mathcal{P}'$ when and only when $\sim_{P'} \subset \sim_P$. Conclude that the collection of all partitions of X is a complete lattice with respect to the ordering $\leq$.

(ii) If $\{E_1, \ldots, E_n\}$ is a finite collection of subsets of a set X, then the partition of X *determined* by $\{E_1, \ldots, E_n\}$ is the coarsest partition of X that partitions each of the sets E_i, $i = 1, \ldots, n$. Show that this partition consists of the collection of all nonempty sets of the form

$$A_1 \cap \ldots \cap A_n$$

where each A_i is either E_i of $X \backslash E_i$. (There are 2^n such sequences $\{A_1, \ldots, A_n\}$, but the number of sets in the partition may be smaller, of course.)

T. Let $\mathcal{C}$ be a collection of (not necessarily pairwise disjoint) sets. If for each set E in $\mathcal{C}$ we write $E' = E \times \{E\}$, then $x \to (x, E)$ is a one-to-one mapping of E onto E', and the collection $\mathcal{C}'$ of all sets E', $E \in \mathcal{C}$, is pairwise disjoint.

U. Let $\{A_n\}_{n=1}^{\infty}$ be a sequence of sets. Show that there exists a unique disjoint sequence of sets $\{B_n\}_{n=1}^{\infty}$ with the property that

$$\bigcup_{k=1}^{n} B_k = \bigcup_{k=1}^{n} A_k$$

for every positive integer n. (The sequence $\{B_n\}$ will be referred to as the *disjointification* of the given sequence $\{A_n\}$.) Verify that $B_n \subset A_n$ for each positive integer n, and also that

$$\bigcup_{n=1}^{\infty} B_n = \bigcup_{n=1}^{\infty} A_n.$$

V. A partially ordered set X is said to be *inductive* if every simply ordered subset of X has a supremum in X. Clearly every inductive partially ordered set satisfies the hypotheses of Zorn's lemma, as that result is stated above, so Zorn's lemma is formally stronger than the assertion that every inductive partially ordered set possesses a maximal element. Nonetheless, in many treatments of set theory this latter assertion is called "Zorn's lemma". Justify these terminological variations by showing that these two versions of Zorn's lemma and the maximum principle are all equivalent to one another. (Hint: The collection of all simply ordered subsets of an arbitrary partially ordered set is inductive in the inclusion ordering.)

W. In connection with each and every notion of an abstract mathematical structure there is an appropriate concept of *isomorphism*, meaning, in every case, a one-to-one correspondence between any two examplars X and Y of that structure that *preserves* the structure, that is, with the property that any statement (that is germane to the structure in question) concerning the elements of X is true if and only if it is also true of the corresponding elements of Y. The significance of this idea is simply put: If two examplars of a mathematical structure are *isomorphic*, i.e., if there exists such an isomorphism between them, then they are indistinguishable as far as the theory of the structure in question is concerned. Suppose, for example, that X and Y are order isomorphic partially ordered sets. Then if either X or Y possesses a maximal element, both must. Again, if either is a lattice, or a directed set, or simply ordered, then both must be. Prove these four statements (as tokens of the general fact that any assertion of an order-theoretic nature that is true of either X or Y is true of both). Show also that (1) the identity mapping on any partially ordered set X is an order isomorphism of X onto itself, (2) if ϕ is an order isomorphism of a partially ordered set X onto a partially ordered set Y, then ϕ^{-1} is an order isomorphism of Y onto X, and (3) if ϕ is an order isomorphism of X onto Y and ψ is a second order isomorphism of Y onto a third partially ordered set Z, then $\psi \circ \phi$ is an order isomorphism of X onto Z. (These three properties are common to every notion of isomorphism; in the sequel we shall ordinarily leave it to the reader to formulate and check them on his own.)

2 Number systems

The real number system

In this second preliminary chapter we turn attention to the structure that lies at the heart of all mathematical analysis—the *real number system*, or *real number field*. In so doing we shall not waste a great deal of time on the basic properties of the real numbers. Indeed, the reader has been dealing with the real numbers in one way or another for many years. Rather it is our intention, first, to fix some terminology and notation and, second and more important, to present a fairly detailed discussion of certain aspects of the order-theoretic properties of the real numbers that are sometimes not treated adequately in undergraduate mathematics courses.

To begin with, the real number system—denoted throughout the text by the usual symbol $\mathbb{R}$—has both an algebraic structure and an order-theoretic structure. Algebraically the real numbers are equipped with both an *addition* and a *multiplication*, so that every pair s, t of real numbers has both a *sum* $s + t$ and a *product* st. Both of these binary operations satisfy the *associative* and *commutative laws*—meaning that, if s, t and u are real numbers, then

$$s + (t + u) = (s + t) + u, \quad s(tu) = (st)u, \tag{1}$$

and

$$s + t = t + s, \quad st = ts \tag{2}$$

(cf. Example 1F). From these two laws it follows, via a routine argument which we omit, that if F is an arbitrary nonempty finite set of real numbers, then the *sum* $\sum F$ and *product* $\prod F$ of the numbers in the set F are uniquely and unambiguously defined, independently of the order in which $F = \{s_1, \ldots, s_n\}$ is arranged or how the sum and product of the numbers $s_1, \ldots, s_n$ are grouped. Moreover, the addition and multiplication of real numbers are related by the *distributive law*, which says that

$$s(t+u) = st + su \tag{3}$$

for all real numbers s, t and u.

There are special and distinct *neutral elements* 0 (*zero*) and 1 (*one*) in $\mathbb{R}$ for the operations of addition and multiplication, respectively, so that

$$s + 0 = s \quad \text{and} \quad s1 = s$$

for every real number s. Moreover, each real number s has a (unique) *negative* $-s$ such that $s+(-s) = 0$, which implies that the equation $s+x = t$ has the (unique) solution

$$x = t + (-s),$$

written as $x = t - s$ (t *minus* s). Similarly, if $s \neq 0$, then there exists a (unique) *reciprocal* $1/s$ such that $s(1/s) = 1$, which implies that if $s \neq 0$, then the equation $sx = t$ has the (unique) solution

$$x = t(1/s),$$

written as $x = t/s$ (t *over* s, or t *divided by* s). (The distributive law implies that $s0 = 0$ for every real number s, so the real number 0 cannot have a reciprocal; accordingly, division by zero is impossible in $\mathbb{R}$. See Problem C.)

In the language of abstract algebra, the properties of $\mathbb{R}$ set forth thus far are summarized by saying that the real numbers form a *field*. The order-theoretic structure of the real number system can be summarized almost as succinctly: There is an order relation $<$ given on $\mathbb{R}$ that turns it into a *simply ordered set* (see Problem 1N). In the simply ordered set $\mathbb{R}$ the number 0 plays a special role; indeed, a real number s such that $s > 0$ is *positive*, while a real number s such that $s < 0$ is *negative*. Thus it is a special case of the trichotomy law that every real number is positive, negative or zero, and that no real number falls into any two of these classes.

The algebraic and order-theoretic structures of the real number system are connected with one another in a number of ways, but, as it turns out (see Problems F, G, H, I, J and K), these various connections all follow readily from two facts, which we now state.

(i) If a and b are real numbers such that $a < b$, and if s is any real number, then

$$a + s < b + s.$$

(ii) If a and b are real numbers such that $a < b$, and if p is a *positive* real number, then

$$ap < bp.$$

As a simply ordered set, $\mathbb{R}$ is automatically a lattice in which, for any two real numbers s and t, $s \vee t$ and $s \wedge t$ are simply the larger and smaller, respectively, of s and t (Prob. 1N). In particular, for any real number s the number $s^+ = s \vee 0$ is called the *positive part* of s, while $s^- = -(s \wedge 0)$ is the *negative part* of s. According to these definitions, both s^+ and s^- are nonnegative for every real number s, and $s = s^+ - s^-$. The number $s^+ + s^-$, which is equal to s when s is nonnegative and to $-s$ when s is nonpositive, is called the *absolute value* of s (notation: $|s|$). Thus

$$|s| = \begin{cases} s, & s \geq 0 \\ -s, & s \leq 0. \end{cases}$$

(In this connection, see also Problem L.)

If a and b are real numbers such that $a < b$, then the real numbers between a and b constitute an *interval* bounded by a and b. More specifically, if $a < b$, then the *closed interval* $[a, b]$ is given by

$$[a, b] = \{t \in \mathbb{R}\colon\ a \leq t \leq b\},$$

while the *open interval* (a, b) is

$$(a, b) = \{t \in \mathbb{R}\colon\ a < t < b\}.$$

(In particular, $[0, 1]$ and $(0, 1)$ are the closed and open *unit intervals*, respectively.) In the same spirit we define the *half-open intervals*

$$(a, b] = \{t \in \mathbb{R}\colon\ a < t \leq b\} \quad \text{and} \quad [a, b) = \{t \in \mathbb{R}\colon\ a \leq t < b\}.$$

(If the assumption that $a < b$ is dropped, these definitions still make sense, of course, but the intervals $[a, b], (a, b), (a, b]$ and $[a, b)$ are all empty for $a > b$, and also $(a, a) = [a, a) = (a, a] = \varnothing$, while $[a, a] = \{a\}$.) This notation for intervals is also extended to include *rays*. Thus for any real number a we write

$$[a, +\infty) \quad \text{and} \quad (a, +\infty)$$

for the *closed right ray* $\{t \in \mathbb{R}\colon\ a \leq t\}$ and the *open right ray* $\{t \in \mathbb{R}\colon a < t\}$, determined by a, respectively, and also

$$(-\infty, a] \quad \text{and} \quad (-\infty, a)$$

for the *closed left ray* $\{t \in \mathbb{R}\colon\ a \geq t\}$ and the *open left ray* $\{t \in \mathbb{R}\colon\ a > t\}$ determined by a, respectively. Such rays are frequently regarded as special intervals, as is the entire real number system $\mathbb{R} = (-\infty, +\infty)$ itself.

A field that is simultaneously a simply ordered set in such a way that (i) and (ii) are satisfied is, in the language of modern algebra, called an *ordered field*. That $\mathbb{R}$ is not uniquely determined as an ordered field is clear from the fact that the field of rational numbers—familiar from primary school—is also an ordered field. (There are many other ordered fields as well.) What sets the real number system apart from other ordered fields is a property known as *completeness*.

Axiom of Completeness. *Every nonempty set of real numbers that is bounded above in $\mathbb{R}$ has a supremum in $\mathbb{R}$. (Dually, a nonempty set of real numbers that is bounded below in $\mathbb{R}$ possesses an infimum in $\mathbb{R}$; cf. Problem G.)*

The property of completeness just formulated is important and is worth a paraphrase. Here is another way of saying the same thing. Let A be a set of real numbers, and let M denote the set of all upper bounds of A in $\mathbb{R}$. Then M may be empty (this happens when A is not bounded above in $\mathbb{R}$), and M may coincide with $\mathbb{R}$ (this happens when A is empty), but *in all other cases M is a closed right ray*: $M = [\sup A, +\infty)$. Yet another way to describe this situation is to say that $\mathbb{R}$ is a boundedly complete lattice.

In treatments of the real numbers that touch upon questions of the foundations of mathematical analysis one customarily finds a painstaking and fairly lengthy construction of the real number system based on some other simpler and more familiar number system—sometimes the system of rational numbers (see [12], for example), more usually the very primitive number system set forth in the *Peano postulates* (Prob. O). (See [16] and [14], to name but two sources.) In this book we eschew any consideration of foundation problems, and are content simply to assume the existence of a number system $\mathbb{R}$ possessing all of the heretofore stated properties.

From this more or less postulational point of view the various constructions of $\mathbb{R}$ from more familiar number systems may be seen primarily as *existence theorems*, i.e., as proofs of the existence of a complete ordered field. However, the various constructions of $\mathbb{R}$ also serve to elucidate just how it is related to the other, more primitive, number systems, and these relations are by no means clear from what we have said so far. Accordingly, we close this discussion of the real number system with a brief account of how the more primitive number systems may be identified within $\mathbb{R}$.

Definition. A set J of real numbers is *inductive* if (i) $1 \in J$ and (ii) if $s \in J$, then $s + 1 \in J$. (This notion is to be distinguished from the one introduced in Problem 1V.)

That inductive sets of real numbers exist is obvious; the set $\mathbb{R}$ itself is inductive, as is the set $P = (0, +\infty)$ of all positive real numbers (Prob.

J). Moreover, the intersection of any collection of inductive sets is clearly again an inductive set. Thus there exists a smallest inductive set, viz., the intersection of the collection of *all* inductive sets in $\mathbb{R}$.

Definition. The smallest inductive set of real numbers is known as the set of *natural numbers* and is denoted by $\mathbb{N}$.

According to this definition, if a set S of real numbers is shown to be inductive, then $\mathbb{N} \subset S$. Let us spell this out in greater detail: If for a set S of real numbers it can be shown that (i) $1 \in S$ and (ii) if $s \in S$, then $s + 1 \in S$, then every natural number belongs to S. The more usual formulation of this fact is the following principle.

Principle of Mathematical Induction. *If a set S of natural numbers satisfies the conditions* (i) $1 \in S$ *and* (ii) *for each natural number n, $n \in S$ implies $n + 1 \in S$, then $S = \mathbb{N}$.*

Proposition 2.1. *The real number 1 is the smallest natural number. The set $\mathbb{N}$ is not bounded above in $\mathbb{R}$, and is therefore cofinal in $\mathbb{R}$.*

PROOF. The first assertion of the proposition is an immediate consequence of the definition of $\mathbb{N}$ and the fact that the ray $[1, +\infty)$ is inductive. To see that $\mathbb{N}$ is cofinal in $\mathbb{R}$, suppose that $\mathbb{N}$ is bounded above in $\mathbb{R}$. Then $M = \sup \mathbb{N}$ exists, and $M - 1$ is therefore not an upper bound for $\mathbb{N}$. Hence there exists a natural number n such that $n > M - 1$. But then $n + 1 > M$, which is impossible. □

Example A. As has been noted, every natural number n is positive. It follows that all reciprocals $1/n$ of natural numbers n are also positive (Prob. K). But if ε is an arbitrary positive number, then there exists a natural number n such that $n > 1/\varepsilon$ (by the preceding proposition and Problem K), so $1/n < \varepsilon$. Thus the set R of all reciprocals of the natural numbers is a subset of $\mathbb{R}$ with $\inf R = 0$.

Example B. The function $\phi(t) = t/(1+t)$ is a strictly increasing mapping (and thus an order isomorphism) of the ray $[0, +\infty)$ onto the half-open interval $[0, 1)$, the inverse mapping being given by $\phi^{-1}(t) = t/(1 - t)$, $0 \le t < 1$ (cf. Problems D, F, I, and K). The mapping

$$\widehat{\phi}(t) = \frac{t}{1 + |t|}, \quad t \in \mathbb{R},$$

with inverse

$$\widehat{\phi}^{-1}(t) = \frac{t}{1-|t|}, \quad |t| < 1,$$

is an extension of ϕ providing an order isomorphism of all of $\mathbb{R}$ onto $(-1,+1)$ (cf. Problems G and L).

Proposition 2.2. *If m and n are natural numbers, then their sum $m+n$ and their product mn are also natural numbers.*

PROOF. Consider first the set S of all those natural numbers m with the property that $m+n \in \mathbb{N}$ for every n in $\mathbb{N}$. It is obvious that $1 \in S$, and, if $m \in S$, then $(m+1)+n = m+(n+1) \in \mathbb{N}$ for every n in $\mathbb{N}$, so S is inductive, and therefore $S = \mathbb{N}$. Thus sums of natural numbers are again natural numbers.

Next let T denote the set of all those natural numbers m with the property that $mn \in \mathbb{N}$ for every n in $\mathbb{N}$. It is again obvious that $1 \in T$. Moreover, if $m \in T$ and $n \in \mathbb{N}$, then

$$(m+1)n = mn + n$$

is the sum of natural numbers, and is therefore itself a natural number by what has just been shown. Thus T also is inductive, and the proof is complete. □

Another important property of the set $\mathbb{N}$ of natural numbers is that the next natural number larger than a natural number n is $n+1$. (That is, for each natural number n, there is *no* natural number in the open interval $(n, n+1)$; see Problem O.) By exploiting this fact we are able to establish an apparently stronger version of the principle of mathematical induction (formally stronger because the inductive hypothesis is weaker while the conclusion remains the same, but only apparently stronger since we derive it, in fact, from the original principle of mathematical induction). In the formulation of the next result it will be convenient to denote the set $\{k \in \mathbb{N} : k \leq n\}$ by A_n.

Theorem 2.3. *If a set S of natural numbers satisfies the conditions* (i) $1 \in S$, *and* (ii) *for each natural number n, $A_n \subset S$ implies that $n+1 \in S$, then $S = \mathbb{N}$.*

PROOF. Consider the set T of all those natural numbers n with the property that $A_n \subset S$. Since $A_1 = \{1\}$, we have $1 \in T$. Moreover, if $n \in T$, then $A_n \subset S$ and therefore $n+1 \in S$. But then

$$A_{n+1} = A_n \cup \{n+1\} \subset S,$$

and therefore $n+1 \in T$. Thus $T = \mathbb{N}$, and consequently $S = \mathbb{N}$ too. □

Our next result, while quite important in its own right, is an immediate consequence of Theorem 2.3.

Theorem 2.4. *The set $\mathbb{N}$ of natural numbers is well-ordered. (That is, every nonempty subset of $\mathbb{N}$ possesses a least element; see Chapter 1.)*

PROOF. Let A be a subset of $\mathbb{N}$ that does *not* possess a least element, and set $S = \mathbb{N} \setminus A$. Then $1 \in S$ since A clearly cannot contain the least natural number 1. But also, if $A_n \subset S$, and if $n+1$ belonged to A, then $n+1$ would surely be the least element of A, contrary to hypothesis. Thus $A_n \subset S$ implies $n+1 \in S$, so S is inductive, and therefore $S = \mathbb{N}$. But then A is empty. □

The method of mathematical induction is much more than a device for proving theorems; it also provides a powerful and exceedingly useful tool for giving definitions.

Theorem 2.5 (Principle of Inductive Definition). *Let X be a nonempty set, let $\widetilde{x}$ be an element of X, and let $\mathcal{F}$ denote the collection of all finite sequences $\{x_1, \ldots, x_n\}$ in X. Suppose given a mapping g of $\mathcal{F}$ into X. Then there exists a unique sequence $\{x_n\}_{n=1}^{\infty}$ in X satisfying the conditions*

(1) $x_1 = \widetilde{x}$,
(2) $x_{n+1} = g(\{x_1, \ldots, x_n\}), \quad n \in \mathbb{N}$.

Note. The proof of this theorem is omitted because it is wholly subsumed under the discussion of the principle of definition by *transfinite* induction to be found in Chapter 5. The role of the function g in the above formulation is simply to provide the "inductive step" in the definition, and this rule can ordinarily be set forth quite informally, so that in an actual definition by mathematical induction the function g need never appear explicitly. It is also only fair to point out that in most applications the value assigned by the inductive rule g to a sequence $\{x_1, \ldots, x_n\}$ depends only on x_n and not on $x_k, k = 1, \ldots, n-1$. Thus, however ponderous the machinery of Theorem 2.5 may seem, an actual inductive definition is typically quite brief and wholly perspicuous.

Example C. For each real number t we define $t^1 = t$ and then, for each natural number n, assuming t^n already defined, we set

$$t^{n+1} = tt^n. \tag{4}$$

Thus the power function $t \to t^n$ is defined inductively for all real numbers t and all natural numbers n, according to the inductive rule (4) and the initial requirement

$$t^1 = t, \quad t \in \mathbb{R}. \tag{5}$$

It may be noted that in this example it is entirely immaterial whether one thinks of the definition of t^n as being given for one t_0 at a time (in which case, in the notation of Theorem 2.5, $X = \mathbb{R}, \widetilde{x} = t_0$ and g assigns the number $t_0 x_n$ to a finite sequence $\{x_1, x_2, \ldots, x_n\}$ of real numbers) or as the definition of the function $t \to t^n, t \in \mathbb{R}$ (in which case X becomes the set $\mathbb{R}^{\mathbb{R}}$ of real-valued functions on $\mathbb{R}, \widetilde{x} = \iota_{\mathbb{R}}$, the identity mapping on $\mathbb{R}$ (Ex. 1D) and g becomes the rule assigning to an n-tuple $\{f_1, \ldots, f_n\}$ of functions in $\mathbb{R}^{\mathbb{R}}$ the function $\iota_{\mathbb{R}} f_n$). It may also be noted that the inductive definition of positive integral powers given in this example applies equally well in any system in which an associative product is defined (cf. Example 1F).

Definition. The subset of $\mathbb{R}$ consisting of the natural numbers and the negatives of the natural numbers, along with the number zero, is the set $\mathbb{Z}$ of *integers*.

If we write $-\mathbb{N} = \{t \in \mathbb{R} : \ -t \in \mathbb{N}\}$, then $\mathbb{Z} = \mathbb{N} \cup \{0\} \cup -\mathbb{N}$. We see that the set $\mathbb{N}$ coincides with the set of all *positive integers*, while $-\mathbb{N}$ is precisely the set of all *negative integers* (Prob. G). Also of interest in this context is the set $\mathbb{N}_0 = \{0\} \cup \mathbb{N}$ of *nonnegative integers*. The set of integers has a number of important properties as a subset of $\mathbb{R}$.

Theorem 2.6. *If j and k are integers, then the sum $j + k$, the difference $j - k$, and the product jk are all integers as well.*

PROOF. Since the negative of an integer is clearly an integer, we may, in view of Problem C, assume that neither j nor k is zero. As for products, we already know (Prop. 2.2) that the product of positive integers is a positive integer, and since $jk = (-j)(-k)$, the product of two negative integers is also a positive integer. On the other hand, if one of j, k is positive and the other negative, then $jk = -(-j)k$ is a negative integer.

In proving that the difference $j - k$ is an integer we first note that since $j - k = -((-j) - (-k))$ (Prob. D), we may also assume without loss of generality that j is a positive integer. But then, if k is negative, $j - k = j + (-k)$ is also a positive integer. On the other hand, if j and k are both positive, then either $j \geq k$, in which case $j - k$ is a nonnegative integer (see Problem O), or $j < k$, in which case $j - k = -(k - j)$ is a negative integer.

Finally, to show that the sum $j + k$ is an integer, we write $j + k = j - (-k)$. □

A quick check of cases shows that if k is an integer, then there is no integer between k and $k+1$, and we have the following fact.

Proposition 2.7. *For any integer k the next larger integer is $k+1$, and the next smaller is $k-1$.*

From this last result it follows at once that if j and k are distinct integers, then the half-open intervals $[j, j+1)$ and $[k, k+1)$ are disjoint subsets of $\mathbb{R}$. The following proposition is a sharpening of this observation.

Proposition 2.8. *The intervals $[k, k+1), k \in \mathbb{Z}$, constitute a partition of $\mathbb{R}$.*

PROOF. All that is needed is to show that for any real number t there is an integer k such that $k \leq t < k+1$, and to this end it clearly suffices to treat the case in which t is not itself an integer. If t is positive, there is a least positive integer n such that $t < n$ (Prop. 2.1, Th. 2.4). If $n = 1$, then $0 < t < 1$, while if $n > 1$, then $n-1$ is a positive integer such that $n-1 < t < n$. Thus, in either case, $k = n-1$ satisfies the required condition. Finally, if $t < 0$, and if $k < -t < k+1$, then $-(k+1) < t < -(k+1)+1$. □

Corollary 2.9. *Every nonnegative real number t is uniquely expressible as the sum of a nonnegative integer (called the* integral part *of t and denoted by $[t]$) and a number in the interval $[0,1)$ (called the* fractional part *of t and denoted by (t)).*

Corollary 2.10. *For an arbitrary positive real number d, the intervals $[kd, (k+1)d), k \in \mathbb{Z}$, also constitute a partition of $\mathbb{R}$.*

PROOF. A real number t belongs to $[kd, (k+1)d)$ if and only if t/d belongs to $[k, k+1)$. □

Finally, we consider the field of rational numbers in the context of our postulational development of the real number system.

Definition. The subset of $\mathbb{R}$ consisting of all numbers of the form j/k, where j and k are both integers and $k \neq 0$, is the set $\mathbb{Q}$ of *rational numbers*. (Since $-j/(-k) = j/k$, we may, and usually do, assume that k is positive in the representation j/k of a rational number.)

As has already been noted, the rational numbers constitute a field. What is meant by this assertion is simply that if q and r are two rational numbers, then the sum $q+r$, the difference $q-r$, the product qr, and, if $r \neq 0$, the quotient q/r are all rational numbers too. Without stating a formal proposition to this effect, we can verify these facts by supposing that $q = j/k$ and $r = \ell/m$, where j and ℓ are integers, while k and m are positive integers. Then

$$qr = \frac{j\ell}{km}$$

is a rational number, as is

$$\frac{q}{r} = \frac{jm}{k\ell}$$

provided $r \neq 0$, i.e., provided $\ell \neq 0$. Moreover, we also have

$$q = \frac{jm}{km} \quad \text{and} \quad r = \frac{k\ell}{km},$$

so that

$$q \pm r = \frac{jm \pm k\ell}{km},$$

and these are also rational numbers.

Since properties (i) and (ii) in the definition of an ordered field are clearly enjoyed by $\mathbb{Q}$ as well, it is also true that $\mathbb{Q}$ is an ordered field along with $\mathbb{R}$. The following observation concerns the way the *subfield* $\mathbb{Q}$ is situated in $\mathbb{R}$.

Theorem 2.11. *If a and b are real numbers such that $a < b$, then there are (infinitely many) rational numbers in the interval (a, b).*

PROOF. There exists a positive integer N exceeding the positive number $1/d$, where $d = b - a$ (Prop. 2.1). Let k be the unique integer such that $k(1/N) \leq a < (k+1)(1/N)$ (Cor. 2.10), and set $r = (k+1)/N$. Then $a < r$ while $r - a \leq 1/N < d = b - a$. Hence $r < b$, and therefore r is contained in the interval (a, b). To see that this interval contains, in fact, an infinity of rational numbers, we note that the interval (a, r) must also contain a rational number, etc. □

Definition. By a *base* in $\mathbb{R}$ is meant any positive integer greater than one. If p is a base, then the *digits* with respect to p are the integers $0, 1, \ldots, p-1$. (The most favored bases are $p = 2, 3$ and, of course, $p = 10$, though $p = 8$, 12 and 16 have also found adherents.)

The following result is just a summary of the basic facts concerning the "place holder" system of notation that is universally employed for denoting real numbers. The interested reader will have no difficulty supplying the proof of Theorem 2.12 on the basis of the preceding material.

Theorem 2.12. *Let p be a base in $\mathbb{R}$. Then every positive integer n has a unique expression in the form*

$$n = \varepsilon_1 + \varepsilon_2 p + \ldots + \varepsilon_m p^m$$

where $\varepsilon_1, \ldots, \varepsilon_m$ are digits with respect to p and $\varepsilon_m \neq 0$. This relation is customarily expressed by writing

$$n = \varepsilon_m \varepsilon_{m-1} \ldots \varepsilon_1. \tag{6}$$

Moreover, every number t in $[0, 1)$ may be written as

$$t = \sum_{n=1}^{\infty} \varepsilon_{-n} p^{-n}, \tag{7}$$

where $\{\varepsilon_{-n}\}_{n=1}^{\infty}$ is a sequence of digits with respect to p. This relation is expressed by writing

$$t = 0.\varepsilon_{-1}\varepsilon_{-2} \ldots \varepsilon_{-n} \ldots, \tag{8}$$

or by

$$t = 0.\varepsilon_{-1} \ldots \varepsilon_{-n} \tag{9}$$

when $\varepsilon_{-k} = 0$ for every $k > n$. (Such numbers are rational of the form k/p^n, and are called p-adic fractions.*) This expression for t is not necessarily unique, but if $\{\eta_{-n}\}_{n=1}^{\infty}$ is another, different sequence of digits such that (7) holds, then*

(1) *if n_0 denotes the first index at which ε_{-n} and η_{-n} differ, then $|\varepsilon_{-n_0} - \eta_{-n_0}| = 1$, and, assuming that $\eta_{-n_0} = \varepsilon_{-n_0} + 1$,*

(2) *$\eta_{-n} = 0$ for all $n > n_0$, while $\varepsilon_{-n} = p - 1$ for all $n > n_0$.*

Thus every nonnegative real number t may be written as

$$t = \varepsilon_m \ldots \varepsilon_1 + 0.\varepsilon_{-1}\varepsilon_{-2} \ldots,$$

or, as is more customary,

$$t = \varepsilon_m \ldots \varepsilon_1.\varepsilon_{-1} \ldots \varepsilon_{-n} \ldots, \tag{10}$$

and this representation is unique except for p-adic fractions, which, as noted, admit exactly two such representations.

For $p = 2$ the expansion (10) is called a *binary expansion* of t. When $p = 3$, it is a *ternary expansion*; when $p = 10$, a *denary expansion*. In general, one refers to (10) as a *"p-ary" expansion* of t.

The extended real number system

It is frequently very convenient to enlarge the real number system by admitting to it the "ideal" numbers $+\infty$ and $-\infty$. Thus it is customary to write $\sup E = +\infty$ for any set E of real numbers that is not bounded above in $\mathbb{R}$. In the study of analysis the usefulness of this symbolism is so great that it is desirable to introduce it on a formal basis. Accordingly, we adjoin to $\mathbb{R}$ two new "numbers" $+\infty$ and $-\infty$. The enlarged number system

$$\mathbb{R} \cup \{+\infty, -\infty\}$$

will be called the *extended real numbers* and will be denoted by $\mathbb{R}^\natural$. (In dealing with $\mathbb{R}^\natural$ we will distinguish the elements of $\mathbb{R}$ as *ordinary* real numbers, or as *finite* numbers.) The simple ordering of $\mathbb{R}$ is extended to $\mathbb{R}^\natural$ by defining $-\infty < +\infty$ and

$$-\infty < t < +\infty$$

for every finite real number t. Thus $+\infty$ is the largest element of $\mathbb{R}^\natural$ and $-\infty$ is the smallest. (Note that in $\mathbb{R}^\natural$ it is not a notational convention but a literal fact that $\sup E = +\infty$ for a set E of real numbers that is not bounded above in $\mathbb{R}$.)

As for algebraic operations, we define

$$t + \pm\infty = \pm\infty + t = \pm\infty$$

for every finite real number t, and likewise

$$\pm\infty + \pm\infty = \pm\infty.$$

Similarly, we agree that

$$t - \pm\infty = \mp\infty \quad \text{and} \quad \pm\infty - t = \pm\infty$$

for every finite real number t, and that

$$\pm\infty - \mp\infty = \pm\infty.$$

(The symbols $\pm\infty + \mp\infty$ and $\pm\infty - \pm\infty$ remain undefined.) As regards multiplication, we define

$$t(\pm\infty) = (\pm\infty)t = \begin{cases} \pm\infty & t > 0 \\ 0 & t = 0 \\ \mp\infty & t < 0 \end{cases}$$

for every finite real number t, and likewise

$$(\pm\infty)(\pm\infty) = +\infty \quad \text{and} \quad (\pm\infty)(\mp\infty) = -\infty.$$

It must be admitted, to be sure, that the systematic use of $\mathbb{R}^\natural$ in place of $\mathbb{R}$ brings some inconvenience with it (see, for example, Problem T below). The best that can be said is that, on balance, it is easier to get along with $\pm\infty$ in real analysis than it is to get along without them.

The complex number system

While this book is basically a treatise on real analysis, it is a feature of its development that the complex numbers appear in the sequel with fair frequency. Accordingly, we close this chapter with a brief account of the *complex number system* (notation: $\mathbb{C}$). Here again, as above in the case of the real number system, we shall not bother to construct $\mathbb{C}$ (though this turns out to be a fairly simple thing to do, and would have the advantage of proving the existence of $\mathbb{C}$), but we will rather simply assume that such a number system exists, having properties that serve to determine it uniquely.

The first and most important thing about $\mathbb{C}$ is that it is a field and is algebraically an *extension* of $\mathbb{R}$, so that $\mathbb{C}$ contains $\mathbb{R}$ as a *subfield* (meaning that the ordinary sums and products of real numbers are the same as their sums and products when they are thought of as complex numbers). It follows at once that the elements 0 and 1, neutral with respect to addition and multiplication, respectively, in $\mathbb{C}$, are just the real numbers 0 and 1. Hence the difference $s - t$ and, if $t \neq 0$, the quotient s/t, of any two real numbers s and t are the same as their difference and quotient when they are regarded as complex numbers.

The second, and most characteristic, feature of $\mathbb{C}$ is that the equation

$$x^2 + 1 = 0 \tag{11}$$

has a solution i in $\mathbb{C}$ (so that i is a "square root of minus one"). Clearly no real number can be a root of (11) (Prob. J) so $i \in \mathbb{C}\backslash\mathbb{R}$. Complex numbers of the form ib, $b \in \mathbb{R}$, are said to be *pure imaginary*, and *every* complex number is of the form $\zeta = a + ib$, where a is real and ib is pure imaginary.

It is easily seen that the only complex number that is both real and pure imaginary is 0 $(= 0 + i0)$, and hence that

$$\zeta = a + ib, \qquad a, b \in \mathbb{R},$$

called the *standard form* of ζ, is uniquely determined by ζ. In this standard form the real number a is the *real part* of ζ (notation: Re ζ), while b is the *imaginary part* of ζ (notation: Im ζ) Thus for any complex number ζ both Re ζ and Im ζ are real numbers.

We conclude this brief discussion of the field $\mathbb{C}$ by recalling how the operations of arithmetic are conducted on complex numbers expressed in standard form. To this end let $a + ib$ and $s + it$ be complex numbers in standard form. Then

$$(a + ib) + (s + it) = (a + s) + i(b + t),$$

and

$$-(a+ib) = (-a) + i(-b).$$

Moreover

$$(a+ib)(s+it) = (as - bt) + i(at + bs).$$

Finally, if $a + ib \neq 0$, then

$$\frac{a}{a^2+b^2} - i\frac{b}{a^2+b^2}$$

is readily seen to be the reciprocal of $a + ib$.

Note. The following problems that bear on $\mathbb{R}$ are not intended to acquaint the reader with new properties of the real number system. Indeed, he knows most if not all of these properties already. The real goal here is rather to convince the reader that our description of the real number system is, in fact, full and complete enough to permit the derivation of all of the properties of $\mathbb{R}$. Accordingly, in solving the following problems bearing on $\mathbb{R}$, it is a cardinal principle of the enterprise that all arguments be based—either directly or indirectly—on those assertions concerning $\mathbb{R}$ explicitly set forth in the above text.

Problems

A. Show that if s, t and t' are real numbers, and if $s + t = s + t'$, then $t = t'$, and use this basic uniqueness result to establish the following facts.

(i) The solution of the equation $s + x = t$ is uniquely determined by s and t.

(ii) The real number zero is unique; indeed, if s and t are any two real numbers such that $s + t = s$, then $t = 0$.

(iii) The negative $-t$ of a real number t is uniquely determined by t, and $-(-t) = t$.

B. Show that if t and t' are real numbers and s is a nonzero real number, and if $st = st'$, then $t = t'$, and use this result to establish the following facts.

(i) If $s \neq 0$, then the solution of the equation $sx = t$ is uniquely determined by s and t.

(ii) The real number one is unique; indeed, if s and t are any two real numbers such that $st = s$, then either $t = 1$ or $s = 0$.

(iii) The reciprocal $1/s$ of a nonzero real number s is uniquely determined by s, and the reciprocal of $1/s$ is s.

C. Verify that $0t = 0$ for every real number t. (Hint: $0 + 0 = 0$.) Show, conversely, that if s and t are real numbers such that $st = 0$, then either $s = 0$ or $t = 0$. Show too that for any real numbers s and t, $(-s)t = s(-t) = -(st)$, while $(-s)(-t) = st$.

D. For any two real numbers r and s, $-(r+s) = (-r) + (-s)$ and $-(r-s) = s - r$, while $r - s = (-s) - (-r)$. For any four real numbers r, s, t and u, $(r-s) + (t-u) = (r+t) - (s+u)$. Verify the analogous facts as regards multiplication.

E. According to the text, the sum $\sum F$ of a nonempty finite set F of real numbers is uniquely and unambiguously defined (by virtue of an unstated proof by mathematical induction). Verify that, if $F = F_1 \cup F_2$ is any partition of F into two nonempty subsets, then $\sum F = \sum F_1 + \sum F_2$. Devise a definition of $\sum \varnothing$ that permits the deletion of the word "nonempty" from this formulation. What, if any, are the analogous facts as regards multiplication?

F. If a and b are real numbers such that $a \leq b$, and if s is any real number, then $a+s \leq b+s$. Similarly, if $a \leq b$ and p is any nonnegative real number, then $ap \leq bp$.

G. Show that if a and b are real numbers, then $a < b$ if and only if $b - a$ is positive, and conclude that $a < b$ if and only if $-b < -a$. In particular, a real number t is positive [negative] if and only if $-t$ is negative [positive].

H. If a and b are real numbers such that $a < b$, and if s is a negative real number, then $as > bs$.

I. The sum and product of positive real numbers are also positive. The product of two negative real numbers is positive. The product of two real numbers is negative if and only if one of the factors is positive and the other is negative.

J. The square $t^2 = tt$ of every real number is nonnegative, while the square of every nonzero real number is positive. In particular, $1 > 0$, and $s < s + 1$ for every real number s.

K. If s is a positive [negative] real number, then $1/s$ is also positive [negative]. If s and t are positive real numbers such that $s < t$, then $1/t < 1/s$.

L. For any real number s the absolute value $|s|$ is nonnegative and equal to $|-s|$, while $|s| = 0$ if and only if $s = 0$. For any real number s we have $s \leq |s|$ and $-s \leq |s|$. Moreover, $|s| \leq t$ if and only if $s \leq t$ and $-s \leq t$.

(i) Verify that for any two real numbers s and t we have

$$|st| = |s||t|$$

and

$$|s+t| \leq |s| + |t|$$

(the *triangle inequality* for real numbers). Use the triangle inequality to show that

$$\Big||s| - |t|\Big| \leq |s - t|$$

for any two real numbers s and t. (Hint: To establish the triangle inequality, consider cases.)

(ii) In the same context verify that, for any two real numbers s and t,

$$\begin{aligned} s + t &= (s \vee t) + (s \wedge t), \\ |s - t| &= (s \vee t) - (s \wedge t), \end{aligned}$$

and

$$\begin{aligned} s \vee t &= [(s + t) + |s - t|]/2, \\ s \wedge t &= [(s + t) - |s - t|]/2. \end{aligned}$$

Conclude that

$$\begin{aligned} s^+ &= (|s| + s)/2, \\ s^- &= (|s| - s)/2. \end{aligned}$$

M. Any mapping of a set X into $\mathbb{R}$ is called a *real-valued function* on X. For any set X the set $\mathcal{F}_{\mathbb{R}}(X)$ of all real-valued functions defined on X is a lattice in which, for each x in X, $(f \vee g)(x)$ is simply the larger of $f(x)$ and $g(x)$, while $(f \wedge g)(x)$ is the smaller of $f(x)$ and $g(x)$ (see Problem 1Q). If 0 denotes the function identically equal to zero on X, then for any f in $\mathcal{F}_{\mathbb{R}}(X)$ the function $f^+ = f \vee 0$ is called the *positive part* of f, while the function $f^- = -(f \wedge 0)$ is the *negative part* of f. Thus for any real-valued function f on $X, f = f^+ - f^-$, while $|f| = f^+ + f^-$ has the value $|f|(x) = |f(x)|$ everywhere on X.

(i) A sublattice of the lattice $\mathcal{F}_{\mathbb{R}}(X)$ is called a *function lattice* on X. Thus if $\mathcal{L}$ is a function lattice on X that contains the function 0, and if $f \in \mathcal{L}$, then $\mathcal{L}$ also contains f^+ and $-f^-$.

(ii) If $\mathcal{L}$ is a nonempty function lattice on X and if $\mathcal{L}$ contains $f - g$ along with any two functions f and g that it contains, then for any function f in $\mathcal{L}$ the functions f^+, f^- and $|f|$ are also in $\mathcal{L}$.

N. If s is a real number and A a set of real numbers, it is customary to write $s + A$ for the set of all real numbers of the form $s + t, t \in A$. Verify that, in this notation, the set $\{1\} \cup (1 + \mathbb{N})$ is inductive.

(i) Conclude that there is no positive integer in the open interval $(1, 2)$, where $2 = 1 + 1$.

(ii) Conclude also that if n is a positive integer such that $n > 1$, then $n - 1$ is also a positive integer.

O. (i) Show that for each positive integer n there is no positive integer in the open interval $(n, n+1)$. (Hint: Use mathematical induction.)

(ii) Show that for each positive integer n the set of positive integers greater than n is $n + \mathbb{N}$. (Thus if, as above, $A_n = \{m \in \mathbb{N} : m \leq n\}$, then $\mathbb{N} = A_n \cup (n + \mathbb{N})$ for each positive integer n.) Conclude that if m and n are any two positive integers such that $m < n$, then $n - m$ is a positive integer.

If we define the *successor* n^+ of each positive integer n to be $n+1$, then according to Problem N the *successor function*, mapping each positive integer n onto n^+, is a one-to-one mapping of $\mathbb{N}$ onto $\mathbb{N}\backslash\{1\}$. This fact and the principle of mathematical induction (appropriately reformulated in terms of successors) together constitute the *Peano postulates*. The best known and most entertaining development of $\mathbb{R}$ from the Peano postulates is to be found in [16].

P. For each real number a let $\mathbb{Q}_a$ be the set of all those rational numbers q such that $q < a$. Show that $a = \sup \mathbb{Q}_a$ for every real number a.

Q. Verify that the function $f(t) = t^2$ is a strictly monotone increasing (that is, monotone increasing and one-to-one; cf. Chapter 1) mapping of the ray $[0, +\infty)$ into itself. Show too that f maps the open unit interval $I = (0,1)$ into itself, and that, in fact, $0 < f(t) < t$ for t in I.

(i) Show next that if x and y belong to I, and if $x^2 > y$, then there exist real numbers x' such that $0 < x' < x$ and such that $x'^{\,2}$ also exceeds y. (Hint: If η is a positive real number such that $x - \eta$ is also positive, then $(x-\eta)^2 > x^2 - 2\eta$; choose η such that $2\eta < x^2 - y$.)

(ii) Show also that if x and y belong to I, and if $x^2 < y$, then there exist real numbers x'' such that $x < x'' < 1$, and such that $x''^{\,2} < y$. (Hint: If η is a positive real number such that $x + \eta < 1$, then $(x+\eta)^2 < x^2 + 3\eta$ $(3 = 2+1)$; choose η such that $3\eta < y - x^2$.)

(iii) Use (i) and (ii) to show that if y belongs to I, then there exists a (unique) x in I such that $y = x^2$. (Hint: Let $A = \{x \in I\colon x^2 < y\}$ and prove that $y = (\sup A)^2$.)

(iv) Conclude that a real number y is positive if and only if $y = x^2$ for some positive real number x (cf. Problem J). The number x is the positive *square root* of y (notation: $\sqrt{y}$).

(v) Generalize the foregoing argument to show that, for each positive integer n, every nonnegative real number t has a unique nonnegative nth *root* $t^{\frac{1}{n}}$, i.e., a solution of the equation $x^n = t$. (Hint: According to the binomial theorem, if x and $x + \eta$ are both in I, then

$$|(x+\eta)^n - x^n| < 2^n|\eta|.)$$

R. Suppose given a mapping ϕ of $\mathbb{R}$ into itself that *preserves* both *sums* and *products*, i.e., has the property that

$$\varphi(s+t) = \varphi(s) + \varphi(t)$$

and

$$\varphi(st) = \varphi(s)\varphi(t)$$

for every pair of real numbers s and t, and suppose also that φ is *not identically zero*.

(i) Show that $\varphi(0) = 0$ and also that $\varphi(1) = 1$. Conclude that φ is one-to-one.

(ii) Verify that $\varphi(n) = n$ for every positive integer n, and conclude that, in fact, $\varphi(q) = q$ for every rational number q.

(iii) Show too that if p is any positive real number, then $\varphi(p)$ is also positive, and conclude that φ is strictly monotone increasing.

(iv) Conclude finally that φ is necessarily the identity mapping on $\mathbb{R}$. (Hint: Recall Problem P.)

S. A strictly increasing finite sequence $\{x_0 < x_1 < \ldots < x_N\}$ of real numbers is called a *partition* of the interval $[x_0, x_N]$. (This notion of partition is to be distinguished from the one introduced in Chapter 1.) If $\mathcal{P} = \{a = x_0 < \ldots < x_N = b\}$ is such a partition of $[a, b]$, then the numbers $x_i, i = 0, \ldots, N$, are the *points* of $\mathcal{P}$, and the closed interval $[x_{i-1}, x_i], i = 1, \ldots, N$, is the ith *subinterval* of $\mathcal{P}$. If $\mathcal{P}' = \{a = y_0 < \ldots < y_M = b\}$ is another partition of $[a, b]$, and if every point of $\mathcal{P}$ is also a point of $\mathcal{P}'$, then $\mathcal{P}'$ *refines* $\mathcal{P}$, or is a *refinement* of $\mathcal{P}$ (notation: $\mathcal{P} \leq \mathcal{P}'$). Verify that the collection of all partitions of $[a, b]$ is a lattice with respect to the ordering $\leq$.

(i) For any partition $\mathcal{P} = \{a = x_0 < \ldots < x_N = b\}$ of $[a, b]$ the maximum length of a subinterval of $\mathcal{P}$, i.e., the largest of the numbers $x_i - x_{i-1}, i = 1, \ldots, N$, is the *mesh* $\mu_\mathcal{P}$ of $\mathcal{P}$. Verify that $\mathcal{P} \to \mu_\mathcal{P}$ is a monotone decreasing function on the lattice of partitions of $[a, b]$.

(ii) Let f be a real-valued function defined and bounded on an interval $[a, b]$. If $\mathcal{P} = \{x_0 < \ldots < x_N\}$ is a partition of $[a, b]$, we set

$$M_i = \sup\{f(x) : x_{i-1} \leq x \leq x_i\}, \quad m_i = \inf\{f(x) : x_{i-1} \leq x \leq x_i\}$$

for the supremum and infimum, respectively, of f over the ith subinterval of $\mathcal{P}, i = 1, \ldots, N$, and form the sums

$$D_\mathcal{P}(f) = \sum_{i=1}^{N} M_i(x_i - x_{i-1}), \quad d_\mathcal{P}(f) = \sum_{i=1}^{N} m_i(x_i - x_{i-1}),$$

known as the *upper* and *lower Darboux sums*, respectively, of the function f based on the partition $\mathcal{P}$. In this way we associate with f the two nets $\{D_{\mathcal{P}}(f)\}$ and $\{d_{\mathcal{P}}(f)\}$ indexed by the lattice of partitions of $[a,b]$. Concerning these nets it is obvious that $d_{\mathcal{P}}(f) \leq D_{\mathcal{P}}(f) \leq D_{\mathcal{P}}(|f|)$ for every $\mathcal{P}$ and every f. Show that the net $\{D_{\mathcal{P}}(f)\}$ is monotone decreasing and the net $\{d_{\mathcal{P}}(f)\}$ is monotone increasing for any one fixed function f.

T. As has been noted, the fact that the addition and multiplication of complex numbers are associative and commutative ensures that if $\{\zeta_1, \ldots, \zeta_N\}$ is a finite set of complex numbers, then the sum $\sum_{i=1}^N \zeta_i$ and product $\prod_{i=1}^N \zeta_i$ are well-defined (that is, independent of order or grouping). In particular, if $\{\zeta_\gamma\}_{\gamma\in\Gamma}$ is an indexed family of complex numbers and D is a finite subset of Γ, then we may define, without ambiguity, the *finite sum*

$$s_D = \sum_{\gamma\in D} \zeta_\gamma.$$

Since for an arbitrary index set Γ the collection $\mathcal{D}$ of all finite subsets of Γ is directed (upward) under inclusion, we obtain in this way the *net of finite sums* $\{s_D\}_{D\in\mathcal{D}}$ of the given indexed family $\{\zeta_\gamma\}_{\gamma\in\Gamma}$.

(i) If $\{\zeta'_\gamma\}_{\gamma\in\Gamma}$ and $\{\zeta''_\gamma\}_{\gamma\in\Gamma}$ are similarly indexed families of complex numbers with corresponding nets of finite sums $\{s'_D\}$ and $\{s''_D\}$, respectively, and if for each finite set D of indices we write

$$s_D = \sum_{\gamma\in D} \left(\alpha\zeta'_\gamma + \beta\zeta''_\gamma\right)$$

for the corresponding finite sum of the family $\{\alpha\zeta'_\gamma + \beta\zeta''_\gamma\}$, where α and β denote complex numbers, then

$$s_D = \alpha s'_D + \beta s''_D.$$

(ii) For any indexed family $\{\zeta_\gamma\}_{\gamma\in\Gamma}$ of complex numbers, if we write $\{s_D\}$ and $\{\widetilde{s}_D\}$ for the nets of finite sums of the families $\{\zeta_\gamma\}$ and $\{|\zeta_\gamma|\}$, respectively, then $|s_D| \leq \widetilde{s}_D$ for every D. Likewise, if D_1 and D_2 are two finite sets of indices, then

$$|s_{D_1} - s_{D_2}| \leq \widetilde{s}_{D_1 \triangledown D_2}.$$

(iii) If $\{a_1, \ldots, a_N\}$ is a finite set of extended real numbers, the product $\prod_{i=1}^N a_i$ is a well-defined extended real number independent of order and grouping. Likewise the sum $\sum_{i=1}^N a_i$ is well-defined provided *one at most* of the numbers $\pm\infty$ appears in the list $a_1, \ldots, a_N$. (If both $+\infty$ and $-\infty$ are among the numbers $a_1, \ldots, a_N$, then the sum $a_1 + \ldots + a_N$ is undefined.) Hence if $\{a_\gamma\}_{\gamma\in\Gamma}$ is an indexed family of extended real numbers such that $-\infty < a_\gamma \leq +\infty$ for each index γ (or such that $-\infty \leq a_\gamma < +\infty$ for each γ), then the finite sum

$$s_D = \sum_{\gamma \in D} a_\gamma$$

is well-defined for each finite set D of indices, and we obtain, once again, the net $\{s_D\}_{D \in \mathcal{D}}$ of finite sums of $\{a_\gamma\}$. What may be said of the results of (i) and (ii) in the context of extended real numbers? (For the purposes of this exercise let us agree to write $|\pm\infty| = +\infty$.)

(iv) If $\{a_\gamma\}_{\gamma \in \Gamma}$ is an indexed family of complex numbers (or a family of extended real numbers among which at least one of the numbers $\pm\infty$ does not appear), the corresponding net $\{s_D\}$ of finite sums is monotone increasing if and only if the numbers a_γ are all (real and) nonnegative.

U. For any complex number $\zeta = s + it$ in standard form the complex number $s - it$ is called the *complex conjugate* of ζ (notation: $\overline{\zeta}$). Show that *complex conjugation*, i.e., the mapping carrying ζ to $\overline{\zeta}$, preserves both sums and products, and is its own inverse. Show too that ζ is real if and only if $\zeta = \overline{\zeta}$, while ζ is pure imaginary if and only if $\zeta = -\overline{\zeta}$.

V. Let ψ be a mapping of the complex number system $\mathbb{C}$ into itself that preserves both sums and products, and that also preserves the real numbers, so that $\psi(\mathbb{R}) \subset \mathbb{R}$. Prove that ψ is either identically zero, or the identity map on $\mathbb{C}$, or complex conjugation.

W. (i) For any complex number ζ the sum $\zeta + \overline{\zeta}$ and the product $\zeta\overline{\zeta}$ are both real. Indeed,

$$\text{Re } \zeta = \frac{\zeta + \overline{\zeta}}{2}$$

while

$$\text{Im } \zeta = \frac{\zeta - \overline{\zeta}}{2i}.$$

Moreover, if $\zeta = s + it$ in standard form, then $\zeta\overline{\zeta}$ is the nonnegative number $s^2 + t^2$.

(ii) Any mapping of a set X into $\mathbb{C}$ is called a *complex-valued function* on X. If f is a complex-valued function on X, we define $\overline{f}$ (the *complex conjugate* of f) by setting $\overline{f}(x) = \overline{f(x)}, x \in X$. Likewise we define Re$f$ and Imf (the *real* and *imaginary parts* of f) pointwise by setting

$$(\text{Re} f)(x) = \text{Re} f(x), (\text{Im} f)(x) = \text{Im} f(x), \; x \in X.$$

Verify that for any complex-valued function f on X, we have

$$f = \text{Re} f + i \text{ Im} f, \quad \text{Re} f = \frac{1}{2}(f + \overline{f}), \quad \text{Im} f = \frac{1}{2i}(f - \overline{f}).$$

X. For each complex number ζ the nonnegative real number $\sqrt{\zeta\bar{\zeta}}$ (see Problem Q) is called the *absolute value* or *modulus* of ζ (notation: $|\zeta|$; clearly this definition agrees with the one given earlier for the absolute value of a real number).

(i) Verify that if $\zeta \neq 0$, then $1/\zeta = \bar{\zeta}/|\zeta|^2$.

(ii) Show also that if α and β are any two complex numbers, then

$$|\alpha\beta| = |\alpha||\beta| \quad \text{and} \quad |\alpha + \beta| \leq |\alpha| + |\beta|.$$

(Hint: To verify the inequality, still called the *triangle inequality*, show first that

$$|as + bt| \leq \sqrt{a^2 + b^2}\sqrt{s^2 + t^2},$$

where $\alpha = a + ib$ and $\beta = s + it$ in standard form.)

Y. For any complex number $\zeta \neq 0$ the complex number

$$\frac{\zeta}{|\zeta|} = a + ib$$

has real and imaginary parts a and b such that $a^2 + b^2 = 1$. Hence there exists a real number θ (unique up to integral multiples of 2π) such that

$$\frac{\zeta}{|\zeta|} = \cos\theta + i\sin\theta,$$

and therefore such that

$$\zeta = \rho(\cos\theta + i\sin\theta),$$

where $\rho = |\zeta|$. (This representation of a nonzero complex number, called its *polar representation*, is not unique, in that the *polar angle* θ is determined only up to integral multiples of 2π.) Use trigonometric identities to show that if $\zeta = \rho(\cos\theta + i\sin\theta)$, then

$$\zeta^n = \rho^n(\cos n\theta + i\sin n\theta)$$

for each positive integer n. Use this observation to show (*De Moivre's theorem*) that for any complex number $\zeta \neq 0$ and any positive integer n, the equation

$$x^n = \zeta$$

has exactly n distinct solutions in $\mathbb{C}$, viz.,

$$x = \rho^{1/n}\left(\cos\frac{\theta + 2k\pi}{n} + i\sin\frac{\theta + 2k\pi}{n}\right), \qquad k = 0, \ldots, n-1.$$

3 Linear algebra

We shall assume that the reader is familiar with the rudiments of linear algebra. In particular, he should be acquainted with the notion of a *linear space*, or *vector space*, and the elementary concepts associated with linear spaces. In this chapter we review these ideas, largely to fix terminology and notation. Readers wishing to improve their acquaintance with any part of linear algebra, or to pursue in greater depth any of the topics discussed below, might consult [10]. Another excellent source is [13].

Definition. Let F be a field (as defined in Chapter 2). A *vector space* or *linear space over* F is a set $\mathcal{E}$ of elements (called *vectors*) satisfying the following postulates.

(A) The set $\mathcal{E}$ is equipped with an associative and commutative binary operation, called *addition* and denoted by $+$, in such a way that the following two conditions are satisfied: (i) there is a neutral element 0 (called the *origin* of $\mathcal{E}$) with respect to addition; (ii) for each vector x in $\mathcal{E}$ there exists a vector $-x$ in $\mathcal{E}$ (called the *negative* of x) such that $x + (-x) = 0$.

(M) There is also given a mapping of $F \times \mathcal{E}$ into $\mathcal{E}$ assigning to each element α of F and vector x in $\mathcal{E}$ a vector αx, called the *product* of α and x, in such a way that the following four conditions are satisfied for all elements α and β of F and all vectors x and y: (i) $\alpha(\beta x) = (\alpha\beta)x$; (ii) $\alpha(x + y) = \alpha x + \alpha y$; (iii) $(\alpha + \beta)x = \alpha x + \beta x$; (iv) $1x = x$.

Note. The elements of the field F are customarily called *scalars* to distinguish them from the *vectors* that are the elements of $\mathcal{E}$, and F is accordingly known as the scalar field of $\mathcal{E}$. In all that follows, the *scalar field of each and every vector space to be considered will be either the field* $\mathbb{R}$ *of real numbers or the field* $\mathbb{C}$ *of complex numbers*. (A vector space over $\mathbb{R}$ is a *real* vector space; a vector space over $\mathbb{C}$ is a *complex* vector space.) Furthermore, the following convention will be in force throughout the book: *In any statement, proposition, or definition concerning linear spaces, if no specific distinction is made, then it is understood that the scalar field may be either* $\mathbb{R}$ *or* $\mathbb{C}$. In the *product* αx of a scalar α times a vector x the left factor is always the scalar, the right factor always the vector. (For that reason what we have here called a *vector space* is sometimes known as a *left vector space*.) It is perhaps worth noting that the product of a scalar times a vector in a vector space is not (ordinarily) a binary operation in the sense of Example 1F.

The following proposition is nothing more than a summary of the most immediate consequences of the above definition, and is included here solely for the sake of completeness.

Proposition 3.1. *Let $\mathcal{E}$ be a linear space, and let x_0 be a vector in $\mathcal{E}$. If for any one vector y in $\mathcal{E}$ it is the case that $x_0 + y = y$ (or $y + x_0 = y$), then $x_0 = 0$. Consequently, $0x = 0$ and $(-1)x = -x$ for every vector x in $\mathcal{E}$. Likewise, $\alpha 0 = 0$ for every scalar α, whence it follows that a product αx is equal to 0 if and only if either $\alpha = 0$ or $x = 0$.*

Example A. The field $\mathbb{R}$ of real numbers is a real vector space if we agree to define vector addition in $\mathbb{R}$ to be the ordinary addition of real numbers, and the product of a real number by a real number the ordinary product of two real numbers. Similarly, the field $\mathbb{C}$ of complex numbers is a complex vector space.

Example B. If $\mathcal{E}$ is a complex linear space, then $\mathcal{E}$ is also a real linear space—we have but to retain the given vector addition in $\mathcal{E}$ and simply refuse to multiply by any but real scalars. Thus, in particular, $\mathbb{C}$ itself is a real linear space in which vector addition is the ordinary addition of complex numbers, and the product $t\alpha, t \in \mathbb{R}, \alpha \in \mathbb{C}$, is the ordinary product of complex numbers.

Example C. It is an immediate consequence of the above definition that if x, y and z are vectors in a linear space $\mathcal{E}$, then $x + y + z$ is a well-defined vector in $\mathcal{E}$, independent of either the ordering or grouping of the three vectors x, y, z. Similarly, if $\{x_1, \ldots, x_n\}$ is an arbitrary finite sequence of vectors in a vector space $\mathcal{E}$, then the sum $x_1 + \cdots + x_n$ exists as a vector in $\mathcal{E}$ independently of all possible permutations or groupings of the vectors in the sequence. Thus, if $\{x_\delta\}_{\delta\in\Delta}$ is a finite indexed family of vectors in a vector space $\mathcal{E}$, we may and do define

$$\sum_{\delta\in\Delta} x_\delta = x_{\delta_1} + \cdots + x_{\delta_n},$$

where $\{\delta_1, \ldots, \delta_n\}$ represents any one enumeration of the index set Δ.

More generally, if Γ is an arbitrary index set, and if $\{x_\gamma\}_{\gamma\in\Gamma}$ is an indexed family of vectors in a linear space $\mathcal{E}$ with the property that there exists a *finite* subset Δ_0 of Γ such that $x_\gamma = 0$ for all γ in $\Gamma\backslash\Delta_0$, then we define, unambiguously,

$$\sum_{\gamma\in\Gamma} x_\gamma = \sum_{\gamma\in\Delta} x_\gamma$$

where Δ denotes an arbitrary finite subset of Γ such that $\Delta \supset \Delta_0$. (It is clear that this definition is independent of the choice of Δ since $x_\gamma = 0$ for all γ in $\Delta\backslash\Delta_0$ by construction.)

If $\mathcal{E}$ is a linear space and if M_1 and M_2 are arbitrary subsets of $\mathcal{E}$, we shall write $M_1 + M_2$ for the set of sums

$$\{x_1 + x_2 : x_i \in M_i, \quad i = 1, 2\}.$$

Similarly, if A denotes a set of scalars and M a set of vectors in $\mathcal{E}$, we shall write AM for the set $\{\alpha x : \alpha \in A, x \in M\}$.

In any linear space $\mathcal{E}$ a vector x is a *linear combination* of vectors $y_1, \ldots, y_n$ in $\mathcal{E}$ if there exist scalars $\alpha_1, \ldots, \alpha_n$ such that $x = \alpha_1 y_1 + \cdots + \alpha_n y_n$. A nonempty subset $\mathcal{M}$ of $\mathcal{E}$ is a *linear manifold* in $\mathcal{E}$ (or a *linear submanifold* of $\mathcal{E}$) if, for every pair of vectors x, y in $\mathcal{M}$ and every pair of scalars α, β, the linear combination $\alpha x + \beta y$ belongs to $\mathcal{M}$. If $\mathcal{M}$ is a linear submanifold of $\mathcal{E}$, then it is easily seen that every linear combination of vectors in $\mathcal{M}$ belongs to $\mathcal{M}$. Among the linear submanifolds of $\mathcal{E}$ are the space $\mathcal{E}$ itself and the *trivial* submanifold (0) consisting of the single vector 0.

If $\{\mathcal{M}_\gamma\}_{\gamma\in\Gamma}$ is an arbitrary indexed family of linear submanifolds of a linear space $\mathcal{E}$, we write $\sum_{\gamma\in\Gamma} \mathcal{M}_\gamma$ for the *vector sum* of the family, which consists, by definition, of all sums of the form $x = \sum_{\gamma\in\Gamma} x_\gamma$ where $x_\gamma \in \mathcal{M}_\gamma$ for each index γ and $x_\gamma = 0$ except for some finite set of indices (which may vary with x; recall Example C). If, in particular, $\Gamma = \{1, \ldots, n\}$, we write $\mathcal{M}_1 + \cdots + \mathcal{M}_n$ for the vector sum of the family $\{\mathcal{M}_i\}_{i=1}^n$. (For $n = 2$ this notion of vector sum agrees with the more general notion of the vector sum of two *sets* of vectors introduced above.) The vector sum of an arbitrary indexed family of linear submanifolds of a linear space $\mathcal{E}$ is itself a linear submanifold of $\mathcal{E}$ containing each of the given submanifolds $\mathcal{M}_\gamma$. In particular, if $\mathcal{M}$ and $\mathcal{N}$ are linear submanifolds of $\mathcal{E}$, then $\mathcal{M} + \mathcal{N}$ is a linear submanifold of $\mathcal{E}$ containing both $\mathcal{M}$ and $\mathcal{N}$.

For any set M of vectors in a linear space $\mathcal{E}$ there exists a smallest linear submanifold $\mathcal{M}$ of $\mathcal{E}$ that contains M. If $M = \varnothing$, then $\mathcal{M} = (0)$; otherwise $\mathcal{M}$ consists of all linear combinations of elements of M. We say that $\mathcal{M}$ is *generated* (*algebraically*) by M, or that M is an *algebraic system of generators* for $\mathcal{M}$, and we write $\mathcal{M} = \langle M \rangle$. If $\mathcal{M}$ and $\mathcal{N}$ are given linear manifolds in $\mathcal{E}$, then the linear manifold $\langle \mathcal{M} \cup \mathcal{N} \rangle$ generated algebraically by $\mathcal{M} \cup \mathcal{N}$ is $\mathcal{M} + \mathcal{N}$. More generally, if $\{\mathcal{M}_\gamma\}_{\gamma \in \Gamma}$ is an indexed family of linear manifolds in $\mathcal{E}$, then the vector sum $\sum_{\gamma \in \Gamma} \mathcal{M}_\gamma$ is the linear manifold $\langle \bigcup_\gamma \mathcal{M}_\gamma \rangle$ generated algebraically by the union $\bigcup_\gamma \mathcal{M}_\gamma$.

A nonempty finite set $J = \{x_1, \ldots, x_n\}$ in a vector space $\mathcal{E}$ is *linearly independent* if the only way in which 0 can be expressed as a linear combination $0 = \alpha_1 x_1 + \cdots + \alpha_n x_n$ is with $\alpha_1 = \cdots = \alpha_n = 0$. An arbitrary subset J of $\mathcal{E}$ is *linearly independent* if every nonempty finite subset of J is linearly independent. A linearly independent set of vectors in $\mathcal{E}$ that is at the same time an algebraic system of generators for $\mathcal{E}$ is a *Hamel basis* for $\mathcal{E}$. Every vector space has a Hamel basis (Prob. B). If $\{x_\gamma\}_{\gamma \in \Gamma}$ is an indexed Hamel basis for a linear space $\mathcal{E}$, then for each vector y in $\mathcal{E}$ there exists a uniquely determined (similarly indexed) family of scalars $\{\lambda_\gamma\}_{\gamma \in \Gamma}$ such that $\lambda_\gamma = 0$ for all but a finite number of indices γ (the set of which depends, in general, on y) and such that $y = \sum_{\gamma \in \Gamma} \lambda_\gamma x_\gamma$. The scalars λ_γ are called the *coordinates* of y with respect to the indexed basis $\{x_\gamma\}_{\gamma \in \Gamma}$.

A linear space possessing a *finite* Hamel basis is called *finite dimensional*; a linear space that is not finite dimensional is *infinite dimensional*. If $X = \{x_1, \ldots, x_n\}$ is a Hamel basis for a finite dimensional linear space $\mathcal{E}$, then every Hamel basis for $\mathcal{E}$ contains exactly n vectors (Prob. C). A Hamel basis for a finite dimensional linear space $\mathcal{E}$ is called simply a *basis* for $\mathcal{E}$, and the number of vectors in any one basis (and therefore in all bases) for $\mathcal{E}$ is the *dimension* of $\mathcal{E}$, denoted by $\dim \mathcal{E}$. The vector space (0) consisting of the vector 0 alone is finite dimensional and has dimension 0.

Example D. If X is a Hamel basis for a complex vector space $\mathcal{E}$, and if $\mathcal{E}_0$ denotes the space $\mathcal{E}$ viewed as a real vector space (Ex. B), then a Hamel basis for $\mathcal{E}_0$ is provided by the set $X \cup iX$. Thus if $\mathcal{E}$ is finite dimensional, then $\dim \mathcal{E}_0 = 2 \dim \mathcal{E}$. In particular, $\mathbb{C}$ is a two dimensional vector space over the real field $\mathbb{R}$, a basis for which is given by the pair $\{1, i\}$.

Example E. The system $\mathbb{R}[x]$ of all real polynomials in the indeterminate x is a real vector space if we agree to make vector addition in $\mathbb{R}[x]$ the ordinary addition of polynomials, and define the product of a real number b and a real polynomial $p(x) = a_0 + a_1 x + \cdots + a_N x^N$ to be

$$bp(x) = ba_0 + ba_1 x + \cdots + ba_N x^N \tag{1}$$

(which is the same thing as the product of the constant polynomial b with $p(x)$). Similarly, the system $\mathbb{C}[x]$ of all complex polynomials is a complex vector space. An indexed Hamel basis for both $\mathbb{R}[x]$ and $\mathbb{C}[x]$ is provided by the infinite sequence $\{x^n\}_{n=0}^{\infty}$. Thus $\mathbb{R}[x]$ and $\mathbb{C}[x]$ are infinite dimensional.

Let N be a nonnegative integer and let $\mathbb{R}_N[x]$ denote the subset of $\mathbb{R}[x]$ consisting of the zero polynomial together with the set of all those nonzero elements of $\mathbb{R}[x]$ having degree no greater than N. (Recall that the degree of a polynomial $p(x)$ is m if $p(x) = a_0 + a_1x + \cdots + a_m x^m$ with $a_m \neq 0$. Thus, in particular, the degree of each nonzero scalar is zero. The degree of the polynomial 0 is either $-\infty$ or undefined, according to one's point of view.) Then, as is readily seen, $\mathbb{R}_N[x]$ is a real linear space for which the system of polynomials $\{1, x, \ldots, x^N\}$ is a basis, so that $\dim \mathbb{R}_N[x] = N+1$. Similarly, the zero polynomial together with the set of all complex polynomials having degree no greater than N is a complex vector space $\mathbb{C}_N[x]$ having dimension $N+1$.

If $\mathcal{E}_1, \ldots, \mathcal{E}_n$ are linear spaces, all over the same scalar field, we write $\mathcal{E} = \mathcal{E}_1 + \ldots + \mathcal{E}_n$ for the set of all n-tuples of the form $(x_1, \ldots, x_n)$ where $x_i \in \mathcal{E}_i, i = 1, \ldots, n$. Then $\mathcal{E}$ is a linear space with respect to the *linear operations*

$$(x_1, \ldots, x_n) + (y_1, \ldots, y_n) = (x_1 + y_1, \ldots, x_n + y_n)$$

and

$$\alpha (x_1, \ldots, x_n) = (\alpha x_1, \ldots, \alpha x_n)$$

for all $(x_1, \ldots, x_n)$ and $(y_1, \ldots, y_n)$ in $\mathcal{E}$ and all scalars α. The space $\mathcal{E}$ is the (*linear space*) *direct sum* of the spaces $\{\mathcal{E}_i\}_{i=1}^n$. (Clearly the origin in $\mathcal{E}_1 + \ldots + \mathcal{E}_n$ is the n-tuple $(0, \ldots 0)$, and $-(x_1, \ldots, x_n) = (-x_1 \ldots, -x_n)$.) If $\mathcal{E}_1, \ldots, \mathcal{E}_n$ are all finite dimensional, and if $\mathcal{E} = \mathcal{E}_1 + \ldots + \mathcal{E}_n$, then $\mathcal{E}$ is also finite dimensional, with

$$\dim \mathcal{E} = \dim \mathcal{E}_1 + \ldots + \dim \mathcal{E}_n.$$

In the case $\mathcal{E}_1 = \ldots = \mathcal{E}_n = \mathcal{F}, \mathcal{E}$ is called the direct sum of n *copies* of $\mathcal{F}$.

Example F. The direct sum of n copies of $\mathbb{R}$ (viewed, as in Example A, as a real vector space) consists of the set of all real n-tuples added and multiplied by real scalars *termwise* or *coordinatewise*. We shall denote this space by $\mathbb{R}^n$. Similarly, we denote by $\mathbb{C}^n$ the direct sum of n copies of $\mathbb{C}$ (viewed as a complex vector space). The n-tuples $e_i = (\delta_{i1}, \ldots, \delta_{in}),\ i = 1, \ldots, n$ (where, by definition, δ_{ij} is the *Kronecker delta*

$$\delta_{ij} = \begin{cases} 0, & i \neq j, \\ 1, & i = j, \end{cases}$$

for all integers, i, j), constitute a basis for both $\mathbb{R}^n$ and $\mathbb{C}^n$, sometimes called the *natural basis*. (In dealing with $\mathbb{R}^n$ and $\mathbb{C}^n$, it is this natural basis that is ordinarily assumed to be in use.) Clearly $\mathbb{R}^n$ and $\mathbb{C}^n$ are n-dimensional.

Example G. The collection $\mathcal{F}_{\mathbb{R}}(X)[\mathcal{F}_{\mathbb{C}}(X)]$ of all real-valued functions [complex-valued functions] on an arbitrary set X is a real [complex] linear space with respect to the *pointwise* linear operations defined by

$$(f+g)(x) = f(x) + g(x) \quad \text{and} \quad (\alpha f)(x) = \alpha f(x), \quad x \in X, \alpha \in \mathbb{R}[\mathbb{C}].$$

(The origin in these spaces is the zero function, and $(-f)(x) = -f(x)$. The spaces $\mathbb{R}^n$ and $\mathbb{C}^n$ are clearly special cases of this construction.) Whenever, in the sequel (as in the following example), we refer to a "real [complex] linear space of functions" on a set X, it is always some linear submanifold of $\mathcal{F}_{\mathbb{R}}(X)[\mathcal{F}_{\mathbb{C}}(X)]$ that is meant.

Example H. Suppose given a linear space $\mathcal{E}$. A scalar-valued function f defined on $\mathcal{E}$ is a *linear functional* on $\mathcal{E}$ if $f(\alpha x + \beta y) = \alpha f(x) + \beta f(y)$ for all vectors x, y in $\mathcal{E}$ and all scalars α, β. It is a triviality to verify that a linear combination of linear functionals on $\mathcal{E}$ is again a linear functional on $\mathcal{E}$, and hence that the collection of all linear functionals on $\mathcal{E}$ forms a linear submanifold of the space of all scalar-valued functions on $\mathcal{E}$. The linear space of all linear functionals on $\mathcal{E}$ (which is real or complex according as $\mathcal{E}$ is real or complex) will be called the *full algebraic dual* of $\mathcal{E}$.

If $\mathcal{E}$ is a linear space and $\mathcal{M}$ is a linear submanifold of $\mathcal{E}$, then the relation $\sim$ on $\mathcal{E}$ defined by setting $x \sim y$ if x is *congruent* to y *modulo* $\mathcal{M}$, that is, if $x - y \in \mathcal{M}$, is an equivalence relation on $\mathcal{E}$. The equivalence class $[x] = x + \mathcal{M}$ of a vector x will be called the *coset* of x modulo $\mathcal{M}$. The set of all cosets $[x]$ modulo $\mathcal{M}$ (i.e., the quotient space $E/\sim$) is turned into a new linear space by the definitions $[x] + [y] = [x + y]$ and $\alpha[x] = [\alpha x]$. This space, which is real or complex according as $\mathcal{E}$ itself is real or complex, is denoted by $\mathcal{E}/\mathcal{M}$, and is called the *quotient space* of $\mathcal{E}$ *modulo* $\mathcal{M}$. The mapping π of $\mathcal{E}$ onto $\mathcal{E}/\mathcal{M}$ defined by $\pi(x) = [x]$ is the natural projection of $\mathcal{E}$ onto $\mathcal{E}/\mathcal{M}$. If we take for $\mathcal{M}$ the trivial submanifold (0), then the natural projection π of $\mathcal{E}$ onto $\mathcal{E}/\mathcal{M}$ is a one-to-one mapping which may be used to identify $\mathcal{E}$ with $\mathcal{E}/\mathcal{M}$. Dually, the linear space $\mathcal{E}/\mathcal{E}$ is the trivial space (0).

Example I. Let $p_0(x)$ be a fixed nonconstant real polynomial, and let $(p_0(x))$ denote the collection of all real polynomials of the form $p_0(x)p(x)$, $p(x) \in \mathbb{R}[x]$. Then $(p_0(x))$ is a linear submanifold of $\mathbb{R}[x]$ having the property that every element of $(p_0(x))$ is either 0 or has degree at least N, where N denotes the degree of $p_0(x)$. Consequently, if $[1], [x], \ldots, [x^{N-1}]$ denote the cosets of $1, x, \ldots, x^{N-1}$ in $\mathbb{R}[x]/(p_0(x))$, and if

$$\alpha_0[1] + \cdots + \alpha_{N-1}[x^{N-1}] = [\alpha_0 + \alpha_1 x + \cdots + \alpha_{N-1}x^{N-1}] = 0$$

in $\mathbb{R}[x]/(p_0(x))$, then $\alpha_0 + \cdots + \alpha_{N-1}x^{N-1} \in (p_0(x))$ and therefore $\alpha_0 + \cdots + \alpha_{N-1}x^{N-1} = 0$ in $\mathbb{R}[x]$, which shows that all of the α_i are zero and hence that $[1], \ldots, [x^{N-1}]$ are linearly independent in $\mathbb{R}[x]/(p_0(x))$. On the other hand, if $p(x)$ denotes an arbitrary real polynomial, then, according to the familiar division algorithm, there exist real polynomials $q(x)$ and $r(x)$ such that the degree of $r(x)$ is less than N and such that

$$p(x) = p_0(x)q(x) + r(x),$$

and $[p(x)] = [r(x)]$ therefore belongs to the linear manifold in $\mathbb{R}[x]/(p_0(x))$ generated by the cosets $[1], \ldots, [x^{N-1}]$. Thus $\{[1], \ldots, [x^{N-1}]\}$ is a basis for $\mathbb{R}[x]/(p_0(x))$. In particular, this quotient space is N-dimensional. (If $p_0(x)$ is allowed to be constant, then either $p_0(x) = a \neq 0$, in which case $(p_0(x)) = \mathbb{R}[x]$, or else $p_0(x) = 0$, in which case $(p_0(x)) = (0)$. Thus, in either case, we are reduced to one of the trivial situations considered above.) Needless to say, all of these facts remain valid if $\mathbb{R}[x]$ is systematically replaced by the complex space $\mathbb{C}[x]$.

If $\mathcal{E}$ is a real vector space, then the *complexification* $\mathcal{E}^+$ of $\mathcal{E}$ is the complex vector space consisting of the Cartesian product of $\mathcal{E}$ with itself with addition defined by $(x_1, y_1)+(x_2, y_2) = (x_1+x_2, y_1+y_2)$ and multiplication by a complex scalar $\alpha = s+it$ defined by $\alpha(x, y) = (sx-ty, tx+sy)$. (Thus $\mathbb{C}$ may be viewed as the complexification of $\mathbb{R}$.) If the mapping $x \to (x, 0)$ is used to identify $\mathcal{E}$ with a real linear manifold in $\mathcal{E}^+$ regarded as a real space, then, since $i(x, 0) = (0, x)$, every vector in $\mathcal{E}^+$ has a unique expression in the form $x + iy$, where x and y belong to $\mathcal{E}$. (Recall (Ex. B) that a complex linear space may always be regarded as a real space simply by refusing to multiply by any but real scalars.)

Example J. If X is a Hamel basis for a real linear space $\mathcal{E}$, and if, as above, we identify $\mathcal{E}$ with the real submanifold $\mathcal{E} \times (0)$ of its complexification $\mathcal{E}^+$, then X is also a basis for $\mathcal{E}^+$. Thus if $\mathcal{E}$ is finite dimensional, the dimension of the complex space $\mathcal{E}^+$ is the same as that of the real space $\mathcal{E}$. (On the other hand, if the dimension of $\mathcal{E}$ is n and $\mathcal{E}^+$ is regarded as a real space, then the dimension of $\mathcal{E}^+$ is $2n$; cf. Example D.)

Example K. If $\mathcal{F}$ is a complex linear space of complex-valued functions on a set X and if $\mathcal{F}_{\mathbb{R}}$ denotes the set of all real-valued functions in $\mathcal{F}$, then it is clear that $\mathcal{F}_{\mathbb{R}}$ is a real linear space of functions on X and that the complexification $(\mathcal{F}_{\mathbb{R}})^+$ may be identified with the linear submanifold of $\mathcal{F}$ consisting of functions of the form $f + ig, f, g \in \mathcal{F}_{\mathbb{R}}$. This submanifold, however, does not coincide with $\mathcal{F}$ in general. Indeed, it is readily seen that a necessary and sufficient condition for this to be so is that $\mathcal{F}$ contain the complex conjugate $\overline{f}$ of each function f in $\mathcal{F}$, a condition that is customarily expressed by saying that $\mathcal{F}$ is *self-conjugate*. Thus we may identify a self-conjugate linear space $\mathcal{F}$ of complex-valued functions with the complexification of the real linear space of real-valued functions in $\mathcal{F}$. If $\mathcal{F}$ is a self-conjugate linear space of complex-valued functions on a set X, and if ϕ is a linear functional defined on $\mathcal{F}$, then ϕ is said to be *self-conjugate* if $\phi(\overline{f}) = \overline{\phi(f)}$ for every function f in $\mathcal{F}$. It can be seen that ϕ is self-conjugate if and only if the restriction $\phi \mid \mathcal{F}_{\mathbb{R}}$ is a real linear functional on $\mathcal{F}_{\mathbb{R}}$ or, equivalently, if and only if $\phi(Ref) = Re\phi(f)$ for every f in $\mathcal{F}$ (Prob. I).

Example L. If a and b are real numbers, $a < b$, we shall denote by $\mathcal{C}_{\mathbb{R}}((a,b)) = \mathcal{C}_{\mathbb{R}}^{(0)}((a,b))$ the collection of all continuous real-valued functions on the open interval (a,b). Clearly $\mathcal{C}_{\mathbb{R}}((a,b))$ is a real linear space. Similarly one sees, using the rules of elementary calculus, that the collection $\mathcal{C}_{\mathbb{R}}^{(n)}((a,b))$ of n times continuously differentiable real functions on (a,b), i.e., the collection of those real functions f on (a,b) with the property that the nth derivative $f^{(n)}$ exists and is continuous on (a,b), is a real linear space. Moreover, if $0 \leq m \leq n$, then $\mathcal{C}_{\mathbb{R}}^{(n)}((a,b))$ is a linear submanifold of $\mathcal{C}_{\mathbb{R}}^{(m)}((a,b))$.

In the same spirit, the collection $\mathcal{C}_{\mathbb{C}}^{(n)}((a,b))$ of all n times continuously differentiable complex-valued functions on (a,b) is a complex vector space that we may identify with the complexification of $\mathcal{C}_{\mathbb{R}}^{(n)}((a,b))$. (Here, as always, we identify $\mathcal{C}_{\mathbb{C}}^{(0)}((a,b))$ with the space $\mathcal{C}_{\mathbb{C}}((a,b))$ of all continuous complex-valued functions on (a,b). Recall that if f is a complex-valued function defined on (a,b), then $\frac{df(t)}{dt} = \frac{d}{dt}Ref(t) + i\frac{d}{dt}Imf(t)$.) If $\mathcal{P}_{\mathbb{R}}(a,b)[\mathcal{P}_{\mathbb{C}}(a,b)]$ denotes the space of all real [complex] polynomial functions on (a,b), then $\mathcal{P}_{\mathbb{R}}(a,b)[\mathcal{P}_{\mathbb{C}}(a,b)]$ is a linear submanifold of $\mathcal{C}_{\mathbb{R}}^{(n)}((a,b))$ $[\mathcal{C}_{\mathbb{C}}^{(n)}((a,b))]$ for every n. It follows that the vector spaces $\mathcal{C}_{\mathbb{R}}^{(n)}((a,b))$ and $\mathcal{C}_{\mathbb{C}}^{(n)}((a,b))$ are all infinite dimensional.

Example M. To define analogs of the spaces of Example L for functions on a *closed* interval, special arrangements must be made regarding the endpoints of the interval. We shall say that a complex-valued function f on a closed interval $[a,b]$ $(a < b)$ is *differentiable* on that interval if (i) f is differentiable on the open interval (a,b) and (ii) the one-sided derivatives $f'_+(a)$ and $f'_-(b)$ exist, and we denote by f' the function

$$f'(t) = \begin{cases} f'_+(a), & t = a, \\ f'(t), & a < t < b, \\ f'_-(b), & t = b. \end{cases}$$

We then declare $\mathcal{C}_{\mathbb{C}}([a,b]) = \mathcal{C}_{\mathbb{C}}^{(0)}([a,b])$ to be the complex linear space of all continuous complex-valued functions on $[a,b]$, and define $\mathcal{C}_{\mathbb{C}}^{(n)}([a,b])$ inductively for positive integers n by setting $\mathcal{C}_{\mathbb{C}}^{(n)}([a,b])$ equal to the collection of all those differentiable complex-valued functions f defined on $[a,b]$ with the property that f' belongs to $\mathcal{C}_{\mathbb{C}}^{(n-1)}([a,b])$. Here again it is not hard to see that each $\mathcal{C}_{\mathbb{C}}^{(n)}([a,b])$ is an infinite dimensional complex vector space, that $\mathcal{C}_{\mathbb{C}}^{(n)}([a,b])$ is a linear submanifold of $\mathcal{C}_{\mathbb{C}}^{(m)}([a,b])$ when and only when $m \le n$, and that, if $\mathcal{C}_{\mathbb{R}}^{(n)}([a,b])$ denotes the set of real-valued functions in $\mathcal{C}_{\mathbb{C}}^{(n)}([a,b])$, then $\mathcal{C}_{\mathbb{R}}^{(n)}([a,b])$ is a real vector space whose complexification is $\mathcal{C}_{\mathbb{C}}^{(n)}([a,b])$.

If $\mathcal{E}$ and $\mathcal{F}$ are linear spaces over the same scalar field, and if T is a mapping defined on $\mathcal{E}$ and taking its values in $\mathcal{F}$, then T is a *linear transformation* of $\mathcal{E}$ into $\mathcal{F}$ provided $T(\alpha x + \beta y) = \alpha Tx + \beta Ty$ for all x, y in $\mathcal{E}$ and all scalars α, β. (When $\mathcal{E} = \mathcal{F}$ we refer to T as a linear transformation *on* $\mathcal{E}$. A linear transformation of a linear space $\mathcal{E}$ into its scalar field is a linear functional on $\mathcal{E}$ (Ex. H).) The range of a linear transformation T will be denoted by $\mathcal{R}(T)$. Likewise the *kernel* or *null space* of T, that is, the set of vectors mapped into 0 by T, will be denoted by $\mathcal{K}(T)$.

Proposition 3.2. *Let $\mathcal{E}$ and $\mathcal{F}$ be linear spaces over the same scalar field, and let T be a linear transformation of $\mathcal{E}$ into $\mathcal{F}$. Then $\mathcal{K}(T)$ and $\mathcal{R}(T)$ are submanifolds of $\mathcal{E}$ and $\mathcal{F}$, respectively. Moreover, T is a one-to-one mapping of $\mathcal{E}$ into $\mathcal{F}$ if and only if $\mathcal{K}(T) = (0)$, and, if $\mathcal{K}(T) = (0)$, so that T is one-to-one, then the (set-theoretic) inverse of T is itself a linear transformation (of $\mathcal{R}(T)$ onto $\mathcal{E}$).*

PROOF. If $x, x' \in \mathcal{K}(T)$ and α, β are scalars, then $T(\alpha x + \beta x') = \alpha\cdot 0 + \beta \cdot 0 = 0$, so $\alpha x + \beta x' \in \mathcal{K}(T)$. Similarly, if $y, y' \in \mathcal{R}(T)$, and if $y = Tx, y' = Tx'$, and α, β are scalars, then $\alpha y + \beta y' = T(\alpha x + \beta x') \in \mathcal{R}(T)$. Moreover, it is clear that $\mathcal{K}(T) = (0)$ if T is one-to-one. On the other hand, if $\mathcal{K}(T) = (0)$ and if $Tx = Tx'$ for some vectors x, x' in $\mathcal{E}$, then $T(x - x') = Tx - Tx' = 0$, so $x - x' \in \mathcal{K}(T)$, and therefore $x = x'$. Thus T is one-to-one.

Suppose now that $\mathcal{K}(T) = (0)$, so that T is one-to-one, and let y and y' be elements of $\mathcal{R}(T)$, so that there exist unique vectors x and x' in $\mathcal{E}$ such that $Tx = y$ and $Tx' = y'$. Then for arbitrary scalars α and β we have

$$T(\alpha x + \beta x') = \alpha y + \beta y',$$

so $T^{-1}(\alpha y + \beta y') = \alpha x + \beta x' = \alpha T^{-1} y + \beta T^{-1} y'$. Thus T^{-1} is a linear transformation. □

Example N. For any vector space $\mathcal{E}$ and any fixed scalar α the mapping $x \to \alpha x, x \in \mathcal{E}$, is a linear transformation on $\mathcal{E}$, which we denote by α, or, when necessary in order to avoid confusion, by $\alpha_{\mathcal{E}}$. In particular, the identity mapping 1 and zero mapping 0 are linear transformations.

Example O. Let $\mathcal{E}$ be a linear space and let $\mathcal{M}$ be a linear submanifold of $\mathcal{E}$. Then the natural projection π of $\mathcal{E}$ onto the quotient space $\mathcal{E}/\mathcal{M}$ is a linear transformation. Moreover, if T is any linear transformation of $\mathcal{E}$ into a linear space $\mathcal{F}$, then there exists a linear transformation $\widehat{T}$ of $\mathcal{E}/\mathcal{M}$ into $\mathcal{F}$ such that $T = \widehat{T} \circ \pi$ if and only if $\mathcal{M} \subset \mathcal{K}(T)$. (Briefly: linear transformations T on $\mathcal{E}$ with $T(\mathcal{M}) = (0)$ can be *factored through* $\mathcal{E}/\mathcal{M}$. Conversely, of course, if T can be factored through $\mathcal{E}/\mathcal{M}$, then $T(\mathcal{M}) = (0)$.)

Example P. If $\mathcal{E}$ is a real linear space and T is a linear transformation of $\mathcal{E}$ into a complex linear space $\mathcal{F}$ (regarded as a real space; see Example B), then

$$T^+(x + iy) = Tx + iTy, \quad x, y \in \mathcal{E},$$

defines a linear transformation T^+ of the complexification $\mathcal{E}^+$ into $\mathcal{F}$. The linear transformation T^+ is called the *complexification* of T.

Example Q. Let $\mathcal{E}$ and $\mathcal{F}$ be finite dimensional linear spaces of dimension n and m, and let $X = \{x_1, \ldots, x_n\}$ and $Y = \{y_1, \ldots, y_m\}$ be ordered bases in $\mathcal{E}$ and $\mathcal{F}$, respectively. If T is a linear transformation of $\mathcal{E}$ into $\mathcal{F}$, then the equations

$$Tx_j = \sum_{i=1}^{m} \alpha_{ij} y_i, \quad j = 1, \ldots, n, \tag{2}$$

define an $m \times n$ matrix

$$A = \begin{pmatrix} \alpha_{11} & \ldots & \alpha_{1n} \\ \vdots & & \vdots \\ \alpha_{m1} & \ldots & \alpha_{mn} \end{pmatrix}$$

called the *matrix of T with respect to X and Y*. (When $\mathcal{E} = \mathcal{F}$ and $X = Y$, A is called the *matrix of T with respect to X*.)

A linear transformation of a linear space $\mathcal{E}$ into a linear space $\mathcal{F}$ is determined by its action on a Hamel basis for $\mathcal{E}$, but is otherwise quite arbitrary.

Proposition 3.3. *Let $\mathcal{E}$ and $\mathcal{F}$ be linear spaces over the same scalar field, and let X be a Hamel basis for $\mathcal{E}$. Then for each mapping ϕ of X into $\mathcal{F}$ there exists a unique linear transformation $T_\phi : \mathcal{E} \to \mathcal{F}$ such that $T_\phi | X = \phi$.*

PROOF. We may suppose X to be indexed as $X = \{x_\gamma\}_{\gamma\in\Gamma}$, so each vector y in $\mathcal{E}$ has a unique representation

$$y = \sum_{\gamma\in\Gamma} \lambda_\gamma x_\gamma$$

where $\{\lambda_\gamma\}_{\gamma\in\Gamma}$ is a similarly indexed family of scalars, all but some finite number of which are zero. If S and T are two linear transformations of $\mathcal{E}$ into $\mathcal{F}$ such that $S|X = T|X$, then it is clear that

$$S\left(\sum_\gamma \lambda_\gamma x_\gamma\right) = T\left(\sum_\gamma \lambda_\gamma x_\gamma\right)$$

for any such indexed family $\{\lambda_\gamma\}$, and hence that $S = T$. Thus T_ϕ is unique if it exists. On the other hand, simply setting

$$T_\phi\left(\sum_\gamma \lambda_\gamma x_\gamma\right) = \sum_\gamma \lambda_\gamma \phi(x_\gamma)$$

defines the desired linear transformation. □

Definition. A one-to-one linear transformation of a linear space $\mathcal{E}$ onto a linear space $\mathcal{F}$ is a (*linear space*) *isomorphism* of $\mathcal{E}$ onto $\mathcal{F}$. Two linear spaces are *isomorphic* if there exists a linear space isomorphism of one onto the other.

Example R. If $\mathcal{E}$ is an n-dimensional real [complex] linear space and $X = \{x_1, \ldots, x_n\}$ is an ordered basis for $\mathcal{E}$, then the mapping

$$\sum_{i=1}^n \alpha_i x_i \xrightarrow{\eta} (\alpha_1, \ldots, \alpha_n)$$

that assigns to each vector in $\mathcal{E}$ its n-tuple of coordinates with respect to X is a linear space isomorphism of $\mathcal{E}$ onto $\mathbb{R}^n[\mathbb{C}^n]$.

If $\mathcal{E}$ and $\mathcal{F}$ are linear spaces over the same scalar field, and if S and T are two linear transformations of $\mathcal{E}$ into $\mathcal{F}$, then the sum $S + T$ is defined by pointwise addition: $(S + T)x = Sx + Tx$ for all x in $\mathcal{E}$. Likewise, for each scalar α, the mapping αS is defined by $(\alpha S)x = \alpha(Sx)$ for all x in $\mathcal{E}$. Clearly $S+T$ and αS are also linear transformations of $\mathcal{E}$ into $\mathcal{F}$. Moreover, these definitions turn the set of all linear transformations of $\mathcal{E}$ into $\mathcal{F}$ into a new linear space—the *full space of linear transformations of* $\mathcal{E}$ *into* $\mathcal{F}$. The zero element of this linear space is the linear transformation 0 defined by $0x = 0$ for all x in $\mathcal{E}$. (The full space of linear transformations of $\mathcal{E}$ into its scalar field coincides with the full algebraic dual of $\mathcal{E}$ (Ex. H).)

Suppose now that $\mathcal{E}, \mathcal{F}$, and $\mathcal{G}$ are all vector spaces over the same scalar field. Let T be a linear transformation of $\mathcal{E}$ into $\mathcal{F}$ and let S be a linear transformation of $\mathcal{F}$ into $\mathcal{G}$. Then the composition $S \circ T$ is a linear transformation of $\mathcal{E}$ into $\mathcal{G}$ called the *product* of S and T and denoted by ST. The multiplication of linear transformations satisfies the following relations whenever the various products are defined:

$$\begin{aligned} &\text{(a) } R(ST) = (RS)T, \\ &\text{(b) } R(S+T) = RS + RT; \; (R+S)T = RT + ST, \\ &\text{(c) } \alpha(ST) = (\alpha S)T = S(\alpha T). \end{aligned} \tag{3}$$

In particular, if R, S and T denote linear transformations of a linear space $\mathcal{E}$ into itself, then all of these products are defined, and the relations (3) hold without exception.

Conditions (3) are the main ingredients in the definition of a *linear algebra.*

Definition. A real [complex] vector space $\mathcal{A}$ on which is given a product satisfying the conditions

$$\begin{aligned} &\text{(a)} \quad x(yz) = (xy)z, \\ &\text{(b)} \quad x(y+z) = xy + xz; \; (x+y)z = xz + yz, \\ &\text{(c)} \quad \alpha(xy) = (\alpha x)y = x(\alpha y), \end{aligned}$$

for all elements x, y, z of $\mathcal{A}$ and all scalars α is a real [complex] (*associative, linear*) *algebra.* If $\mathcal{A}$ possesses an element 1 such that $1x = x1 = x$ for every x in $\mathcal{A}$, then 1 is the *identity* or *unit* of $\mathcal{A}$ (such an element is obviously unique if it exists and will, if necessary, be denoted by $1_{\mathcal{A}}$), and $\mathcal{A}$ is said to be a *unital* algebra or an algebra *with identity* or *unit.* If $\mathcal{A}$ is a unital algebra with identity 1, and if λ is a scalar, we shall simply write λ for $\lambda 1$ when no confusion can result. Likewise, if $\mathcal{A}$ is a unital algebra and if x is an element of $\mathcal{A}$, then an element y of $\mathcal{A}$ is the *inverse* of x in $\mathcal{A}$ if $xy = yx = 1$. (The inverse of an element x is obviously unique if it exists and is denoted by x^{-1}.) An element of $\mathcal{A}$ that possesses an inverse in $\mathcal{A}$ is said to be *invertible* (in $\mathcal{A}$).

Thus the full space of linear transformations on a linear space $\mathcal{E}$ is a unital algebra in which the transformation $1_{\mathcal{E}}$ is the identity element and in which an element S is invertible if and only if $\mathcal{R}(S) = \mathcal{E}$ and $\mathcal{K}(S) = (0)$ (Prop. 3.2). Similarly, the field $\mathbb{R}$ is a real algebra, while the field $\mathbb{C}$ is both a real and a complex algebra. Indeed, according to the general principle laid down in Example B, every complex linear algebra is also a real algebra. Other examples of important and useful algebras abound.

Example S. The linear space $\mathbb{R}[x]$ of real polynomials (Ex. E) is a real algebra with respect to ordinary multiplication of polynomials. Similarly, $\mathbb{C}[x]$ is a complex algebra.

Example T. On any set X the real [complex] linear space $\mathcal{F}_{\mathbb{R}}(X)[\mathcal{F}_{\mathbb{C}}(X)]$ of all real-valued [complex-valued] functions on X (Ex. G) is a real [complex] algebra with respect to pointwise multiplication:

$$(fg)(x) = f(x)g(x), \ x \in X, \quad f, g \in \mathcal{F}_{\mathbb{R}}(X)[\mathcal{F}_{\mathbb{C}}(X)].$$

These algebras are unital, having for identity the constant function identically equal to one, and an element of either algebra is invertible if and only if it never assumes the value zero.

Thus far we have considered only functions of one vector variable, but we shall also be interested in certain kinds of functions of two variables. Suppose that $\mathcal{E}$ and $\mathcal{F}$ are linear spaces over the same scalar field, and let $\phi = \phi(x, y)$ be a scalar-valued mapping defined on the direct sum $\mathcal{E} + \mathcal{F}$. If for each fixed y_0 in $\mathcal{F}$ the function $\phi(x, y_0)$ is a linear functional on $\mathcal{E}$, and, for each fixed x_0 in $\mathcal{E}, \phi(x_0, y)$ is a linear functional on $\mathcal{F}$, then ϕ is a *bilinear functional on* $\mathcal{E} + \mathcal{F}$. If $\mathcal{E} = \mathcal{F}$, then ϕ is a *bilinear functional on* $\mathcal{E}$. The set of all bilinear functionals on $\mathcal{E} + \mathcal{F}$ is a linear space with linear operations defined pointwise. In particular, the set of all bilinear functionals on $\mathcal{E}$ is a linear space. A bilinear functional on a linear space $\mathcal{E}$ is *symmetric* if $\phi(x, y) = \phi(y, x)$ for every pair of vectors x, y in $\mathcal{E}$.

When $\mathcal{E}$ and $\mathcal{F}$ are complex, there is a notion closely related to that of a bilinear functional on $\mathcal{E} + \mathcal{F}$ that we shall have occasion to use. A mapping $\psi : \mathcal{E} + \mathcal{F} \to \mathbb{C}$ is said to be a *sesquilinear functional* on $\mathcal{E} + \mathcal{F}$ if $\psi(x, y_0)$ is a linear functional on $\mathcal{E}$ for each y_0 in $\mathcal{F}$ and $\overline{\psi(x_0, y)}$ is a linear functional on $\mathcal{F}$ for each x_0 in $\mathcal{E}$. (Another way to state the second of these conditions is to say that $\psi(x_0, \alpha y_1 + \beta y_2)$ is equal to $\overline{\alpha}\psi(x_0, y_1) + \overline{\beta}\psi(x_0, y_2)$ for all complex numbers α, β and all vectors y_1, y_2 in $\mathcal{F}$; such a functional is said to be *conjugate linear*.) When $\mathcal{E} = \mathcal{F}, \psi$ is called a sesquilinear functional *on* $\mathcal{E}$. A sesquilinear functional ψ on $\mathcal{E}$ is said to be *Hermitian symmetric* if $\psi(x, y) = \overline{\psi(y, x)}$ for all x and y in $\mathcal{E}$.

Problems

A. Verify that the intersection of an arbitrary nonempty collection of linear submanifolds of a vector space $\mathcal{E}$ is a linear submanifold of $\mathcal{E}$. Use this fact to show that for any subset M of $\mathcal{E}$ the linear manifold $\langle M \rangle$ coincides with the intersection of the collection of all those linear submanifolds of $\mathcal{E}$ that contain M. Conclude also that the system of all linear submanifolds of $\mathcal{E}$ is a complete lattice (in the inclusion ordering). (Hint: Recall Problem 1K.)

B. Let $\mathcal{E}$ be a vector space and let $\mathcal{J}$ denote the partially ordered set consisting of all linearly independent subsets of $\mathcal{E}$ (in the inclusion ordering).

(i) Verify that the following conditions are equivalent for an element J of $\mathcal{J}$: (a) J is maximal in $\mathcal{J}$, (b) $\langle J\rangle = \mathcal{E}$, (c) J is a Hamel basis for $\mathcal{E}$.

(ii) Prove that if $\mathcal{J}_0$ is any simply ordered subset of $\mathcal{J}$, then $\bigcup \mathcal{J}_0 \in \mathcal{J}$.

(iii) Conclude that if J_0 is an arbitrary element of $\mathcal{J}$ and M is any subset of $\mathcal{E}$ that contains J_0, then there exists a Hamel basis X for $\langle M\rangle$ such that $J_0 \subset X \subset M$. In particular, $\mathcal{E}$ itself possesses a Hamel basis.

C. Let $X = \{x_1, \ldots, x_m\}$ and $Y = \{y_1, \ldots, y_n\}$ be linearly independent sets of vectors in a linear space $\mathcal{E}$, and suppose X is contained in the linear submanifold $\mathcal{M} = \langle Y\rangle$. Show that $m \le n$ and that it is possible to select a set Z of exactly $n - m$ vectors from Y so that $X \cup Z$ is also a basis for $\mathcal{M}$. (In particular, if $m = n$, then the set X is itself a basis for $\mathcal{M}$.) Conclude that if a vector space $\mathcal{E}$ possesses a finite basis, then any two bases for $\mathcal{E}$ contain the same number of vectors, and that, if $\dim \mathcal{E} = n$, then no linearly independent subset of $\mathcal{E}$ contains more than n vectors and $\mathcal{E}$ itself is the only n-dimensional linear submanifold of $\mathcal{E}$. (Hint: The heart of the matter is that if a vector x belongs to $\mathcal{M}$, and if, in the expression $x = \sum_{i=1}^{n} \lambda_i y_i$, some $\lambda_{i_0} \neq 0$, then the set $\widetilde{Y}$ consisting of x and $\{y_i \in Y : i \neq i_0\}$ is linearly independent, and is therefore another basis for $\mathcal{M}$.)

D. If $\mathcal{M}$ is a linear manifold in a linear space $\mathcal{E}$ and $\mathcal{N}$ is another linear manifold in $\mathcal{E}$ such that $\mathcal{M} \cap \mathcal{N} = (0)$ and $\mathcal{M} + \mathcal{N} = \mathcal{E}$, then $\mathcal{N}$ is a *complement* of $\mathcal{M}$ in $\mathcal{E}$. Show that every linear manifold in a linear space $\mathcal{E}$ has a complement in $\mathcal{E}$. (Hint: Recall Problem B.)

E. Two subsets M and N of a linear space $\mathcal{E}$ are *independent* if $\langle M\rangle \cap \langle N\rangle = (0)$. More generally, an indexed family $\{M_\gamma\}_{\gamma\in\Gamma}$ of subsets of $\mathcal{E}$ is *independent* if $M_{\gamma'}$ and $\bigcup_{\gamma\neq\gamma'} M_\gamma$ are independent for each index γ'.

(i) Show that a set J of nonzero vectors in $\mathcal{E}$ is linearly independent if and only if for every partition of J into the union $J' \cup J''$ of two subsets the sets J' and J'' are independent.

(ii) Prove that if $\{J_\gamma\}_{\gamma\in\Gamma}$ is an independent family of linearly independent subsets of $\mathcal{E}$, then $J = \bigcup_{\gamma\in\Gamma} J_\gamma$ is a Hamel basis for the linear manifold $\sum_{\gamma\in\Gamma}\langle J_\gamma\rangle$.

F. Let $\mathcal{E}$ and $\mathcal{F}$ be linear spaces over the same scalar field, and let T be a linear transformation of some linear submanifold $\mathcal{M}$ of $\mathcal{E}$ into $\mathcal{F}$. Show that there exists a linear transformation $\widetilde{T}$ of $\mathcal{E}$ into $\mathcal{F}$ such that $T = \widetilde{T}|\mathcal{M}$. (Hint: Use Problem B or Problem D.)

G. If T is a linear transformation of a linear space $\mathcal{E}$ into a linear space $\mathcal{F}$, and if its range $\mathcal{R}(T)$ is finite dimensional, then $\dim \mathcal{R}(T)$ is called the *rank* of T (notation: rank T). Likewise, if the null space $\mathcal{K}(T)$ is finite dimensional, then $\dim \mathcal{K}(T)$ is the *nullity* of T. Show that if $\mathcal{E}$ is finite dimensional, then for an arbitrary linear transformation $T : \mathcal{E} \to \mathcal{F}$ we have

$$(\text{rank } T) + (\text{nullity of } T) = \dim \mathcal{E}.$$

Conclude that if S is a linear transformation on an n-dimensional linear space $\mathcal{E}$, then S is invertible if either $\mathcal{K}(S) = (0)$ or $\mathcal{R}(S) = \mathcal{E}$, i.e., if either the nullity of S vanishes or the rank of S equals n. Show finally that if there exists a linear transformation R on $\mathcal{E}$ such that either $RS = 1$ or $SR = 1$, then S is invertible and $R = S^{-1}$.

H. The space $\mathbb{R}_{m,n}[\mathbb{C}_{m,n}]$ of all real [complex] $m \times n$ matrices is a real [complex] linear space when equipped with the usual (entrywise) linear operations. Let $\mathcal{E}$ and $\mathcal{F}$ be finite dimensional real [complex] linear spaces and, as in Example Q, let $X = \{x_1, \ldots, x_n\}$ and $Y = \{y_1, \ldots, y_m\}$ be ordered bases in $\mathcal{E}$ and $\mathcal{F}$, respectively. Show that the mapping Φ that assigns to each linear transformation of $\mathcal{E}$ into $\mathcal{F}$ its matrix with respect to X and Y is a linear space isomorphism of the full space of linear transformations of $\mathcal{E}$ into $\mathcal{F}$ onto $\mathbb{R}_{m,n}[\mathbb{C}_{m,n}]$.

I. Let $\mathcal{F}$ be a self-conjugate linear space of complex-valued functions on a set X, and let ϕ be a linear functional on $\mathcal{F}$.

(i) Verify that ϕ is self-conjugate if and only if its restriction $\phi \mid \mathcal{F}_{\mathbb{R}}$ to the real-valued functions in $\mathcal{F}$ takes real values.

(ii) The functional ϕ is called *positive* if $\phi(f) \geq 0$ whenever f is a non-negative-valued function belonging to $\mathcal{F}$. Show that if $\mathcal{F}$ has the property that the system $\mathcal{F}_{\mathbb{R}}$ of real-valued functions in $\mathcal{F}$ is a function lattice, and if ϕ is positive, then ϕ is automatically self-conjugate. (Hint: Recall Problem 2M.)

J. Let $\mathcal{A}$ be a linear algebra. A linear manifold $\mathcal{S}$ in $\mathcal{A}$ that is closed with respect to products is a *subalgebra* of $\mathcal{A}$. If $\mathcal{S}$ has the additional property that, for every s in $\mathcal{S}$ and x in $\mathcal{A}$, sx belongs to $\mathcal{S}$ [xs belongs to $\mathcal{S}$], then $\mathcal{S}$ is a *right* [*left*] *ideal* in $\mathcal{A}$. If $\mathcal{S}$ is both a left and a right ideal, then $\mathcal{S}$ is a *two-sided ideal*, or, more simply, an *ideal* in $\mathcal{A}$.

(i) Show that if $\mathcal{S}$ is a two-sided ideal in $\mathcal{A}$, then the quotient vector space $\mathcal{A}/\mathcal{S}$ equipped with the product $[x][y] = [xy]$ forms an algebra. This algebra $\mathcal{A}/\mathcal{S}$ is called the *quotient algebra* of $\mathcal{A}$ *modulo* $\mathcal{S}$.

(ii) The linear spaces $\mathcal{C}^{(m)}_{\mathbb{R}}((a,b))$ and $\mathcal{C}^{(m)}_{\mathbb{R}}([a,b])$ of Examples L and M are subalgebras of the real algebras $\mathcal{F}_{\mathbb{R}}((a,b))$ and $\mathcal{F}_{\mathbb{R}}([a,b])$, respectively. Similarly, $\mathcal{C}^{(m)}_{\mathbb{C}}((a,b))$ and $\mathcal{C}^{(m)}_{\mathbb{C}}([a,b])$ are subalgebras of the complex algebras $\mathcal{F}_{\mathbb{C}}((a,b))$ and $\mathcal{F}_{\mathbb{C}}([a,b])$, respectively. Are any of these subalgebras ideals?

K. Let $\mathcal{A}$ and $\mathcal{B}$ be algebras over the same scalar field, and suppose ϕ is a linear transformation of $\mathcal{A}$ into $\mathcal{B}$ such that ϕ *preserves products* (i.e., such that $\phi(xy) = \phi(x)\phi(y)$ for every pair x, y of elements of $\mathcal{A}$). Then ϕ is called an (*algebra*) *homomorphism* of $\mathcal{A}$ into $\mathcal{B}$. (If $\mathcal{A}$ and $\mathcal{B}$ are both unital algebras, then the homomorphism ϕ is said to be *unital* if $\phi(1_{\mathcal{A}}) = 1_{\mathcal{B}}$.) If ϕ is a vector space isomorphism of $\mathcal{A}$ onto $\mathcal{B}$ that is also an algebra homomorphism, then ϕ is an *algebra isomorphism* of $\mathcal{A}$ onto $\mathcal{B}$, and if such an isomorphism exists, $\mathcal{A}$ and $\mathcal{B}$ are said to be *isomorphic* (*as*) *algebras*.

(i) Let ϕ be a linear transformation of $\mathcal{A}$ into $\mathcal{B}$, and let $X = \{x_\gamma\}$ be a Hamel basis for $\mathcal{A}$ (regarded as a linear space). Verify that ϕ is an algebra homomorphism if and only if $\phi(x_\gamma x_{\gamma'}) = \phi(x_\gamma)\phi(x_{\gamma'})$ for every pair $x_\gamma, x_{\gamma'}$ of elements of the basis X.

(ii) Verify that if $\mathcal{J}$ is an ideal in $\mathcal{A}$, then the natural projection π of $\mathcal{A}$ onto the quotient algebra $\mathcal{A}/\mathcal{J}$ is a homomorphism with kernel $\mathcal{J}$. Show in the converse direction that if ϕ is an arbitrary homomorphism of $\mathcal{A}$ into $\mathcal{B}$, then the kernel $\mathcal{K}(\phi)$ is an ideal in $\mathcal{A}$, the range $\mathcal{R}(\phi)$ is a subalgebra of $\mathcal{B}$, and the result of factoring ϕ through $\mathcal{K}(\phi)$ (Ex. O) is an algebra isomorphism of the quotient algebra $\mathcal{A}/\mathcal{K}(\phi)$ onto $\mathcal{R}(\phi)$.

(iii) Show that if $\mathcal{E}$ and $\mathcal{F}$ are vector spaces, and if η is a linear space isomorphism of $\mathcal{E}$ onto $\mathcal{F}$, then $\phi(T) = \eta T \eta^{-1}$ defines an algebra isomorphism ϕ of the full space of linear transformations on $\mathcal{E}$ onto the full space of linear transformations on $\mathcal{F}$. (The isomorphism ϕ is said to be *spatially implemented* by η.)

L. Let $\mathcal{E}$ be a finite dimensional real [complex] linear space, and let $X = \{x_1, \dots, x_n\}$ be an ordered basis for $\mathcal{E}$. Verify that the linear space isomorphism of the full space of linear transformation on $\mathcal{E}$ onto $\mathbb{R}_{n,n}[\mathbb{C}_{n,n}]$ obtained by assigning to each linear transformation on $\mathcal{E}$ its matrix with respect to X (Ex. Q, Prob. H) becomes an algebra isomorphism when $\mathbb{R}_{n,n}[\mathbb{C}_{n,n}]$ is equipped with the standard *row-by-column* product, according to which the product AB of two $n \times n$ matrices $A = (\alpha_{ij})$ and $B = (\beta_{ij})$ is the matrix $C = (\gamma_{ij})$ where

$$\gamma_{ij} = \sum_{k=1}^{n} \alpha_{ik}\beta_{kj}, \quad i, j = 1, \dots, n.$$

M. Let $\mathcal{A}$ be a real [complex] unital algebra with identity 1, and let x be an element of $\mathcal{A}$. Prove that there exists a unique unital homomorphism ϕ_x of the algebra $\mathbb{R}[\lambda]$ $[\mathbb{C}[\lambda]]$ of all polynomials in the indeterminate λ into $\mathcal{A}$ such that $\phi_x(\lambda) = x$. (Hint: The sequence $\{1, \lambda, \lambda^2, \dots\}$ constitutes a Hamel basis for $\mathbb{R}[\lambda]$ and $\mathbb{C}[\lambda]$ (Ex. E); use Problem K(i).) The image

$\phi_x(p)$ of a polynomial p under the homomorphism ϕ_x is denoted by $p(x)$ and is referred to as the result of *evaluating* p at x.

N. The subspace $(p_0(x))$ of the algebra $\mathbb{R}[x]$ constructed in Example I is an ideal in $\mathbb{R}[x]$, so that, according to Problem J, the quotient space $\mathbb{R}[x]/(p_0(x))$ is actually a real N-dimensional algebra where N denotes the degree of p_0. Show that if one takes for p_0 the quadratic polynomial x^2+1, then the quotient algebra $\mathbb{R}[x]/(p_0(x))$ is, in a natural way, isomorphic as a real algebra to the complex field $\mathbb{C}$.

O. Let $\mathcal{E}$ and $\mathcal{F}$ be linear spaces over the same scalar field, let $\mathcal{M}$ and $\mathcal{N}$ be linear submanifolds of $\mathcal{E}$ and $\mathcal{F}$, respectively, and let ϕ be a bilinear functional on $\mathcal{M}+\mathcal{N}$. Show that ϕ can be extended to a bilinear functional on $\mathcal{E}+\mathcal{F}$. (Hint: Recall Problem D.)

P. (i) By the *quadratic form* of a bilinear functional ϕ on a vector space $\mathcal{E}$ is meant the functional

$$\widehat{\phi}(x) = \phi(x,x), \qquad x \in \mathcal{E}.$$

Verify that ϕ and $\widehat{\phi}$ are related by the identity

$$\phi(x,y) + \phi(y,x) = \frac{1}{2}\left[\widehat{\phi}(x+y) - \widehat{\phi}(x-y)\right], \quad x,y \in \mathcal{E},$$

and conclude that if ϕ is *symmetric*, then, in fact,

$$\phi(x,y) = \frac{1}{4}\left[\widehat{\phi}(x+y) - \widehat{\phi}(x-y)\right]$$

for all x and y in $\mathcal{E}$. Thus a bilinear functional is uniquely determined by its quadratic form if it is symmetric. Show, however, that two different nonsymmetric bilinear functionals may have the same quadratic form.

(ii) Similarly, by the *quadratic form* of a sesquilinear functional ψ on a complex vector space $\mathcal{E}$ is meant the functional

$$\widehat{\psi}(x) = \psi(x,x), \qquad x \in \mathcal{E}.$$

Verify that ψ and $\widehat{\psi}$ are related by the identity

$$\psi(x,y) = \frac{1}{4}\left\{\left[\widehat{\psi}(x+y) - \widehat{\psi}(x-y)\right] + i\left[\widehat{\psi}(x+iy) - \widehat{\psi}(x-iy)\right]\right\}, \tag{4}$$

valid for every pair of vectors x and y in $\mathcal{E}$. Thus on a complex linear space *every* sesquilinear functional is uniquely determined by its quadratic form. (The identity (4) is known as the *polarization identity*.) Show also that a sesquilinear functional is Hermitian symmetric if and only if its quadratic form is real-valued.

Q. Let $\mathcal{E}$ be a complex vector space and let ψ be a sesquilinear functional on $\mathcal{E}$. Then ψ is *positive semidefinite* if $\widehat{\psi} \geq 0$, where $\widehat{\psi}$ denotes the quadratic form of ψ. (If $\widehat{\psi}(x) > 0$ for all $x \neq 0$, then ψ is *positive definite*.)

(i) Verify that if ψ is positive semidefinite, then ψ is automatically Hermitian symmetric, and also satisfies the inequality

$$|\psi(x, y)|^2 \leq \widehat{\psi}(x)\widehat{\psi}(y), \qquad x, y \in \mathcal{E}. \tag{5}$$

(Hint: Inequality (5) is unchanged if y is multiplied by a complex number γ such that $|\gamma| = 1$. Hence it suffices to treat the case $\psi(x, y) \geq 0$. Consider the quadratic function

$$\widehat{\psi}(x + ty) = \widehat{\psi}(x) + 2t\psi(x, y) + t^2\widehat{\psi}(y)$$

of a *real* variable t. Setting $a = \widehat{\psi}(y), b = \psi(x, y)$, and $c = \widehat{\psi}(x)$, one has

$$at^2 + 2bt + c \geq 0$$

for all values of t, and therefore $b^2 \leq ac$ by elementary algebra.) Show also that if ψ is a positive semidefinite sesquilinear functional on $\mathcal{E}$ and if $\widehat{\psi}(z) = 0$, then $\psi(x, z) = 0$ for every x in $\mathcal{E}$, and conclude that the set $\mathcal{Z} = \{z \in \mathcal{E} : \widehat{\psi}(z) = 0\}$ is a linear submanifold of $\mathcal{E}$.

(ii) Similarly, if $\mathcal{E}$ is a real vector space and ϕ is a bilinear functional on $\mathcal{E}$, then ϕ is said and to be *positive semidefinite* [*positive definite*] if ϕ is symmetric and $\widehat{\phi} \geq 0$ [$\widehat{\phi}(x) > 0$ for all $x \neq 0$]. Verify that a positive semidefinite bilinear functional on a real vector space $\mathcal{E}$ satisfies the inequality

$$|\phi(x, y)|^2 \leq \widehat{\phi}(x)\widehat{\phi}(y), \qquad x, y \in \mathcal{E}. \tag{6}$$

Show also that the assumption of symmetry cannot be dropped in the real case by giving an example of a (nonsymmetric) bilinear functional ϕ on a real vector space that fails to satisfy (6) even though $\widehat{\phi} \geq 0$. (Hint: Construct a 2×2 matrix.)

R. Let $\mathcal{E}$ be a real vector space. If $x_0, \ldots, x_m$ and x are vectors in $\mathcal{E}$, then x is an *affine combination* of the vectors $x_0, \ldots, x_m$ if there exist scalars $s_0, \ldots, s_m$ such that $x = s_0x_0 + \ldots + s_mx_m$ and $s_0 + \ldots + s_m = 1$. A subset A of $\mathcal{E}$ is an *affine variety* in $\mathcal{E}$ if $x \in A$ whenever x is an affine combination of vectors belonging to A.

(i) If x_0, x_1 are distinct vectors in $\mathcal{E}$, then the *line* through x_0 and x_1 is the set L of all affine combinations of x_0 and x_1. Verify that a line is an affine variety in $\mathcal{E}$, and that a subset A of $\mathcal{E}$ is an affine variety if and only if A contains the entire line L through x_0 and x_1 whenever x_0 and x_1 are distinct vectors in A.

(ii) Show that if $\mathcal{A}$ is an arbitrary nonempty collection of affine varieties in $\mathcal{E}$, then $\bigcap \mathcal{A}$ is also an affine variety in $\mathcal{E}$. Conclude that if M is any subset of $\mathcal{E}$, then there is a smallest affine variety $A(M)$ (called the affine variety *spanned* by M) in $\mathcal{E}$ that contains M. Verify that the affine variety spanned by a finite set of vectors $\{x_0, \ldots, x_m\}$ coincides with the set of all affine combinations $s_0x_0 + \ldots + s_mx_m$.

(iii) Show that if A is an affine variety in $\mathcal{E}$ and $a \in \mathcal{E}$, then the translate $a + A$ is also an affine variety. Verity that an affine variety in $\mathcal{E}$ is a linear submanifold of $\mathcal{E}$ if and only if $0 \in A$. Conclude that the nonempty affine varieties in $\mathcal{E}$ coincide with the various cosets of linear submanifolds of $\mathcal{E}$.

S. Let $\mathcal{E}$ be a real vector space. If $x_0, \ldots, x_m$ and x are vectors in $\mathcal{E}$, then x is a *convex combination* of the vectors $x_0, \ldots, x_m$ if there exist *nonnegative* scalars $s_0, \ldots, s_m$ such that $x = s_0x_0 + \ldots + s_mx_m$ and $s_0 + \ldots + s_m = 1$. A subset C of $\mathcal{E}$ is *convex* if $x \in C$ whenever x is a convex combination of vectors $x_0, \ldots, x_m$ belonging to C.

(i) If x_0 and x_1 are vectors in $\mathcal{E}$, then the *line segment* $\ell(x_0, x_1)$ joining x_0 to x_1 is the set of all convex combinations of x_0 and x_1. Verify that $\ell(x_0, x_1)$ is convex, and that a subset C of $\mathcal{E}$ is convex if and only if C contains $\ell(x_0, x_1)$ whenever $x_0, x_1 \in C$.

(ii) If $\mathcal{C}$ is an arbitrary nonempty collection of convex sets in $\mathcal{E}$, then $\bigcap \mathcal{C}$ is also convex. Hence if M is an arbitrary subset of $\mathcal{E}$, there exists a smallest convex set $C(M)$ in $\mathcal{E}$ (called the *convex hull* of M) that contains M. Verify that the convex hull of a finite set of vectors $\{x_0, \ldots, x_m\}$ coincides with the set of all convex combinations $s_0x_0 + \ldots + s_mx_m$.

(iii) Let C be a convex set in $\mathcal{E}$. If $a \in \mathcal{E}$, then the translate $a + C$ is also convex. If $T : \mathcal{E} \to \mathcal{F}$ is a linear transformation of $\mathcal{E}$ into another real vector space $\mathcal{F}$, then $T(C)$ is convex in $\mathcal{F}$.

Cardinal numbers 4

If X and Y are finite sets, containing, say, m and n elements, respectively, then it is obvious that there exists a one-to-one mapping of X onto Y if and only if $m = n$. A quite natural extension of this use of numbers to classify sets according to their size was made by Georg Cantor [6], who introduced the "cardinal number" of any set X, finite or not, to represent the number of elements in X. This goes as follows: A symbol, called the *cardinal number* of X (notation: card X), is associated with each set X according to the rule that card X and card Y shall be equal if and only if there exists a one-to-one mapping of X onto Y.

There is a small logical difficulty here. We have said when two cardinal numbers are equal without saying precisely what a "cardinal number" is. This difficulty could easily be overcome if it were worth the trouble to do so; for our purposes such a discussion would not be fruitful, and we therefore omit it. The reader is encouraged to consult [8] or [11], where such matters are discussed in detail. It should be noted that this agreement on when two sets have the same cardinal number ensures that card $X =$ card Y if and only if card $Y =$ card X, that card $X =$ card X for every set X, and likewise that if card $X =$ card Y and card $Y =$ card Z, where Z is some third set, then card $X =$ card Z (see Problem 1B). Thus equality of cardinal numbers is an equivalence relation.

Example A. As noted above, two finite sets have the same cardinal number if and only if they possess the same number of elements. Thus if X is a finite set containing n elements, we may and do take for the cardinal number of X the number n of elements in X (card $X = n$). In particular, we take card $\varnothing$ to be 0.

Example B. The function $f(n) = n+1$ provides a one-to-one mapping of the set $\mathbb{N}_0$ of all nonnegative integers onto the (strictly smaller) set $\mathbb{N}$ of all positive integers. Thus card $\mathbb{N}_0 =$ card $\mathbb{N}$, even though $\mathbb{N}$ is a proper subset of $\mathbb{N}_0$. Employing universally accepted notation, we write card $\mathbb{N} =$ card $\mathbb{N}_0 = \aleph_0$ (in English: *aleph naught*). A set X with card $X = \aleph_0$—i.e., one that can be placed in one-to-one correspondence with the system $\mathbb{N}$ of natural numbers—will be called *countably infinite*.

Example C. The linear function $f(t) = (b-a)t + a$ provides a one-to-one mapping of the closed unit interval $[0,1]$ onto the closed interval $[a,b]$ provided $a < b$. It follows at once that any two nondegenerate closed intervals of real numbers have the same cardinal number.

Example D. The same argument as in the preceding example also shows that any two nonempty *open* intervals of real numbers have the same cardinal number. Moreover, the function $\widehat{\phi}(t) = t/(1+|t|), t \in \mathbb{R}$, provides a one-to-one correspondence between all of $\mathbb{R}$ and the open interval $(-1,+1)$ (Ex. 2B). Hence for any two real numbers a and b such that $a < b$,

$$\text{card } (a,b) = \text{ card } \mathbb{R}.$$

This cardinal number will, in the sequel, be denoted by $\aleph$ (*aleph*) and will also be called the *power of the continuum*. (The cardinal number of a set was sometimes referred to by Cantor and others as the *power* of that set.)

Definition. If X and Y are sets, we say that card X is *less than or equal to* card Y (notation: card $X \leq$ card Y) if there exists a one-to-one mapping of X into Y, or equivalently, if there exists a subset Y_0 of Y such that card $X =$ card Y_0. Note that if card $X' =$ card X and card $Y' =$ card Y, and if card $X \leq$ card Y, then card $X' \leq$ card Y'. Thus $\leq$ is actually a relation between cardinal numbers and is independent of the sets representing those cardinal numbers.

Example E. If X is a set and A is a subset of X, then the inclusion mapping of A into X (Ex. 1E) provides a one-to-one mapping of A into X. Thus $A \subset X$ implies card $A \leq$ card X. On the other hand, if X is a finite set and if card $A =$ card X for some subset A of X, then $A = X$. (That this conclusion may fail to hold when X is an infinite set was seen in Example B.)

Example F. Let X be an infinite set, so that there does not exist any nonnegative integer n such that X consists of exactly n elements, and let

s be a mapping that assigns to each subset E of X other than X itself an element of $X \backslash E$ (see Example 1N). Set $\phi(1) = s(\varnothing)$, and suppose ϕ is defined and one-to-one on the set of integers $\{1, \ldots, n\}$. Then the set $E_n = \phi(\{1, \ldots, n\})$ contains exactly n elements and therefore cannot coincide with X, so $s(E_n)$ exists, and we define $\phi(n+1) = s(E_n)$. Thus by mathematical induction we obtain a one-to-one mapping ϕ of all of $\mathbb{N}$ into X. This shows, of course, that card $X \geq \aleph_0$, and hence that $\aleph_0$ is the *smallest infinite cardinal number*, that is, the smallest of the cardinal numbers c such that $n < c$ for every positive integer n. A set X such that card $X \leq \aleph_0$, i.e., a set that is either finite or countably infinite, is said to be *countable*. (This same construction shows also that every infinite set has the property that it has the same cardinal number as some proper subset of itself.)

The distinction between those sets that are countable, and therefore can be arranged somehow into a sequence indexed either by $\mathbb{N}$ or by some segment $\{1, 2, \ldots, n\}$ of $\mathbb{N}$, and those sets that are not countable (so-called *nondenumerable* or *uncountably infinite* sets) is of fundamental significance throughout all of mathematical analysis. The following example is drawn from the theory of ordered sets.

Example G. A simply ordered set X is said to be *densely ordered* or, more simply, *dense* if it contains at least two elements and if for each pair x, y of distinct elements of X there are elements of X strictly between x and y. Suppose Y is a dense simply ordered set that possesses neither a greatest element nor a least element (that is, is unbounded both above and below). Let X be another simply ordered set and let ϕ be an order isomorphism of some nonempty finite subset F of X into Y. If F is enumerated as $F = \{x_1, \ldots, x_n\}$ in such a way that $x_1 < \ldots < x_n$, and if $y_i = \phi(x_i), i = 1, \ldots, n$, then $y_1 < \ldots < y_n$, of course, since ϕ is an order isomorphism. Moreover, if x^+ denotes an arbitrary element of X not belonging to F, then one has (a) $x^+ < x_1$, or (b) $x^+ > x_n$, or (c) there is a unique positive integer $i (1 \leq i < n)$ such that $x_i < x^+ < x_{i+1}$. But according to our assumptions concerning Y, in any of these cases there is an element y^+ of Y that stands in the same relation to the subset $\{y_1, \ldots, y_n\}$; that is, we can select $y^+ < y_1$ in case (a), $y^+ > y_n$ in case (b) and $y_i < y^+ < y_{i+1}$ in case (c). But then, setting $\phi^+(x) = \phi(x)$ for each x in F and $\phi^+(x^+) = y^+$, we extend the order isomorphism ϕ to an order isomorphism ϕ^+ of $F \cup \{x^+\}$ into Y.

It follows at once from the foregoing discussion, by mathematical induction, that an ordered set such as Y contains subsets order isomorphic to every set of integers of the form $\{1, \ldots, n\}$ (and hence to every finite simply ordered set, cf. Example 1X). But more can be said. Suppose that the ordered set X is countably infinite, and that X is enumerated as a sequence

$\{x_n\}_{n=1}^{\infty}$. (The given simple ordering of X need not have anything to do with the ordering of the subscripts in this enumeration; in particular, it is not assumed that X is order isomorphic to $\mathbb{N}$.) We first select an arbitrary element y_1 of Y and set $y_1 = \phi_1(x_1)$. Then if ϕ_n is an order isomorphism of $\{x_1, \ldots, x_n\}$ into Y, we set $\phi_{n+1} = \phi_n^+$ as above, with $x^+ = x_{n+1}$. In this way we obtain, by mathematical induction, an order isomorphism of X into Y. Thus an ordered set such as Y contains subsets order isomorphic to any given countable simply ordered set.

Finally, this construction can be further improved upon in significant fashion when the sets X and Y both possess the properties ascribed to Y, that is, are both dense and unbounded above and below, and are also both countable. Indeed, suppose X is enumerated as an infinite sequence $\{x_n\}_{n=1}^{\infty}$ and that Y is also enumerated as $\{y_n\}_{n=1}^{\infty}$. We begin by setting $\phi_{\{x_1\}}(x_1) = y_1$, and suppose next that ϕ_F is an order isomorphism of some nonempty finite subset F of X into Y. We then extend ϕ_F to an order isomorphism ϕ_{F^+} of a larger set F^+ into Y as follows: If the number n of elements in F is even, we set x^+ equal to the first term of the sequence $\{x_n\}$ that does not belong to F, define $F^+ = F \cup \{x^+\}$, and set $\phi_{F^+} = (\phi_F)^+$ as above; if, on the other hand, the number n is odd, we set y^+ equal to the first term of the sequence $\{y_n\}$ that does not belong to $\phi_F(F)$, extend the order isomorphism ϕ_F^{-1} to an order isomorphism $(\phi_F^{-1})^+$ of $\phi_F(F) \cup \{y^+\}$ into X just as above (but with the roles of X and Y reversed), and then define $\phi_{F^+} = ((\phi_F^{-1})^+)^{-1}$. In this way we obtain, once again by mathematical induction, an order isomorphism of all of X *onto* all of Y. *Thus any two countable simply ordered sets that are both dense and unbounded above and below are order isomorphic.* It follows at once, of course, that two countable, dense, simply ordered sets are order isomorphic if and only if they agree in possessing or not possessing a greatest and/or a least element.

It is easily verified that the relation $\leq$ between cardinal numbers is reflexive and transitive. Thus to prove that $\leq$ is a partial ordering, it suffices to prove the following theorem.

Theorem 4.1 (Cantor-Bernstein Theorem). *If X and Y are any two sets such that* $\operatorname{card} X \leq \operatorname{card} Y$ *and* $\operatorname{card} Y \leq \operatorname{card} X$, *then* $\operatorname{card} X = \operatorname{card} Y$.

PROOF. We are given that there exists a one-to-one mapping ϕ of X into Y and a one-to-one mapping ψ of Y into X, and our task is to construct a one-to-one mapping of X onto Y. For each set A in 2^X define $\Phi(A) = X \backslash \psi(Y \backslash \phi(A))$. Clearly if $A_1 \subset A_2$ in 2^X, then $\Phi(A_1) \subset \Phi(A_2)$, so Φ is a monotone increasing mapping of 2^X into itself with respect to the inclusion ordering. Hence, by the Banach-Knaster-Tarski lemma (Prop. 1.2), there exists a set A_0 in 2^X such that $\Phi(A_0) = A_0$. Write $B_0 = \phi(A_0)$ and consider the mapping ω with domain X obtained by setting $\omega(x) = \phi(x)$ for each x in A_0 and $\omega(x) = \psi^{-1}(x)$ for each x in $X \backslash A_0 = \psi(Y \backslash B_0)$. Direct calculation shows that ω is a one-to-one mapping of X onto Y, and the proof is complete. □

Readers interested in questions concerning the foundations of mathematics, and more particularly in the various uses of the axiom of choice, may be surprised to note that, while the axiom of choice was used in Example F, and that, indeed, the facts established there cannot be proved without the use of some version of the axiom of choice, the apparently deeper and more complicated Cantor-Bernstein theorem is proved without any recourse, direct or indirect, to the axiom of choice.

Example H. If a and b are real numbers such that $a < b$, then the open interval (a, b) is contained in the closed interval $[a, b]$, while if $a' < a$ and $b < b'$, then $[a, b] \subset (a', b')$. Hence by the Cantor-Bernstein theorem the cardinal number of an arbitrary nondegenerate closed interval $[a, b]$ is $\aleph$—the power of the continuum. (Similarly, of course, $\aleph$ is the cardinal number of the half-open intervals $[a, b)$ and $(a, b]$.)

Example I. Let us denote by S the set of all infinite sequences $\{\varepsilon_n\}_{n=1}^{\infty}$ in which each term ε_n is either zero or one. The mapping ϕ assigning to each sequence $\{\varepsilon_n\}$ in S the number

$$0.\eta_1\eta_2 \cdots \eta_n \cdots, \text{ where } \eta_n = 2\varepsilon_n, \quad n \in \mathbb{N},$$

in ternary notation (so that $\phi(\{\varepsilon_n\}) = \sum_{n=1}^{\infty} \eta_n/3^n$) is a one-to-one mapping of S into the unit interval $[0, 1]$ (Th. 2.12), whence it follows that card $S \leq \aleph$. On the other hand, the mapping ψ assigning to each sequence $\{\varepsilon_n\}$ in S the number

$$0.\varepsilon_1\varepsilon_2 \cdots \varepsilon_n \cdots$$

in *binary* notation (so that $\psi(\{\varepsilon_n\}) = \sum_{n=1}^{\infty} \varepsilon_n/2^n$)is a mapping of S *onto* $[0, 1]$ (once again, see Theorem 2.12), and it follows (Prob. A) that card $S \geq \aleph$. Thus card $S = \aleph$ by the Cantor-Bernstein theorem.

As a matter of fact, the relation $\leq$ is a simple ordering on any set of cardinal numbers.

Theorem 4.2. *For any two cardinal numbers c and d, either $c \leq d$ or $d \leq c$.*

PROOF. Let X and Y be sets such that card $X = c$ and card $Y = d$, and consider the collection $\mathcal{Z}$ of all one-to-one mappings of various subsets of X into Y. Then $\mathcal{Z}$ is a partially ordered set in the extension ordering (Prob.

1L). Moreover, if $\mathcal{K}$ is any simply ordered subset of $\mathcal{Z}$, then the supremum $\bigcup\mathcal{K}$ is again an element of $\mathcal{Z}$ (Prob. 1M). Thus $\mathcal{Z}$ is a partially ordered set with the property that every simply ordered subset of $\mathcal{Z}$ is bounded above in $\mathcal{Z}$. Hence, by Zorn's lemma, $\mathcal{Z}$ contains a maximal element Φ_0. Let A and B denote the domain and range, respectively, of Φ_0, so that A is a subset of X, B a subset of Y, and Φ_0 is a one-to-one mapping of A onto B. If both $X\backslash A$ and $Y\backslash B$ are nonempty, we may choose an element x_0 of $X\backslash A$ and an element y_0 of $Y\backslash B$ and then extend Φ_0 to a mapping Φ_1 of $A \cup \{x_0\}$ into Y by writing

$$\Phi_1(x) = \begin{cases} \Phi_0(x), & x \in A, \\ y_0, & x = x_0. \end{cases}$$

Since it is clear that Φ_1 belongs to $\mathcal{Z}$ and properly extends Φ_0, this contradicts the supposed maximality of Φ_0. Hence, either $A = X$ or $B = Y$, and the theorem follows. □

Example J. Just as in any simply ordered set, it follows at once from the foregoing result that in any finite set $\{c_1, \ldots, c_n\}$ of cardinal numbers there exists a *largest* element $c_1 \vee \ldots \vee c_n$ and a *smallest* element $c_1 \wedge \ldots \wedge c_n$.

According to Example F (in the notation introduced in Examples A and B), the list of cardinal numbers begins (in natural order) as follows:

$$0, 1, 2, \ldots, n, \ldots, \aleph_0.$$

That the list of cardinal numbers does not *end* with $\aleph_0$ is another fundamental observation that we owe to the genius of Cantor.

Theorem 4.3 (Cantor's Theorem). *For any set X,* card $2^X >$ card X.

Proof. The mapping $x \to \{x\}$ carrying each element x of X to the singleton on x is a one-to-one mapping of X into 2^X, which shows that card $X \leq$ card 2^X. Hence to complete the proof it suffices to show that card $X \neq$ card 2^X. Suppose, on the contrary, that there exists a one-to-one mapping ϕ of X onto 2^X. Let $A = \{x \in X : x \notin \phi(x)\}$, and denote by x_0 that element of X for which $\phi(x_0) = A$. If x_0 belongs to A, then x_0 belongs to $\phi(x_0)$, so x_0 does not belong to A. On the other hand, if x_0 does not belong to A, then x_0 fails to belong to $\phi(x_0)$, so x_0 belongs to A. Thus neither $x_0 \in A$ nor $x_0 \notin A$ is possible, and we have reached a contradiction. □

If X is a finite set containing n elements, then it is an elementary combinatorial fact that the power class on X contains 2^n elements. In other words, X has exactly 2^n subsets. For this reason, and for others to be elucidated shortly (see Problem D), it is customary to write $2^{\operatorname{card} X}$ for the cardinal number of the power class on an arbitrary set X. Thus, in particular, $2^{\aleph_0}$ is the cardinal number of the power class on $\mathbb{N}$. This cardinal number is the first *uncountable* cardinal number we have encountered, but it is easy to construct others. Indeed, if we write $\aleph^{(1)} = 2^{\aleph_0}, \aleph^{(2)} = 2^{\aleph^{(1)}}$, etc., we obtain a strictly increasing sequence $\aleph^{(1)}, \aleph^{(2)}, \ldots, \aleph^{(n)}, \ldots$ of cardinal numbers, each of which is uncountable.

Example K. For an arbitrary set X it is clear that the correspondence between the subsets of X and their characteristic functions (Prob. 1R) is one-to-one, and hence that for any cardinal number c the cardinal number 2^c may also be regarded as the cardinal number of the collection of all characteristic functions on some set X of cardinal number c. In particular, $2^{\aleph_0}$ is the cardinal number of the collection S of all sequences $\{\varepsilon_n\}_{n=1}^{\infty}$ of zeros and ones. Thus (Ex. I) *the cardinal numbers* $\aleph$ *and* $2^{\aleph_0}$ *coincide*:

$$\aleph = 2^{\aleph_0}.$$

The question whether the cardinal number $\aleph = 2^{\aleph_0}$ is the smallest cardinal number that is larger than $\aleph_0$ is known as the "continuum problem". More generally, it may be asked whether, for any infinite cardinal number c, there exist any cardinal numbers between c and 2^c. These questions pertain to the foundations of mathematics, and we refer the reader to [8] for a treatment of them.

If X and Y are finite sets containing m and n elements, respectively, then it is evident that the cardinal number of the product $X \times Y$ is mn. Furthermore, if X and Y are disjoint, then card $(X \cup Y) = m + n$. These elementary observations lead in a natural way to the definition of the sum and product of any two cardinal numbers.

Definition. Let c and d be cardinal numbers, and let X and Y be sets such that card $X = c$ and card $Y = d$. Then we define $cd =$ card $(X \times Y)$. Moreover, if X and Y are disjoint (such disjoint representations of c and d always exist; see Problem 1T), then we define $c + d =$ card $(X \cup Y)$.

Once again it is easily verified that these definitions are independent of the particular sets X and Y employed, and hence that addition and multiplication are well-defined operations on cardinal numbers. (Note that it follows from these definitions that $c + 0 = 0 + c = c$ and $c1 = 1c = c$ for every cardinal number c.)

Example L. According to Example B we have $\aleph_0 + 1 = \aleph_0$. More generally, it is easy to see (mathematical induction) that $\aleph_0 + n = n + \aleph_0 = \aleph_0$ for every positive integer n. Much more is true however.

Example M. The function $f(n) = 2n$ provides a one-to-one mapping of the set $\mathbb{N}$ of positive integers onto the subset E of all even positive integers. Similarly, the function $g(n) = 2n-1$ yields a one-to-one mapping of $\mathbb{N}$ onto the subset O of all odd positive integers. Since $\mathbb{N} = E \cup O$ and $E \cap O = \varnothing$, this shows that

$$\aleph_0 + \aleph_0 = \aleph_0.$$

More generally, for every positive integer n the sum of $\aleph_0$ with itself n times is $\aleph_0$:

$$\overbrace{\aleph_0 + \ldots + \aleph_0}^{n} = \aleph_0.$$

The equation $\aleph_0 + \aleph_0 = \aleph_0$, simple as it may be, is of historical significance. It seems to have been officially noted first by Galileo, who concluded that "... the attributes 'equal', 'greater', and 'less', are not applicable to infinite, but only to finite, quantities [7]. Such uneasiness concerning the ancient "paradoxes of infinity" clearly contributed to the resistance that Cantor's ideas had to overcome.

Example N. The cardinal number of the set $\mathbb{N}_0$ of nonnegative integers is $\aleph_0$ (Ex. B), as is the cardinal number of the set of all negative integers. Thus card $\mathbb{Z} = \aleph_0 + \aleph_0 = \aleph_0$.

Example O. According to Example F, if c is an infinite cardinal number, then there exists a cardinal number a such that $c = a + \aleph_0$. But then, of course, $c + \aleph_0 = (a + \aleph_0) + \aleph_0 = a + (\aleph_0 + \aleph_0) = a + \aleph_0 = c$. (The associativity here employed is a special case of the general associative law for the addition of cardinal numbers; see Problem B.) It follows at once from this fact that if c is an infinite cardinal number and n is any nonnegative integer, then $c + n = c$.

The multiplicative counterpart of the additive formula $\aleph_0 + \aleph_0 = \aleph_0$ is given by the following proposition.

Proposition 4.4. *The product $\aleph_0\aleph_0$ is also equal to $\aleph_0$.*

PROOF. The mapping p of $\mathbb{N} \times \mathbb{N}$ into $\mathbb{N}$ defined by

$$p(m, n) = \frac{(m+n-1)(m+n-2)}{2} + m, \quad m, n \in \mathbb{N},$$

is easily seen to be one-to-one and onto $\mathbb{N}$. □

Example P. According to Proposition 4.4 and Example N, the cardinal number of the product $P = \mathbb{Z} \times \mathbb{N}$ is $\aleph_0$. Since the mapping $(j, k) \to j/k$ maps P onto the set $\mathbb{Q}$ of all rational numbers, it follows (Prob. A) that this latter set is also countable. Since it is clear that $\mathbb{Q}$ is infinite, this shows that card $\mathbb{Q} = \aleph_0$.

A particularly useful form of Proposition 4.4 is given by the following corollary.

Corollary 4.5. *Any countable union of countable sets is countable. That is, if Γ is a countable index set, and if each of the sets X_γ in the indexed family $\{X_\gamma\}_{\gamma\in\Gamma}$ is countable, then $U = \bigcup_{\gamma\in\Gamma} X_\gamma$ is also countable.*

PROOF. We may suppose Γ to be infinite (for otherwise the desired result may be obtained without recourse to Proposition 4.4; see Example M). Hence we may suppose that the given indexed family is simply an infinite sequence $\{X_n\}_{n=1}^\infty$. Let $\{Y_n\}_{n=1}^\infty$ be the disjointification of this sequence (Prob. 1U), and for each positive integer n let g_n be a one-to-one mapping of Y_n into the countably infinite set $\{(m, n) : m \in \mathbb{N}\}$. Then the mapping $g = \bigcup_{n\in\mathbb{N}} g_n$ (Prob. 1M) is a one-to-one mapping of U into $\mathbb{N} \times \mathbb{N}$, so card $U \leq \aleph_0$. □

The notions of sum and product extend without difficulty to arbitrary collections of cardinal numbers.

Definition. Let $\{c_\gamma\}_{\gamma\in\Gamma}$ be an indexed family of cardinal numbers, and let $\{X_\gamma\}_{\gamma\in\Gamma}$ be a similarly indexed family of sets such that card $X_\gamma = c_\gamma, \gamma \in \Gamma$. Then we define

$$\prod_{\gamma\in\Gamma} c_\gamma = \text{card} \prod_{\gamma\in\Gamma} X_\gamma,$$

and, in the event that the sets X_γ are pairwise disjoint (such pairwise disjoint representatives of the cardinal numbers c_γ always exist; see Problem 1T), we likewise define

$$\sum_{\gamma\in\Gamma} c_\gamma = \text{card} \bigcup_{\gamma\in\Gamma} X_\gamma.$$

(Here again it is seen at once that addition and multiplication are well-defined operations on the cardinal numbers c_γ and do not depend on the representing sets X_γ. If $\{c_1, \ldots, c_n\}$ is a finite system of cardinal numbers, so that the index family is just the set $\{1, \ldots, n\}$, we also write $c_1 + \ldots + c_n$ and $c_1 \ldots c_n$ for the sum and product, respectively.) Clearly, these definitions agree with our earlier ones when the index family is the doubleton $\{1, 2\}$.

Example Q. Let Γ be an index set, let X be a set, and let $c = \text{card } X$. The product $X \times \Gamma$ is the disjoint union of the indexed family $\{X_\gamma\}_{\gamma \in \Gamma}$, where $X_\gamma = X \times \{\gamma\}, \gamma \in \Gamma$. It follows that if $c_\gamma = c$ for each γ in Γ, then

$$\sum_{\gamma \in \Gamma} c_\gamma = c(\text{card } \Gamma).$$

Thus the multiplication of cardinal numbers may be viewed as the result of repeated additions, just as in the case of the multiplication of natural numbers.

The following result provides a useful generalization of Proposition 4.4.

Proposition 4.6. *If c is an arbitrary infinite cardinal number, then $c\aleph_0 = c$.*

PROOF. Let X be a set such that card $X = c$, and consider the collection $\mathcal{Z}$ of all disjoint collections of countably infinite subsets of X. Then $\mathcal{Z}$ is a partially ordered set in the inclusion ordering. Moreover, if $\mathcal{K}$ is any simply ordered set of such collections, then $\mathcal{C} = \bigcup \mathcal{K}$ is also a disjoint collection of countably infinite subsets of X, so $\mathcal{C} \in \mathcal{Z}$. Thus $\mathcal{Z}$ is a partially ordered set with the property that every simply ordered subset of $\mathcal{Z}$ is bounded above in $\mathcal{Z}$, and, according to Zorn's lemma, $\mathcal{Z}$ has a maximal element $\mathcal{C}_0$. The collection $\mathcal{C}_0$ contains at least one countably infinite set A_0 (Ex. F), and the set $Y = \bigcup \mathcal{C}_0$ is a subset of X such that $X \backslash Y$ is finite. (Indeed, if $X \backslash Y$ were infinite, then it would contain a countably infinite subset A (Ex. F), and $\mathcal{C}_0 \cup \{A\}$ would be an element of $\mathcal{Z}$ dominating $\mathcal{C}_0$, thus contradicting the assumed maximality of $\mathcal{C}_0$.) Hence, if we set $A_1 = A_0 \cup (X \backslash Y)$, then card $A_1 = \aleph_0$ (Ex. L). Consequently, if $\mathcal{C}_1$ denotes the collection of sets obtained by replacing A_0 by A_1 in $\mathcal{C}_0$, then $\mathcal{C}_1$ is a partition of X into countably infinite subsets, and it follows, as in the preceding example, that $c = \text{card } X = d\aleph_0$, where d denotes the cardinal number of $\mathcal{C}_1$. But then, of course, $c\aleph_0 = (d\aleph_0)\aleph_0 = d(\aleph_0\aleph_0) = d\aleph_0 = c$ by Proposition 4.4. (The associativity here used is a special case of the general associative law for the multiplication of cardinal numbers; see Problem B.) □

PROBLEMS

A. Let X and Y be sets and suppose that there exists a mapping of X onto Y. Show that card $Y \leq$ card X. (Hint: Recall the axiom of choice.) Show, conversely, that if $0 < \text{card } Y \leq \text{card } X$, then there exists a mapping of X onto Y.

B. Let $\{c_\gamma\}_{\gamma\in\Gamma}$ be an indexed family of cardinal numbers and let $\{\Gamma_\delta\}_{\delta\in\Delta}$ be an arbitrary indexed partition of the index set Γ. Verify the extended associative laws

$$\sum_{\gamma\in\Gamma} c_\gamma = \sum_{\delta\in\Delta}\sum_{\gamma\in\Gamma_\delta} c_\gamma \text{ and } \prod_{\gamma\in\Gamma} c_\gamma = \prod_{\delta\in\Delta}\prod_{\gamma\in\Gamma_\delta} c_\gamma.$$

Formulate and prove similarly extended commutative laws for the addition and multiplication of cardinal numbers. (Hint: Recall Problem 1G.) Show also that the distributive law

$$a\left(\sum_{\gamma\in\Gamma} c_\gamma\right) = \sum_{\gamma\in\Gamma} ac_\gamma$$

holds, where a is an arbitrary cardinal number and $\{c_\gamma\}$ is an arbitrary indexed family of cardinal numbers.

C. Prove that if C is an arbitrary set of cardinal numbers, then there exist cardinal numbers strictly larger than any element of the set C.

Problem C establishes the somewhat disconcerting fact that the notion of a "set of all cardinal numbers" is simply unthinkable—*all* cardinal numbers cannot be encompassed in any one *set*. This observation, known historically as *Cantor's paradox*, is a good example of the new class of "paradoxes of the infinite"—serious problems apparently pertaining to the very heart of our powers of conceptualization. The modern point of view, roughly speaking, is that, among all of the various collections of things (sometimes called *classes*), some are privileged to be *sets*, while others are not. Thus, the *class* of all cardinal numbers is perfectly thinkable, but this is one of the classes that (as we have just seen) cannot possibly be a set.

D. For any two sets X and Y we define $(\text{card } Y)^{\text{card } X}$ to be the cardinal number of the set Y^X of all mappings of X into Y (cf. Example 1L). Verify that this definition of c^d for cardinal numbers c and d depends only on c and d, and not on the representing sets, and that this notation extends the notation 2^c introduced earlier for the cardinal number of the power class on a set X of cardinal number c. Show also that

$$1^c = 1 \text{ and } c^1 = c$$

for every cardinal number c. Show finally that if $\{c_\gamma\}_{\gamma\in\Gamma}$ is an indexed family of cardinal numbers such that $c_\gamma = c$ for each index γ, where c

denotes some one fixed cardinal number, then

$$\prod_{\gamma\in\Gamma} c_\gamma = c^{\text{card }\Gamma}.$$

(Thus the exponentiation of cardinal numbers may be thought of as resulting from repeated multiplications, just as in the case of the exponentiation of natural numbers.)

E. Show that

$$a^{b+c} = a^b a^c,$$
$$(a^b)^c = a^{bc},$$

and

$$(ab)^c = a^c b^c$$

for arbitrary cardinal numbers a, b and c.

F. Verify that $\aleph^2 = \aleph^3 = \ldots = \aleph^{\aleph_0} = \aleph$, and conclude that the cardinal number of every nontrivial finite dimensional linear space is $\aleph$. Show also that $\aleph^\aleph = 2^\aleph$, and conclude that $c^\aleph = 2^\aleph, 2 \leq c \leq \aleph$.

G. Show that $\aleph_0^n = \aleph_0$ for every positive integer n, and also that $\aleph_0^{\aleph_0} = \aleph$.

H. Find $\sum_{n\in\mathsf{N}} n$ and $\prod_{n\in\mathsf{N}} n$.

I. Let X be a set and let $\mathcal{C}$ be a covering of X. Suppose each set A in $\mathcal{C}$ satisfies the inequality card $A \leq c$, where c denotes some one fixed cardinal number. Verify that card $X \leq c(\text{card } \mathcal{C})$. Conclude, in particular, that card $X \leq$ card $\mathcal{C}$ if $c \leq \aleph_0$ and $\mathcal{C}$ is infinite. (Hint: One may assume the sets in $\mathcal{C}$ to be subsets of X; recall Example Q and Problem A.)

J. (i) Let X and Y be infinite sets. Suppose that to each element y of Y there corresponds some finite subset F_y of X, and suppose also that the family $\{F_y\}_{y\in Y}$ covers X. Verify that card $X \leq$ card Y.

(ii) Let $\mathcal{E}$ be an infinite dimensional linear space, and let X and Y denote two Hamel bases for $\mathcal{E}$ (see Chapter 3 for definitions). Use (i) to show that card X = card Y. (Thus *all Hamel bases for any one infinite dimensional linear space* $\mathcal{E}$ *have the same cardinal number*. This common cardinal number is known as the *Hamel dimension* of $\mathcal{E}$. The Hamel dimension of a finite dimensional linear space is simply its dimension.) (Hint: If y is any vector in Y, then there exist a unique, finite, nonempty subset F_y of X and unique nonzero scalars $\{\lambda_x\}_{x\in F_y}$ such that $y = \sum_{x\in F_y} \lambda_x x$.)

K. A partially ordered set X is said to be *countably determined* if there exists a countable cofinal subset of X (Prob. 1I). Give an example of a directed set that is not countably determined. Show that if Λ is a countably determined directed set, then there exists a monotone increasing sequence $\{\lambda_n\}_{n=1}^\infty$ in Λ that is cofinal in Λ.

L. Show that the set $\mathbb{Q}[x]$ of all rational polynomials is countably infinite. (Hint: Use Problem G to show that the set of all rational polynomials of degree less than n is countable for each positive integer n, and employ Corollary 4.5.)

M. A real number t is called *algebraic* if it is a *root* of some rational polynomial, i.e., if there exists a rational polynomial $p(x)$ such that $p(t) = 0$. Verify that the set of all algebraic numbers is countably infinite, and use this fact to conclude that there exist real numbers that are *transcendental*, that is, not algebraic. Show that, in fact, the cardinal number of the set of all transcendental real numbers is $\aleph$. (Hint: A rational polynomial of degree n has at most n distinct roots, real or complex.)

> This simple, elegant proof of the existence of transcendental numbers was one of the early triumphs of cardinal number theory and helped assure that the theory would survive in spite of the many difficulties (both real and imagined) that beset it.

N. Show that $nc = c$ for every infinite cardinal number c and positive integer n.

O. If $c_1, \ldots, c_n$ are cardinal numbers, and if $c_i \leq c, i = 1, \ldots, n$, where c is some infinite cardinal number, then

$$c_1 + \ldots + c_n \leq c.$$

In particular, if $c_1 \vee \ldots \vee c_n$ is infinite, then

$$c_1 + \ldots + c_n = c_1 \vee \ldots \vee c_n.$$

P. Let $\{c_\gamma\}_{\gamma \in \Gamma}$ and $\{d_\gamma\}_{\gamma \in \Gamma}$ be similarly indexed families of cardinal numbers such that $c_\gamma \leq d_\gamma, \gamma \in \Gamma$.

(i) Verify that $\sum_{\gamma \in \Gamma} c_\gamma \leq \sum_{\gamma \in \Gamma} d_\gamma$ and $\prod_{\gamma \in \Gamma} c_\gamma \leq \prod_{\gamma \in \Gamma} d_\gamma$. Show by example that $c_\gamma < d_\gamma, \gamma \in \Gamma$, does not imply either $\sum_{\gamma \in \Gamma} c_\gamma < \sum_{\gamma \in \Gamma} d_\gamma$ or $\prod_{\gamma \in \Gamma} c_\gamma < \prod_{\gamma \in \Gamma} d_\gamma$.

(ii) Show that $\sum_{\gamma \in \Gamma} c_\gamma \leq \prod_{\gamma \in \Gamma} d_\gamma$ provided $d_\gamma > 1$ for every index γ. (Hint: It suffices, in view of (i) to verify that $\sum_{\gamma \in \Gamma} d_\gamma \leq \prod_{\gamma \in \Gamma} d_\gamma$. Suppose first that d_γ is infinite for each index γ, and let $\{X_\gamma\}$ be a similarly indexed family of sets such that card $X_\gamma = d_\gamma, \gamma \in \Gamma$. Then

$\Pi = \prod_{\gamma\in\Gamma} X_\gamma \neq \varnothing$ by the axiom of products (Chap. 1). Fix an element $\widetilde{x}$ of Π and for each index γ let A_γ denote the set of all those elements of Π that agree with $\widetilde{x}$ at every index except γ. Then $\{A_\gamma\backslash\{\widetilde{x}\}\}_{\gamma\in\Gamma}$ is a disjoint family of subsets of Π and card $A_\gamma\backslash\{\widetilde{x}\} = d_\gamma, \gamma \in \Gamma$.)

Q. (König's Theorem [15]) Let $\{c_\gamma\}_{\gamma\in\Gamma}$ and $\{d_\gamma\}_{\gamma\in\Gamma}$ be similarly indexed families of cardinal numbers such that $c_\gamma < d_\gamma, \gamma \in \Gamma$.

(i) Let $\{X_\gamma\}_{\gamma\in\Gamma}$ be a similarly indexed family of sets such that card $X_\gamma = d_\gamma, \gamma \in \Gamma$, let

$$\Pi = \prod_{\gamma\in\Gamma} X_\gamma,$$

and suppose that, for some index γ, A_γ is a subset of Π such that card $A_\gamma = c_\gamma$. Show that if $B_\gamma = \pi_\gamma(A_\gamma)$, then $X_\gamma\backslash B_\gamma \neq \varnothing$, and conclude that if $\{A_\gamma\}_{\gamma\in\Gamma}$ is an arbitrary collection of subsets of Π (indexed by the same set Γ) such that card $A_\gamma = c_\gamma, \gamma \in \Gamma$, then $\bigcup_{\gamma\in\Gamma} A_\gamma \neq \Pi$. (Hint: If $B_\gamma = \pi_\gamma(A_\gamma), \gamma \in \Gamma$, then $\prod_{\gamma\in\Gamma}(X_\gamma\backslash B_\gamma) \neq \varnothing$ by the axiom of products.)

(ii) Prove that $\sum_{\gamma\in\Gamma} c_\gamma < \prod_{\gamma\in\Gamma} d_\gamma$.

R. Let X be an infinite set, and let $\mathcal{Z}$ denote the collection of all one-to-one mappings ϕ of various subsets of X into $X \times X$ having the property that, if the domain of ϕ is A, then the range of ϕ coincides with $A \times A (\subset X \times X)$.

(i) The set $\mathcal{Z}$ is a partially ordered set in the extension ordering (see Problem 1M). Show that if $\mathcal{K}$ is an arbitrary simply ordered subset of $\mathcal{Z}$, then the supremum $\bigcup \mathcal{K}$ belongs to $\mathcal{Z}$, and employ Zorn's lemma to show that $\mathcal{Z}$ contains a maximal element.

(ii) Let ϕ_0 be a maximal element of $\mathcal{Z}$ as in (i), let A_0 denote the domain of definition of ϕ_0 (so that $\phi_0(A_0) = A_0 \times A_0$), and suppose $a =$ card $A_0 <$ card X. Show that the difference $X\backslash A_0$ contains a subset A_1 such that card $A_1 = a$, and also that if we set $A = A_0 \cup A_1$, then card $((A \times A)\backslash(A_0 \times A_0)) = a$. (Hint: Recall Problem O and use the fact that card $(A_0 \times A_0) = a$.)

(iii) Conclude that, in fact, $a =$ card X, and hence that card $(X \times X) =$ card X.

S. If $c_1, \ldots, c_n$ are cardinal numbers, and if $c_i \leq c, i = 1, \ldots, n$, where c is some infinite cardinal number, then

$$c_1 \ldots c_n \leq c.$$

In particular, if $c_1 \vee \ldots \vee c_n$ is infinite, then

$$c_1 \ldots c_n = c_1 \vee \ldots \vee c_n.$$

T. Let $\mathcal{E}$ be a linear space, $\mathcal{E} \neq (0)$, and let X be a Hamel basis for $\mathcal{E}$. Then for each vector $u \neq 0$ in $\mathcal{E}$ there exists a unique nonempty finite subset F_u of X and corresponding unique nonzero scalars $\{\lambda_x\}_{x \in F_u}$ such that $u = \sum_{x \in F_u} \lambda_x x$. Prove that, for every positive integer n, the set $S_n = \{u \in \mathcal{E} : \text{ card } F_u = n\}$ has cardinal number $d\aleph$ where $d = \text{card } X$ denotes the Hamel dimension of $\mathcal{E}$, and conclude that

$$\text{card } \mathcal{E} = d\aleph = d \vee \aleph.$$

(Hint: $\{S_n\}_{n=1}^{\infty}$ is a partition of $\mathcal{E} \backslash \{0\}$.)

U. Let $\mathcal{E}$ be a linear space, let X be a Hamel basis for $\mathcal{E}$, and let $d = \text{card } X$ be the Hamel dimension of $\mathcal{E}$.

(i) Let $\mathcal{E}'$ denote the full algebraic dual of $\mathcal{E}$ (Ex. 3H). If $\mathcal{E}$ is a real [complex] vector space, then $\mathcal{E}'$ can be placed in one-to-one correspondence with the set $\mathbb{R}^X\,[\mathbb{C}^X]$. Hence card $\mathcal{E}' = \aleph^d$. (Hint: Recall Proposition 3.3.)

(ii) Prove that if $\mathcal{E}$ is infinite dimensional, then the Hamel dimension of $\mathcal{E}'$ is greater than or equal to $\aleph$, and use Problem T to conclude that if $\mathcal{E}$ is an arbitrary infinite dimensional linear space, then the Hamel dimension of $\mathcal{E}'$ is precisely $\aleph^d$. (Hint: Let X_0 be a countably infinite subset of X, let $X_1 = X \backslash X_0$, and arrange X_0 in an infinite sequence $\{x_n\}_{n=0}^{\infty}$. For each positive real number t there is a unique element f_t of $\mathcal{E}'$ such that

$$f_t(x_n) = t^n, \quad n \in \mathbb{N}_0,$$
$$f_t(x) = 0, \quad x \in X_1.$$

Show that the set of linear functionals $f_t, 0 < t < +\infty$, is linearly independent.)

5 Ordinal numbers

Recall from Chapter 1 that a partially ordered set W is said to be well-ordered if every nonempty subset of W possesses a least element. As has been noted, a well-ordered set W is simply ordered, and it is easily seen that an element w of W has an immediate successor provided w is not the greatest element of W.

Example A. If W is well-ordered and nonempty, then W itself has a least element w_0, and if W contains more than one element, then w_0 has an immediate successor w_1. Continuing in this way, one sees that if W is a finite well-ordered set containing n elements, then W has the form $W = \{w_0, \ldots, w_{n-1}\}$ with $w_0 < \ldots < w_{n-1}$. In other words, W is order isomorphic to the set $\{0, \ldots, n-1\}$ of the first n nonnegative integers (in its natural order, cf. Example 1X).

Example B. The sets $\mathbb{N}$ and $\mathbb{N}_0$ (in their natural order) are order isomorphic infinite well-ordered sets (Ex. 1W).

Example C. The set of all countable cardinal numbers is a well-ordered set

$$0, 1, \ldots, n, \ldots, \aleph_0$$

possessing the maximum element $\aleph_0$. In this well-ordered set the element $\aleph_0$ has no immediate predecessor, that is, $\aleph_0$ is not the successor of any element of the set. Thus while every element of a well-ordered set (except the maximum element if there is one) has a successor, there may well be elements (other than the least) that are not themselves successors.

Example D. If W_1 and W_2 are disjoint well-ordered sets, then there is a unique ordering of the union $W_1 \cup W_2$ that extends the given orderings of both W_1 and W_2 and that possesses the property that every element of W_1 is less than each element of W_2. It is readily verified that this ordering turns $W_1 \cup W_2$ into a well-ordered set. (If A is a nonempty subset of $W_1 \cup W_2$ and if A contains elements of W_1, then the least element of $A \cap W_1$ in W_1 is the least element of A in $W_1 \cup W_2$; if $A \subset W_2$, then the least element of A in W_2 is also its least element in $W_1 \cup W_2$.) The well-ordered set thus constructed will be denoted by $W_1 + W_2$ (to be distinguished from $W_2 + W_1$). Thus the well-ordered set in Example C is simply $\mathbb{N}_0 + \{\aleph_0\}$. Similarly, if $\mathbb{N}_0'$ is some well-ordered set that is order isomorphic to $\mathbb{N}_0$ and disjoint from the latter, then $\mathbb{N}_0 + \mathbb{N}_0'$ is a well-ordered set that is not order isomorphic to any of the examples presented above.

To facilitate the study of well-ordered sets, and for other purposes as well, we associate with each well-ordered set W a symbol, to be called the *ordinal number* of W (notation: ord W) according to the rule that two well-ordered sets are to have the same ordinal number if and only if they are order isomorphic.

Note that we do not give a definition of the term "ordinal number", just as we did not define "cardinal number" earlier. As before, this is done because to do otherwise would take us too far afield and would serve no useful purpose. The reader is invited to consult [8] and [11] for a discussion of these matters. It is appropriate to point out, however, that if W_1, W_2 and W_3 are any well-ordered sets, then ord $W_1 =$ ord W_2 when and only when ord $W_2 =$ ord W_1, ord $W_1 =$ ord W_1, and also ord $W_1 =$ ord W_2 and ord $W_2 =$ ord W_3 imply ord $W_1 =$ ord W_3.

Example E. As has already been noted, every well-ordered set containing exactly n elements is order isomorphic to the set $\{0, 1, \ldots, n-1\}$ of nonnegative integers, and we take the ordinal number of such a set to be n. (In particular, the ordinal number of the empty set is 0.) Thus for finite well-ordered sets the same symbol does double duty, serving as both cardinal and ordinal number of the set. For infinite well-ordered sets the situation is quite different.

Example F. In keeping with universally accepted notation, we write ω for the ordinal number of $\mathbb{N}_0$. It follows, of course, that $\omega =$ ord $\mathbb{N}$ also.

Example G. If $W_1, \widetilde{W}_1$ and $W_2, \widetilde{W}_2$ are order isomorphic pairs of well-ordered sets, and if $W_1 \cap W_2 = \widetilde{W}_1 \cap \widetilde{W}_2 = \varnothing$, then it is clear that the ordered set $W_1 + W_2$ is order isomorphic to $\widetilde{W}_1 + \widetilde{W}_2$. It follows that the ordinal number $\operatorname{ord}(W_1 + W_2)$ depends only on ord W_1 and ord W_2, and not on the representing sets W_1 and W_2. Accordingly, it makes sense to define

$$\operatorname{ord} W_1 + \operatorname{ord} W_2 = \operatorname{ord}(W_1 + W_2).$$

Thus, in particular, the ordinal number of the well-ordered set

$$0, 1, \ldots, \aleph_0$$

of Example C is $\operatorname{ord} \mathbb{N}_0 + \operatorname{ord} \{\aleph_0\} = \omega + 1$, while the ordinal number of the set $\mathbb{N}_0 + \mathbb{N}_0'$ of Example D is $\omega + \omega$.

If W is any infinite well-ordered set, then W begins

$$w_0 < w_1 < \ldots < w_n < \ldots . \tag{1}$$

In other words, W begins with a copy of $\mathbb{N}_0$. In order to make this idea precise we introduce the following terminology.

Definition. A subset A of a partially ordered set X is an *initial segment* of X if $y \in A$ and $x \leq y$ imply $x \in A$. In particular, if z is an arbitrary element of X, then the set $A_z = \{x \in X : x < z\}$ is an initial segment of X. We shall call A_z the initial segment of X *determined by* z.

It is a special feature of well-ordered sets that (almost) all of their initial segments are determined in this manner.

Lemma 5.1. *If W is a well-ordered set, and if A is an initial segment of W, then either $A = W$ or $A = A_w$ for some (unique) element w of W. The collection $\mathcal{A}$ of all initial segments of a well-ordered set W is well-ordered in the inclusion ordering, and the mapping $w \to A_w$ is an order isomorphism of W onto the complement in $\mathcal{A}$ of the singleton $\{W\}$ (so that* $\operatorname{ord} \mathcal{A} = (\operatorname{ord} W) + 1$).

Proof. If $A \neq W$, then $W \backslash A$ has a least element w, and it is clear that $A_w \subset A$. On the other hand, if x belongs to A, then $x < w$ must hold, since otherwise $x \geq w$ and therefore $w \in A$, contrary to fact. Thus $A \subset A_w$, and the proof of the first assertion of the lemma is complete. The second assertion is an immediate consequence of the first. □

Lemma 5.2. *Suppose that W_1 and W_2 are well-ordered sets, and that ϕ and ψ are both order isomorphisms of W_1 onto initial segments of*

W_2. Then $\phi = \psi$ and consequently $\phi(W_1) = \psi(W_1)$. In particular, no well-ordered set is order isomorphic to any initial segment of itself other than itself.

PROOF. If $\phi \neq \psi$, then the set $\{w \in W_1 : \phi(w) \neq \psi(w)\}$ has a least element w_0. We suppose, without loss of generality, that $\phi(w_0) < \psi(w_0)$. Since the range of ψ is an initial segment of W_2, there must be some $w < w_0$ such that $\psi(w) = \phi(w_0)$. On the other hand, by the definition of w_0, we must have $\psi(w) = \phi(w)$. Thus $\phi(w) = \phi(w_0)$ and therefore $w = w_0$, contrary to hypothesis. (The final assertion of the lemma may also be obtained from Problem 1P.) □

These lemmas enable us to prove one of the central results concerning well-ordered sets.

Theorem 5.3. *If W_1 and W_2 are any two well-ordered sets, then either W_1 is order isomorphic to an initial segment of W_2, or W_2 is order isomorphic to an initial segment of W_1. Moreover, in either case, the order isomorphism is unique, and both conclusions are valid if and only if W_1 and W_2 are order isomorphic to one another.*

PROOF. It suffices, in view of Lemma 5.2, to establish the first assertion of the theorem. Consider the collection $\mathcal{A}_0$ of all those initial segments A of W_1 with the property that there exists an order isomorphism of A onto an initial segment B of W_2. If $A \in \mathcal{A}_0$ then, according to Lemma 5.2, the initial segment B_A of W_2 to which A is order isomorphic, and the order isomorphism ϕ_A of A onto B_A, are both uniquely determined by A. Moreover, if $A \in \mathcal{A}_0$ and if A' is an initial segment of A, then it is easily seen that $\phi_A|A'$ is an order isomorphism of A' onto an initial segment B' of $B_A = \phi_A(A)$, whence it follows that A' belongs to $\mathcal{A}_0$ along with A.

Thus $\mathcal{A}_0$ is not only well-ordered in the inclusion ordering, but is an initial segment in the well-ordered set of all initial segments of W_1, and the mapping $A \to \phi_A$ of $\mathcal{A}_0$ into the partially ordered set of mappings of subsets of W_1 into W_2 (in the extension ordering; see Problem 1M) is an order isomorphism. In particular, the system $\{\phi_A\}_{A \in \mathcal{A}_0}$ is nested. Consider the supremum $\phi_0 = \bigcup_{A \in \mathcal{A}_0} \phi_A$. It is a triviality to check that ϕ_0 is an order isomorphism of the initial segment $A_0 = \bigcup \mathcal{A}_0$ of W_1 onto an initial segment B_0 of W_2, and hence that A_0 is itself an element of $\mathcal{A}_0$, and therefore the greatest element of $\mathcal{A}_0$.

Suppose now that A_0 is not equal to W_1 and B_0 is also not equal to W_2. Then there exist elements w_1 of W_1 and w_2 of W_2 such that $A_0 = A_{w_1}$ and $B_0 = A_{w_2}$. But then the mapping

$$\phi_0^+(w) = \begin{cases} \phi_0(w), & w \in A_0, \\ w_2, & w = w_1, \end{cases}$$

is an order isomorphism of the initial segment $A_1 = A_0 \cup \{w_1\}$ of W_1 onto an initial segment of W_2, in contradiction of the fact that A_0 is the greatest initial segment in $\mathcal{A}_0$. Hence either $A_0 = W_1$ or $B_0 = W_2$. □

Example H. If W_1 and W_2 are order isomorphic well-ordered sets, then by the foregoing result (or Lemma 5.2) the order isomorphism of W_1 onto W_2 is absolutely unique. Thus the order isomorphism $n \to n+1$ of $\mathbb{N}_0$ onto $\mathbb{N}$ is the *only* order isomorphism between these two ordered sets (Ex. 1W).

The preceding theorem shows that any two well-ordered sets are comparable in a certain precise sense. In order to exploit this fact we introduce the following notion.

Definition. Let W_1 and W_2 be well-ordered sets. Then ord $W_1 \leq$ ord W_2 if and only if W_1 is order isomorphic to an initial segment of W_2. (It is obvious that this relation is well-defined and is, in fact, a partial ordering.)

In terms of this notion, Theorem 5.3 translates immediately into the following result.

Corollary 5.4. *If ξ and η are any two ordinal numbers, then either $\xi \leq \eta$ or $\eta \leq \xi$.*

Thus the relation $\leq$ between ordinal numbers is a simple ordering. More than this is true, however. The following idea is of basic importance in the further study of ordinal numbers.

Definition. For any ordinal number α we shall denote by $W(\alpha)$ the *ordinal number segment* consisting of the set of all ordinal numbers ξ such that $\xi < \alpha$. (Thus, for example, $W(0) = \varnothing$ and $W(\omega) = \mathbb{N}_0$.)

If W_0 is a well-ordered set having α for its ordinal number, then the ordinal numbers in $W(\alpha)$ are, by definition, precisely the ordinal numbers of the various initial segments of W_0 (excluding W_0 itself). Thus for each ξ in $W(\alpha)$ there is a unique element w_ξ in W_0 such that $\xi =$ ord A_{w_ξ}. It follows that the mapping $\phi : W(\alpha) \to W_0$ defined by setting $\phi(\xi) = w_\xi$ is an order isomorphism of $W(\alpha)$ onto W_0. Indeed, it is clear that the range of ϕ is W_0, and it is not hard to see that ord $A_{w_\xi} <$ ord A_{w_η} if and only if $\xi < \eta$ in $W(\alpha)$. Thus we have proved the following basic theorem.

Theorem 5.5. *Every well-ordered set W can be indexed in a unique and order preserving manner by means of the ordinal numbers in the ordinal* *number segment $W(\alpha)$ consisting of all ordinal numbers less than $\alpha =$ ord W.*

It is an immediate consequence of this theorem that every ordinal number segment $W(\alpha)$ is well-ordered. Once again, however, more is true.

Corollary 5.6. *Every set of ordinal numbers is well-ordered.*

PROOF. If Θ is a nonempty set of ordinal numbers and if $\alpha \in \Theta$, then either $\Theta \cap W(\alpha)$ is empty, in which case α is the least element of Θ, or $\Theta \cap W(\alpha)$ is nonempty, in which case the least element of $W(\alpha)$ that belongs to Θ is the least element of Θ. □

Corollary 5.6 leads to another historically interesting and logically troubling observation. If there were a set of all ordinal numbers, then this well-ordered set would consist of all those ordinal numbers less than its own ordinal number. But then the latter ordinal number would have to be greater than every ordinal number, a clearly impossible situation known as the *Burali-Forti paradox* [4,5]. This logical difficulty joins the Cantor paradox of cardinal number theory as one of the so-called "antinomies" of set theory. Since these difficulties hold no interest for us, we shall say no more of them, except to remark, once again, that the modern point of view coincides with that of Cantor, viz., that there exist certain (curious) collections that cannot be dealt with as ordinary sets, and that the ordinal numbers and cardinal numbers provide examples of such collections. This unpleasantness does not hinder the study of analysis since in practice one is always able to choose a sufficiently large ordinal number α (or cardinal number c) and conduct business with the set $W(\alpha)$ of all ordinal numbers less than α (or with the set of all cardinal numbers less than c).

Theorem 5.5 provides a very convenient notation for dealing with well-ordered sets. (In fact, since it shows that every well-ordered set is order theoretically indistinguishable from some ordinal number segment $W(\alpha)$, we may, for the most part, deal directly with such number segments whenever well-ordered sets are needed.) Observe that this indexing is consistent with (1); the first element of a well-ordered set W is w_0, the second w_1, etc. If W is infinite and is not exhausted by the initial segment $\{w_n\}_{n \in \mathbb{N}_0}$, then the first element after all of these is w_ω. If still larger indices are required, we use the ordinal numbers $\omega+1, \omega+2, \ldots, \omega+n, \ldots$ that follow ω. Beyond these ordinal numbers we have $\omega+\omega, \omega+\omega+1$, and so on (see Problems D and G).

The question of how large a well-ordered set can be is an important one. In order to formulate the problem precisely, we begin with the trivial observation that if two well-ordered sets are order isomorphic, so that they have the same ordinal number, then they automatically have the same cardinal number as well. Hence it makes sense to define the *cardinal number* of an ordinal number α (notation: card α) to be the cardinal number of any well-ordered set that represents α. (Note that this mapping $\alpha \to$ card α from ordinal numbers to cardinal numbers is far from being one-to-one. Indeed, as we shall see, for each infinite cardinal number c there are infinitely many ordinal numbers α satisfying the equation card $\alpha = c$.) In these more precise terms we ask again: Which cardinal numbers are cardinal numbers of ordinal numbers? The surprising answer to this question is provided by the following theorem.

Theorem 5.7 (Zermelo's Well-ordering Theorem). *Every set can be well-ordered. Equivalently, every cardinal number is the cardinal number of some ordinal number.*

PROOF. Let X be a set and let $\mathcal{W}$ denote the collection of all pairs $(A, \leq_A)$, where A is a subset of X and $\leq_A$ is a well-ordering of A. If $(A, \leq_A)$ and $(B, \leq_B)$ are two elements of $\mathcal{W}$, we define $(A, \leq_A) \prec (B, \leq_B)$ to mean that A (in the ordering $\leq_A$) is contained in B as an initial segment. It is clear (Th. 5.3) that $\prec$ is a partial ordering of $\mathcal{W}$. Moreover, if $\{(A_\gamma, \leq_{A_\gamma})\}_{\gamma\in\Gamma}$ is an indexed subset of $\mathcal{W}$ that is simply ordered in the ordering $\prec$, then the collection of sets $\{A_\gamma\}_{\gamma\in\Gamma}$ is nested, and if we write $A = \bigcup_{\gamma\in\Gamma} A_\gamma$, it is obvious that there exists a unique partial ordering $\leq$ on A that extends the ordering $\leq_{A_\gamma}$ on A_γ for each index γ. We shall show that $(A, \leq)$ is an upper bound for the set $\{(A_\gamma, \leq_{A_\gamma})\}$ in $\mathcal{W}$.

Note first that each set A_γ (in the ordering $\leq_{A_\gamma}$) is an initial segment in A. Indeed, if a belongs to A_γ for some index γ, and if x is an element of A such that $x \leq a$, then $x \in A_{\gamma'}$ for some index γ', and either $A_{\gamma'}$ is an initial segment of A_γ, in which case it is obvious that $x \in A_\gamma$, or else A_γ is an initial segment in $A_{\gamma'}$, in which case, once again, it is clear that $x \in A_\gamma$. Hence, to prove that $(A, \leq)$ is an upper bound in $\mathcal{W}$ for the family $\{(A_\gamma, \leq_{A_\gamma})\}$ it suffices to show that $\leq$ is a well-ordering of A. Suppose B is a nonempty subset of A. If γ is an index such that $B \cap A_\gamma \neq \varnothing$, then the least element of $B \cap A_\gamma$ (in the ordering $\leq_{A_\gamma}$) is also the least element of B (in the ordering $\leq$) since A_γ is an initial segment of A. Thus $\leq$ is a well-ordering of A, so the pair $(A, \leq)$ belongs to $\mathcal{W}$.

We have shown, in summary, that $\mathcal{W}$ is a partially ordered set in which every simply ordered subset is bounded above, and it follows by Zorn's lemma that $\mathcal{W}$ possesses a maximal element—say $(A_0, \leq_0)$—in the ordering $\prec$. The proof of the theorem will be complete if we show that $A_0 = X$. But if $A_0 \neq X$, and if x_0 is any element of $X \backslash A_0$, then the well-ordered set $A_0 + \{x_0\}$ (Ex. D) contains A_0 as a proper initial segment, a manifest contradiction of the maximality of A_0. □

The foregoing argument shows that Zermelo's well-ordering theorem is implied by Zorn's lemma. As a matter of fact, it turns out that Zorn's lemma, Zermelo's well-ordering theorem, and the axiom of choice are all mutually equivalent. By this is meant that, given the other usual elementary apparatus of set theory, any one of these three propositions can be derived from any other. (Thus, from a purely logical point of view, one might as well take Zermelo's well-ordering theorem as one of the axioms of set theory; see Problems R, S, T, U, and V.)

Definition. For each infinite cardinal number c the collection N_c of all those ordinal numbers ξ such that card $\xi = c$ is called the *number class* of c. According to the preceding theorem the set N_c is never empty, and, being well-ordered (Cor. 5.6), possesses a least element called the *initial number* of c.

Example I. The initial number of the cardinal number $\aleph_0$ is ω. By Zermelo's well-ordering theorem there exist uncountable ordinal numbers, i.e., ordinal numbers α such that card $\alpha > \aleph_0$. Hence there is a *first* or *least uncountable ordinal number*. This ordinal number will be denoted by Ω. The number class of $\aleph_0$ (sometimes called the *first number class*) is then $W(\Omega)\backslash W(\omega) = W(\Omega)\backslash\mathbb{N}_0$.

Proposition 5.8. *Every set of cardinal numbers is well-ordered.*

Proof. It suffices to show that if C is a nonempty set of cardinal numbers, then C contains a least element. Let c_0 be a cardinal number belonging to C and let C_0 denote the set $\{c \in C : c \leq c_0\}$. It clearly suffices to show that C_0 has a least element. Let α be an ordinal number such that card $\alpha = c_0$. Then the mapping $c \to \omega_c$ assigning to each element c of C_0 its initial number is an order isomorphism of C_0 into $W(\alpha)$, and the result follows. □

Notation and Terminology. There is a smallest uncountable cardinal number, which we denote by $\aleph_1$ (*aleph one*). There is next a smallest cardinal number greater than $\aleph_1$, which we denote by $\aleph_2$. Continuing in this fashion, and assuming $\aleph_n$ already defined, we define $\aleph_{n+1}$ to be the smallest cardinal number to exceed $\aleph_n$. Then the smallest cardinal number exceeding $\aleph_n$ for every positive integer n we denote by $\aleph_\omega$, etc. In general, for any infinite cardinal number c, the collection of infinite cardinal numbers less than c will be written as

$$\aleph_0, \aleph_1, \ldots, \aleph_n, \ldots \aleph_\omega, \aleph_{\omega+1}, \ldots \aleph_\eta, \ldots,$$

that is, as a family $\{\aleph_\xi\}_{\xi<\alpha}$ of alephs indexed by the ordinal numbers less than a uniquely determined α. In this same spirit, the initial number of the cardinal number $\aleph_\xi$ will also be denoted by ω_ξ. (Whether $\aleph_1 = 2^{\aleph_0}$

is the continuum problem once again; whether $\aleph_{\xi+1} = 2^{\aleph_\xi}$ for all ξ is the *generalized continuum problem.*)

Example J. The initial numbers ω_0 and ω_1 are also known, of course, as ω and Ω, respectively (Ex. I).

Example K. The number class N_ξ of the aleph $\aleph_\xi$ is $W(\omega_{\xi+1})\backslash W(\omega_\xi)$.

Proposition 5.9. *The number class N_ξ of the infinite cardinal number $\aleph_\xi$ has cardinal number $\aleph_{\xi+1}$. (Thus the number class of each aleph is a representative of the next larger aleph.)*

PROOF. By the foregoing example, we have $W(\omega_{\xi+1}) = W(\omega_\xi) + N_\xi$, and therefore card $\omega_{\xi+1} =$ card $\omega_\xi +$ card N_ξ, or, equivalently,

$$\aleph_{\xi+1} = \aleph_\xi + \text{ card } N_\xi,$$

whence the result follows by Problem 4O. □

Corollary 5.10. *For any ordinal number α and any subset Θ of the ordinal number segment $W(\omega_{\alpha+1})$ such that* card $\Theta \leq \aleph_\alpha$*, Θ is bounded above in $W(\omega_{\alpha+1})$.*

PROOF. Let Ξ denote the initial segment $\bigcup_{\xi\in\Theta} W(\xi)$. Since card $W(\xi) =$ card $\xi \leq \aleph_\alpha$ for each ξ in Θ, and card $\Theta \leq \aleph_\alpha$, we have card $\Xi \leq \aleph_\alpha^2 = \aleph_\alpha$ (see Problems 4I and 4S), and it follows by Proposition 5.9 that there exist elements β of $W(\omega_{\alpha+1})$ that dominate every element of Ξ. □

Example L. Every countable subset of $W(\Omega)$ is bounded above in $W(\Omega)$.

We close this chapter with two theorems that encompass most of the usual applications of ordinal numbers to mathematical analysis.

Theorem 5.11 (Principle of Transfinite Induction). *Let α be an ordinal number, and let S be a subset of $W(\alpha)$. If S satisfies the condition that $W(\eta) \subset S$ implies $\eta \in S$ for each ordinal number η in $W(\alpha)$, then $S = W(\alpha)$.*

PROOF. Set $\Theta = W(\alpha)\backslash S$, and suppose $\Theta \neq \varnothing$. Then Θ contains a smallest ordinal number η by Corollary 5.6. But then, $W(\eta) \subset S$ and therefore $\eta \in S$, a contradiction. □

It is easy to see that if $\alpha \leq \omega$, the above theorem reduces to the more familiar principle of mathematical induction (cf. also Problem J). It should come as no surprise that there is likewise a natural extension of the procedure of definition by mathematical induction into the realm of transfinite ordinal numbers. In reading the following theorem, which clearly subsumes Theorem 2.5, the reader should bear in mind that most (if not all) definitions in mathematics may be regarded as definitions of functions.

Theorem 5.12 (Principle of Transfinite Definition). *Let α be an ordinal number, let X be a set, and let $\mathcal{F}$ denote the collection of all mappings f of number segments $W(\xi), \xi < \alpha$, into X. Suppose given a mapping g of $\mathcal{F}$ into X. Then there exists a unique mapping h of $W(\alpha)$ into X satisfying the equation $h(\xi) = g(h|W(\xi))$ for every ξ in $W(\alpha)$.*

PROOF. If $\alpha = 0$ there is nothing to prove. For positive α, consider the set Θ_0 of those ordinal numbers $\eta < \alpha$ such that there exists a mapping f of $W(\eta)$ into X satisfying the equation

$$f(\xi) = g(f|W(\xi)), \quad \xi < \eta. \tag{2}$$

We observe first that if $\eta = 0$, then $W(\eta) = \varnothing$ and the empty mapping satisfies (2), so that 0 belongs to Θ_0. Next, if η belongs to Θ_0 and if f_1 and f_2 are mappings of $W(\eta)$ into X, both of which satisfy (2), then $f_1 = f_2$. (Indeed, if $f_1 \neq f_2$, then there is a least element $\xi_0 < \eta$ such that $f_1(\xi_0) \neq f_2(\xi_0)$. But then $f_1|W(\xi_0) = f_2|W(\xi_0)$, which implies, contrary to hypothesis, that $f_1(\xi_0) = g(f_1|W(\xi_0)) = g(f_2|W(\xi_0)) = f_2(\xi_0)$.) Thus if η is in Θ_0, then there is a unique mapping $f : W(\eta) \to X$—call it f_η—such that (2) holds. Moreover, if ζ belongs to Θ_0 and if $\eta < \zeta$, then $f = f_\zeta|W(\eta)$ clearly satisfies (2), which shows that η also belongs to Θ_0 and that $f_\zeta|W(\eta) = f_\eta$. Thus Θ_0 is an initial segment in $W(\alpha)$, and the family of mappings $\{f_\eta\}_{\eta\in\Theta_0}$ is nested. Denote by k the supremum $k = \bigcup_{\eta\in\Theta_0} f_\eta$. The domain of definition of k is then the set $W = \bigcup_{\eta\in\Theta_0} W(\eta)$, which is an initial segment in $W(\alpha)$, and it is clear that k satisfies (2) for each η in W. (If Θ_0 has a greatest element β, then $W = W(\beta)$; if not, then $W = \Theta_0$.) If $W = W(\alpha)$, then we set $h = k$. On the other hand, if $W \neq W(\alpha)$, then $W = W(\eta_0)$ for some $\eta_0 < \alpha$, so k belongs to $\mathcal{F}$ and $g(k)$ is a well-defined element x_0 of X. Hence we may define a mapping k^+ as follows:

$$k^+(\xi) = \begin{cases} k(\xi), & \xi < \eta_0 \\ x_0, & \xi = \eta_0. \end{cases}$$

Then k^+ satisfies (2) for $\eta = \eta_0 + 1$, so that, if $\eta_0 + 1$ were less than α, then η_0 would belong to W, contrary to definition. Hence, we must have $\eta_0 + 1 = \alpha$, so that in this case we may set $h = k^+$. □

The role of the function g in Theorem 5.12 is to describe exactly how the "inductive step" is taken in a transfinite definition. We have introduced this notion in order to elucidate as clearly as possible the mechanics of an inductive definition. In actual practice, however, it is not customary to set forth a function "g" explicitly. All that is required for a valid transfinite definition is that a clear-cut recipe be given for determining $h(\xi)$ in terms of the behavior of h on $W(\xi)$. It should also be pointed out that what is defined in an inductive definition is most usually thought of as an indexed family $\{x_\xi\}_{\xi<\alpha}$ rather than as a function h; thus in a typical application of Theorem 5.12, neither "g" nor "h" is likely to appear explicitly.

PROBLEMS

A. Show that a simply ordered set S fails to be well-ordered if and only if S contains a subset that is order isomorphic to $\mathbb{N}_0^*$—the set of nonnegative integers in the inverse ordering (Ex. 1P). Conclude that both S and S^* are well-ordered when and only when S is finite.

B. Let W be a well-ordered set and let V be a subset of W. Show that ord $V \leq$ ord W. (Hint: Recall Problem 1P.)

C. Show that addition of ordinal numbers is associative but not commutative. Verify also that the ordinal numbers greater than a given ordinal number α are precisely the ordinal numbers of the form $\alpha + \xi$, $\xi > 0$, and give an example of two nonzero ordinal numbers α and ξ such that $\xi + \alpha = \alpha$. Is it possible for $\xi + \alpha$ to be strictly less than α?

D. According to the preceding problem, the smallest ordinal number greater than a given ordinal number α is $\alpha + 1$, and for each positive integer n the smallest ordinal number greater than $\alpha + n$ is $\alpha + n + 1$. Show that the smallest ordinal number greater than all of the numbers $\alpha + n, n \in \mathbb{N}_0$, is $\alpha + \omega$.

E. Prove that if β and γ are ordinal numbers such that $\beta < \gamma$, then $\alpha + \beta < \alpha + \gamma$ for every ordinal number α, and use this fact to show that if α, β and ξ, η are ordinal numbers such that $\alpha + \beta = \xi + \eta$ and $\beta < \eta$, then $\alpha > \xi$.

F. An ordinal number ρ is a *remainder* of an ordinal number α if there exists an ordinal number ξ such that $\alpha = \xi + \rho$. Show that for an arbitrary ordinal number α the number of distinct remainders of α is finite. Thus, for example, the only remainder of an initial number is that number itself. (Hint: Appeal to Problem A.)

G. An ordinal number that is the successor of some other ordinal number (that is, has an immediate predecessor in the natural ordering of ordinal numbers) is, by Problem D, of the form $\eta = \xi + 1$. Such ordinal numbers are said to be of *type* I. (The ordinal number 0 is also of type I by convention.) All other ordinal numbers are known as *limit ordinals* or *limit numbers*.

Examples of ordinal numbers of type I are $1, \omega+\omega+5$ and $\Omega+9$. Examples of limit ordinals are $\omega, \omega+\omega$, and Ω. Every initial number is a limit number. Every ordinal number α can be written (uniquely) as $\alpha = \lambda + n$ where λ is a limit number and $n \in \mathbb{N}_0$. Show also that $\alpha + \omega$ is the smallest limit number greater than α.

H. Let μ be a limit ordinal.

(i) Show that a limit ordinal λ in $W(\mu)$ is the greatest limit ordinal less than μ if and only if $\mu = \lambda + \omega$.

(ii) If λ is a limit number in $W(\mu)$, then $\lambda + \omega \leq \mu$ and therefore $V_\lambda = W(\lambda + \omega) \backslash W(\lambda) \subset W(\mu)$. Show that the sets V_λ, where λ ranges over the limit numbers in $W(\mu)$, together with $V_0 = W(\omega)$, constitute a partition of $W(\mu)$, and use this fact to give a new proof of Proposition 4.6.

I. Let Γ be a well-ordered index set, let $\{W_\gamma\}_{\gamma \in \Gamma}$ be a disjoint family of well-ordered sets indexed by Γ, and let $\leq_\gamma$ be the ordering of W_γ. Then the union $W = \bigcup_{\gamma \in \Gamma} W_\gamma$ can be ordered as follows: If x and y belong to W_γ and $W_{\gamma'}$, respectively, and if $\gamma \neq \gamma'$, then $x \lneqq y$ if $\gamma \lneqq \gamma'$ while if $\gamma = \gamma'$ then $x \leq y$ if and only if $x \leq_\gamma y$. Show that W is well-ordered by this ordering and that it makes sense to define

$$\sum_{\gamma \in \Gamma} \text{ord } W_\gamma = \text{ ord } W.$$

Show too that for any family $\{\alpha_\gamma\}$ indexed by Γ, we have

$$\text{card}\left(\sum_{\gamma \in \Gamma} \alpha_\gamma\right) = \sum_{\gamma \in \Gamma} \text{ card } \alpha_\gamma.$$

(When the index set is finite, say $\Gamma = \{\gamma_0, \ldots, \gamma_n\}$, we sometimes write $\alpha_{\gamma_0} + \ldots + \alpha_{\gamma_n}$ for the sum $\sum_{\gamma \in \Gamma} \alpha_\gamma$.) Verify that if $\{\alpha_\gamma\}$ is a family of ordinal numbers indexed by Γ, and if $\Delta \subset \Gamma$, then

$$\sum_{\gamma \in \Delta} \alpha_\gamma \leq \sum_{\gamma \in \Gamma} \alpha_\gamma.$$

Prove, finally, that if λ is a limit number, then the sum $\sum_{\xi < \lambda} \alpha_\xi$ of a family $\{\alpha_\xi\}$ of ordinal numbers indexed by $W(\lambda)$ is the least ordinal number greater than or equal to all of the "partial sums" $\sum_{\xi < \eta} \alpha_\xi, \eta < \lambda$. (Thus, in particular, for a sequence of ordinal numbers $\{\alpha_n\}_{n=0}^{\infty}$, the sum

$$\sum_{n=0}^{\infty} \alpha_n = \sum_{n \in \mathbb{N}_0} \alpha_n$$

is the supremum of the set of finite sums $\alpha_0 + \ldots + \alpha_n, n \in \mathbb{N}_0$.)

J. Let $p(\ )$ be a predicate that is either true or false for every ordinal number in some ordinal number segment $W(\alpha), \alpha > 0$. Suppose the following conditions are satisfied:

(1) $p(0)$ is true,
(2) If $p(\xi)$ is true, and if $\xi + 1 < \alpha$, then $p(\xi + 1)$ is true,
(3) If λ is a limit number in $W(\alpha)$, and if $p(\xi)$ is true for every ξ in $W(\lambda)$, then $p(\lambda)$ is true.

Show that $p(\xi)$ is true for every ξ in $W(\alpha)$. (This is another version of Theorem 5.11 having the slight advantage that, when $\alpha \leq \omega$, (3) becomes vacuous, whereupon the result reduces to the familiar principle of mathematical induction.)

K. Call a subset J of a partially ordered set X *inductive* if $y \in X$ and $A_y \subset J$ imply $y \in J$. (This notion of an inductive set should be distinguished from those introduced in Chapters 1 and 2.) The essence of Theorem 5.11 is that if W is a well-ordered set, the only inductive subset of W is W itself. Show, in the converse direction, that if X is simply ordered and if the only inductive subset of X is X itself, then X is well-ordered. Does this assertion remain valid if the assumption that X is simply ordered is dropped?

L. If x is an element of a partially ordered set X, then the *successor* x^+ of x is the least element of the set of elements of X strictly larger than x. (If no such least element exists, then x has no successor.) Suppose X has the following properties:

(1) X has a least element,
(2) Every element of X that is not maximal has a successor,
(3) Every subset of X that is bounded above in X has a supremum in X.

Prove that X is well-ordered. (Give examples to show that no two of these three properties imply the third.)

M. If X and Y are arbitrary simply ordered sets, there is a useful way to order the product $X \times Y$. If $x_i \in X$ and $y_i \in Y$, $i = 1, 2$, we declare (x_1, y_1) to be less than (x_2, y_2) if $x_1 < x_2$ in X or if $x_1 = x_2$ and $y_1 < y_2$ in Y. (This ordering is known as the *lexicographical ordering* of $X \times Y$.) Verify that if V and W are well-ordered sets, then $V \times W$ is well-ordered in the lexicographical ordering, and that it makes sense to define

$$(\text{ord } W)(\text{ord } V) = \text{ord } (V \times W).$$

(Note the reversal in the order of writing this *product* of two ordinal numbers.) Prove that multiplication of ordinal numbers is associative but not commutative. Show also that the distributive law $\alpha(\beta + \gamma) = \alpha\beta + \alpha\gamma$ is valid in the arithmetic of ordinal numbers, but that the equation $(\beta+\gamma)\alpha = \beta\alpha + \gamma\alpha$ may fail to hold. Conclude that if $\alpha > 0$ and $\beta < \gamma$, then $\alpha\beta < \alpha\gamma$, and give an example of ordinal numbers $\alpha > 0$ and $\beta < \gamma$ such that $\beta\alpha = \gamma\alpha$. Is it possible for $\beta\alpha$ to exceed $\gamma\alpha$ when $\beta < \gamma$?

N. Show that if α and β are arbitrary ordinal numbers, then $\alpha\beta = \sum_{\xi<\beta} \alpha_\xi$, where $\alpha_\xi = \alpha$, $\xi < \beta$. (Thus for ordinal numbers, just as for cardinal numbers, multiplication may be viewed as the result of repeated additions; cf. Example 4Q.)

O. The idea of Problem M is readily generalized. Let Γ be a well-ordered index set, and let $\{W_\gamma\}_{\gamma\in\Gamma}$ be a family of well-ordered sets indexed by Γ. The *lexicographical ordering* of $\Pi = \prod_{\gamma\in\Gamma} W_\gamma$ is then defined by setting $\{x_\gamma\} < \{\widetilde{x}_\gamma\}$ if $x_{\gamma_0} < \widetilde{x}_{\gamma_0}$ where γ_0 denotes the least index γ at which $x_\gamma \neq \widetilde{x}_\gamma$. Verify that this ordering turns Π into a well-ordered set, and use this construction to define the *product*

$$\ldots \eta_\xi \ldots \eta_1 \eta_0$$

of a family of ordinal numbers $\{\eta_\xi\}_{\xi<\alpha}$. Prove that card $(\ldots \eta_\xi \ldots \eta_0) = \prod_{\xi<\alpha}$ card η_ξ.

P. There is even a notion of division for ordinal numbers. Show that if α and β are any two ordinal numbers, and if $\alpha > 0$, then there exist uniquely determined ordinal numbers ξ and ρ such that

$$\beta = \alpha\xi + \rho \quad \text{and} \quad \rho < \alpha.$$

(Hint: Let A be a well-ordered set such that ord $A = \alpha$, and let X be a well-ordered set large enough so that ord $(X \times A)$ exceeds β (Th. 5.7).) Show, in particular, that if μ is a limit number, than μ is divisible by ω, by verifying that $\mu = \omega(1 + \lambda)$ where λ denotes the ordinal number of the set of all limit numbers in $W(\mu)$. (Hint: Recall Problem H.)

Q. Let ω_α be an initial number, and let Λ denote the set of limit ordinals in $W(\omega_\alpha)$. Find ord Λ.

R. Prove that Zorn's lemma implies the axiom of products (and hence the axiom of choice; Chapter 1). (Hint: Let $\{X_\gamma\}_{\gamma\in\Gamma}(\Gamma \neq \varnothing)$ be an indexed family of nonempty sets, and consider the collection $\mathcal{M}$ of all mappings ϕ of subsets of the index set Γ into $\bigcup_{\gamma\in\Gamma} X_\gamma$ satisfying the condition $\phi(\gamma) \in X_\gamma$ for every γ in the domain of ϕ. The set $\mathcal{M}$ is partially ordered in the extension ordering (Prob. 1M).)

S. (i) Prove that Zermelo's well-ordering theorem implies the axiom of choice. (Hint: Given a collection $\mathcal{C}$ of nonempty sets, let U denote the union of $\mathcal{C}$, and well-order U.)

(ii) Let X be a partially ordered set with the property that every simply ordered subset of X is bounded above in X, and let W be a well-ordered set. Show that if X has no maximal element, then there exists an order isomorphism of W into X. (Hint: Well-order X and use Theorem 5.12.)

Use this fact to prove that Zermelo's well-ordering theorem also implies Zorn's lemma.

T. Let X be a nonempty partially ordered set with the property that every nonempty simply ordered subset of X has a supremum in X, and let ϕ be a mapping of X into itself such that $\phi(x) \geq x$ for every x in X. We shall call a subset T of X a *ϕ-tower* if (a) $\phi(T) \subset T$ and (b) if S is a nonempty simply ordered subset of T, then the supremum of S (in X) belongs to T. If x is an element of X, then T is a ϕ-tower *over* x if T is a ϕ-tower and $x \in T$.

(i) Show that for each element x of X there exists a unique smallest ϕ-tower t_x over x. Show also that x is the least element of t_x, that an element y of X belongs to t_x if and only if $t_y \subset t_x$, and that $t_x = \{x\} \cup t_{\phi(x)}$.

(ii) For each element x of X and element y of t_x let us write $A_y^{(x)}$ for the initial segment of t_x determined by y : $A_y^{(x)} = \{z \in t_x : z < y\}$. An element y of t_x is called *x-normal* if $t_x = A_y^{(x)} \cup t_y$. Show that an element y of t_x is x-normal if and only if $A_y^{(x)} \cup t_y$ is a ϕ-tower. Show also that if y is an x-normal element of t_x, then (1) y is comparable with every element of t_x, and (2) there is no element z of t_x such that $y < z < \phi(y)$.

(iii) For each element x of X an element y of t_x is *x-hypernormal* if y and all of the elements of $A_y^{(x)}$ are x-normal. Prove that x is x-hypernormal, and that if y is x-hypernormal, then $\phi(y)$ is x-hypernormal also. Complete the proof that the set H_x of all x-hypernormal elements of t_x is a ϕ-tower, and hence that $H_x = t_x$. (Hint: If S is a nonempty simply ordered subset of H_x, and if z_0 denotes the supremum of S in X, then $z_0 \in t_x$ since t_x is a ϕ-tower. Moreover, if $z \in A_{z_0}^{(x)}$, then there exists an element y of S such that $z < y$, and it follows that $\phi(z) \leq y$ since y and z are both x-normal. Show that if S_0 is a nonempty simply ordered subset of $A_{z_0}^{(x)}$, then $\sup S_0 \leq z_0$.)

(iv) Complete the proof that, for each x in X, the ϕ-tower t_x is a well-ordered subset of X. (Hint: Recall Problem L.)

(v) (Bourbaki [2]) For each element x of X the ϕ-tower t_x possesses a greatest element z_0, and z_0 is a fixed point for ϕ. In particular, ϕ has at least one fixed point in X.

U. Use the results of the preceding problem to prove that the axiom of choice implies Zermelo's well-ordering theorem. (Hint: It follows from the axiom of choice (Ex. 1N) that for any set X there is a mapping s of $2^X \backslash \{X\}$ into X such that $s(E) \notin E$ for each E in $2^X \backslash \{X\}$. Set $\phi(E) = E \cup \{s(E)\}$ for each E in $2^X \backslash \{X\}$, define $\phi(X) = X$, and let T be a well-ordered ϕ-tower in 2^X over $\varnothing$. Then for each element x of X there is a smallest set E in T such that $x \in E$. Call this set $E(x)$, and show that the mapping $x \to E(x)$ is one-to-one.)

V. Use the results of Problem T to show that the axiom of choice also implies the maximum principle. (Hint: If X is a partially ordered set, then the simply ordered subsets of X constitute a partially ordered subset of 2^X.)

6 Metric spaces

The root idea in all that follows is that of "closeness" as between two points of a space. This idea turns out to be elusive and difficult to pin down in general, but it presents no difficulty at all if one is given a numerical gauge for measuring how close together two points are, and that is the case of interest in the present chapter.

Definition. If X is an arbitrary set, then a real-valued function ρ on $X \times X$ is a *metric* on X if the following conditions are satisfied for all elements x, y, z of X:

(1) $\rho(x,y) \geq 0$; $\rho(x,y) = 0$ when and only when $x = y$,
(2) $\rho(x,y) = \rho(y,x)$,
(3) $\rho(x,z) \leq \rho(x,y) + \rho(y,z)$.

The value of the metric ρ at a pair (x,y) of points of X is called the *distance* between x and y. A set X equipped with a metric on X is a *metric space.*

Notation and Terminology. In view of this definition it is clear that a metric space is really an ordered pair (X, ρ) in which the first term is a *carrier*—a set of elements—and the second term is a metric on that carrier, and we shall frequently use exactly this notation for metric spaces. In keeping with almost universal practice, however, we shall also speak sometimes of X itself as the metric space, or of the metric space X "equipped with the metric ρ." The inequality (3) in the foregoing definition is known as the *triangle inequality* for metric spaces.

Example A. The function $\rho(s,t) = |s-t|$ is readily seen to be a metric on the real line $\mathbb{R}$ (cf. Problem 2L), and the metric space obtained by equipping $\mathbb{R}$ with this *usual metric* is the most important metric space of all. An immediate generalization is the vector space $\mathbb{R}^n$ of all real n-tuples equipped with the metric

$$\rho(x,y) = \left[\sum_{i=1}^{n}(s_i - t_i)^2\right]^{1/2},$$

where $x = (s_1, \ldots, s_n)$ and $y = (t_1, \ldots, t_n)$. The space $\mathbb{R}^n$ equipped with this *usual* or *Euclidean metric* is called *Euclidean space* of dimension n. Similarly, the function $\rho(\alpha,\beta) = |\alpha - \beta|$ is a metric—the *usual metric*—on the complex plane $\mathbb{C}$ (Prob. 2X). More generally, the function

$$\rho(x,y) = \left[\sum_{i=1}^{n}|\alpha_i - \beta_i|^2\right]^{1/2},$$

where $x = (\alpha_1, \ldots, \alpha_n)$ and $y = (\beta_1, \ldots, \beta_n)$, is a metric on $\mathbb{C}^n$. The space $\mathbb{C}^n$ equipped with this *usual metric* is called n-dimensional *unitary space*. (It is by no means obvious that the usual metrics on $\mathbb{R}^n$ and $\mathbb{C}^n$ are really metrics; see Problem B.) *Whenever in the sequel $\mathbb{R}^n$ or $\mathbb{C}^n$ is regarded as a metric space, the metric on the space will be understood to be the usual metric unless the contrary is expressly stipulated.*

Example B. There are other metrics of interest on the space $\mathbb{R}^n$ besides the Euclidean metric. Thus setting

$$\rho_1(x,y) = \sum_{i=1}^{n}|s_i - t_i|$$

for any two points $x = (s_1, \ldots, s_n)$ and $y = (t_1, \ldots, t_n)$ defines a metric ρ_1 on $\mathbb{R}^n$, and

$$\rho_\infty(x,y) = |s_1 - t_1| \vee \cdots \vee |s_n - t_n|$$

defines yet another metric ρ_∞ on $\mathbb{R}^n$. (These metrics on $\mathbb{R}^n$ are different from the usual metric introduced in Example A except, of course, for the case $n = 1$.) The Euclidean metric on $\mathbb{R}^n$ is sometimes denoted by ρ_2 to emphasize its essentially quadratic natur . These remarks all generalize at once to the complex space $\mathbb{C}^n$.

Example C. Let X be an arbitrary set, and define

$$\rho(x,y) = \begin{cases} 0, & x = y, \\ 1, & x \neq y, \end{cases}$$

for all x and y in X. Then ρ is readily seen to be a metric on X, called the *discrete metric* on X.

If (X, ρ) is a metric space and A is a subset of X, then the restriction $\rho|(A \times A)$ is clearly a metric on A—the *relative metric*. A subset A of (X, ρ) equipped with this relative metric is a *subspace* of X. Whenever in the sequel a subset of a metric space is regarded as a metric space in its own right, it is this relative metric that is in use unless the contrary is expressly stipulated. In particular, each subset A of $\mathbb{R}^n[\mathbb{C}^n]$ becomes a metric space in a natural way as a subspace of $\mathbb{R}^n[\mathbb{C}^n]$ in the usual metric, and it is this subspace of $\mathbb{R}^n[\mathbb{C}^n]$ that will be denoted by A unless some other metric is indicated. (It is also appropriate to note that $\mathbb{R}^n$ in the Euclidean metric is a subspace of $\mathbb{C}^n$ in the usual metric.)

Here is a brief list of some of the most basic concepts pertinent to the theory of metric spaces.

Definition. Let (X, ρ) be a metric space. For each point x_0 of X and each nonnegative number r the set

$$D_r(x_0) = \{x \in X : \rho(x, x_0) < r\}$$

is the *open ball* with *center* x_0 and *radius* r. (If $<$ is replaced by $\leq$ in this definition, the result is the *closed ball* with *center* x_0 and *radius* r. The ball $D_0(x_0)$ with radius zero is empty; otherwise $D_r(x_0)$ always contains at least the center x_0. The closed ball with center x_0 and radius 0 is $\{x_0\}$.) If A is a nonempty subset of X, then the supremum

$$\sup\{\rho(x, y) : x, y \in A\}$$

(taken in $\mathbb{R}^\natural$) is the *diameter* of A (notation: diam A). If A is nonempty and diam $A < +\infty$, then A is *bounded.* (By convention the empty set $\varnothing$ is bounded and diam $\varnothing = 0$.) If A is a nonempty subset of X and y is an arbitrary point of X, then the *distance from* y *to* A (notation: $d(y, A)$) is the infimum

$$\inf\{\rho(x, y) : x \in A\},$$

and the set of all those points y of X such that $d(y, A) < r (r > 0)$ will be denoted by $D_r(A)$. (Thus $D_r(A) = \bigcup\{D_r(x) : x \in A\}$.) Likewise, if A and B are any two nonempty subsets of X, then the *distance between* A and B (notation: $d(A, B)$) is the infimum

$$\inf\{\rho(x, y) : x \in A, y \in B\},$$

and a mapping ϕ of a set Z into X is *bounded* if the range $\phi(Z)$ is a bounded subset of X.

Note. In $\mathbb{R}$ equipped with its usual metric (Ex. A) the open ball with center t and radius r coincides with the open interval $(t-r, t+r)$, and, conversely, the open interval (a, b) is the open ball with center $(a+b)/2$ and radius $(b-a)/2$. We shall use this observation freely in the sequel without further explanation. Likewise, a subset M of $\mathbb{R}$ is bounded as a subset of the metric space $\mathbb{R}$ if and only if M is bounded in $\mathbb{R}$ regarded as an ordered set (see Chapter 1). Thus the notion of a *bounded subset* of $\mathbb{R}$ is unambiguously defined, as is therefore the notion of a *bounded real-valued function* on a given set Z.

The following is nothing more than a convenient summary of some of the obvious relations between the above concepts, and no proof is given.

Proposition 6.1. *Let (X, ρ) be a metric space, and let A be a subset of X. Then A is bounded if and only if there exists some open ball $D_r(x_0)$ containing A. In particular, every open ball in X is a bounded set with* diam $D_r(x_0) \leq 2r$. *The diameter of each singleton $\{x\}$ in X is zero, and the diameter of any subset of X containing two or more distinct points is strictly positive. For any nonempty subset A of X and every point x of $X, d(x, A) = d(\{x\}, A)$. For any two nonempty subsets A and B of X,*

$$d(A, B) = d(B, A) = \inf\{d(x, B) : x \in A\}.$$

Example D. We have defined the distance $d(A, B)$ between two nonempty subsets of a metric space (X, ρ). It will also be convenient to define

$$D(A, B) = \sup\{\rho(x, y) : x \in A, y \in B\}$$

for any two nonempty subsets A and B of X. (Observe that $D(A, B) < +\infty$ when and only when both A and B are bounded, and also that $D(A, A) =$ diam A.) Each of these two functions possesses some of the properties of a metric and fails to possess others. To facilitate the following discussion let us write $\mathcal{B}_0$ for the collection of all bounded nonempty subsets of (X, ρ).

To begin with, the function d on $\mathcal{B}_0 \times \mathcal{B}_0$ is nonnegative and satisfies the conditions $d(A, A) = 0$. Moreover, d is *symmetric* (that is $d(A, B) = d(B, A)$, for all $A, B \in \mathcal{B}_0$). It is not true, however, that $d(A, B) = 0$ implies that $A = B$, and d does not satisfy the triangle inequality (examples?). As regards the function D, it is also symmetric, nonnegative and finite-valued on $\mathcal{B}_0 \times \mathcal{B}_0$, and D *does* satisfy the triangle inequality (proof?), but, as noted above, $D(A, A) > 0$ unless A is a singleton. It is also worthy of note that D is monotone increasing in each variable (where, as usual, $\mathcal{B}_0$ is equipped with the inclusion ordering), while d is monotone decreasing in each variable.

There is another nonnegative real-valued function on $\mathcal{B}_0 \times \mathcal{B}_0$ that is of interest. For each A and B in $\mathcal{B}_0$ we shall write

$$\tau(A, B) = \sup\{d(x, B) : x \in A\}.$$

It is obvious that $d \leq \tau \leq D$ everywhere on $\mathcal{B}_0 \times \mathcal{B}_0$, and also that $\tau(A, A) = 0$ for every A in $\mathcal{B}_0$. Unfortunately, τ is not symmetric, and $\tau(A, B) = 0$ does not imply $A = B$ (examples?). On the other hand, τ *does* satisfy the triangle inequality. Indeed, if r is a positive number and if $\tau(A, B) < r$, then $A \subset D_r(B)$, while if $A \subset D_r(B)$, then $\tau(A, B) \leq r$. Thus if $\tau(A, B) < r$ and $\tau(B, C) < s$, then $A \subset D_r(B)$ and $B \subset D_s(C)$. But then $A \subset D_{r+s}(C)$, so $\tau(A, C) \leq r + s$, and letting r and s tend downward to $\tau(A, B)$ and $\tau(B, C)$, respectively, we obtain the stated result. The function τ is also easily seen to be monotone increasing as a function of its first variable and monotone decreasing as a function of its second variable.

Whenever one metric space is given there are several ways of associating other metric spaces with it. The most important construction of this sort is the following one.

Definition. Let (X, ρ) be a metric space, let Z be a nonempty set, let $\mathcal{B}(Z; X)$ denote the collection of all bounded mappings of Z into X, and let ϕ and ψ be elements of $\mathcal{B}(Z; X)$. Then, as is readily seen, the real-valued function $z \to \rho(\phi(z), \psi(z))$ is bounded above on Z. Hence setting

$$\rho_\infty(\phi, \psi) = \sup\{\rho(\phi(z), \psi(z)) : z \in Z\}$$

for each pair ϕ, ψ of elements of $\mathcal{B}(Z; X)$ defines a nonnegative real-valued function on $\mathcal{B}(Z; X) \times \mathcal{B}(Z; X)$.

Proposition 6.2. *The function ρ_∞ just defined is, in fact, a metric on $\mathcal{B}(Z; X)$. (The reader will note that the notation introduced here squares with that of Example B.)*

PROOF. It is evident that ρ_∞ satisfies both (1) and (2) in the definition of a metric, so it suffices to establish the triangle inequality. To this end we observe that if ϕ, ψ and ω are all bounded mappings of Z into X, then for any element z of Z we have

$$\begin{aligned} \rho(\phi(z), \omega(z)) &\leq \rho(\phi(z), \psi(z)) + \rho(\psi(z), \omega(z)) \\ &\leq \rho_\infty(\phi, \psi) + \rho_\infty(\psi, \omega), \end{aligned}$$

whence it follows at once that $\rho_\infty(\phi, \omega) \leq \rho_\infty(\phi, \psi) + \rho_\infty(\psi, \omega)$. □

Note. In this construction we have thus far excluded the special case $Z = \varnothing$. As it happens, this is both inconvenient and unnecessary. Indeed, if Z is empty, then there is but one mapping of Z into X, viz., the empty mapping $\varnothing$. This mapping is bounded (by convention) and $\mathcal{B}(\varnothing; X)$ is just the singleton $\{\varnothing\}$, which supports only one metric, the trivial *zero metric*. In the sequel it is in this sense that we shall interpret the metric space $(\mathcal{B}(Z;X), \rho_\infty)$ in the event that the set Z is empty.

There is an obvious notion of isomorphism between metric spaces.

Definition. A mapping ϕ of one metric space (X, ρ) into another metric space (Y, ρ') is *isometric* if it preserves distances, i.e., if $\rho'(\phi(x), \phi(y)) = \rho(x, y)$ for all pairs of points x, y in X. An isometric mapping of X onto Y is called an *isometry* of X onto Y. (It is obvious that an isometric mapping is necessarily one-to-one, and that inverses and compositions of isometries are isometries. Moreover, the identity mapping of any metric space onto itself is an isometry on that space.)

Example E. If a is any fixed element of $\mathbb{R}^n$ then the *translation* $x \to x+a$ is an isometry of $\mathbb{R}^n$ onto itself (with respect to any of the metrics of Examples A and B). Similarly, all translations on $\mathbb{C}^n$ are isometries of $\mathbb{C}^n$ onto itself. More generally, the translations on any normed space are isometries with respect to the metric defined by the norm on that space; see Problem A. The mapping assigning to each complex n-tuple $(\xi_1, \ldots, \xi_n)$ the real $2n$-tuple (Re ξ_1, Im $\xi_1, \ldots,$ Re ξ_n, Im ξ_n) is an isometry of $\mathbb{C}^n$ onto $\mathbb{R}^{2n}$ in the Euclidean metric. If X and Y are any two sets of the same cardinal number, then an arbitrary one-to-one correspondence between X and Y is an isometry if X and Y are both equipped with the discrete metric (Ex. C).

Sequences of points provide a very important tool in the theory of metric spaces. In this context the following idea is of basic importance.

Definition. Let (X, ρ) be a metric space, let $\{x_n\}$ be a sequence in X (indexed either by $\mathbb{N}$ or $\mathbb{N}_0$), and let a belong to X. Then the sequence $\{x_n\}$ *converges* to a, or has *limit* a (notation: $\lim_n x_n = a$ or, more simply, $x_n \to a$) if for each positive number ε there exists an index n_0 (depending ordinarily on ε) such that $\rho(a, x_n) < \varepsilon$ whenever $n \geq n_0$ (equivalently, if for each $\varepsilon > 0$ the sequence $\{x_n\}$ belongs to $D_\varepsilon(a)$ eventually).

Note. If $\{x_n\}$ is a sequence in a metric space (X, ρ) that converges to a point a, and if $b \neq a$ in X, then $\{x_n\}$ does not converge to b. (If $d = \rho(a, b) > 0$, then $D_{d/2}(a)$ and $D_{d/2}(b)$ are disjoint open balls in X, and if $\{x_n\}$ converges to a, then an entire tail of $\{x_n\}$ is contained in $D_{d/2}(a)$, and is therefore disjoint from $D_{d/2}(b)$.) Thus *the limit of a convergent sequence in a metric space is unique*.

If $\{t_n\}$ is a sequence of real numbers and $a \in \mathbb{R}$, then $\lim_n t_n = a$ if and only if for each positive number ε there exists a positive integer n_0 such that $|a - t_n| < \varepsilon$ (or, equivalently, such that $a - \varepsilon < t_n < a + \varepsilon$) for all $n \geq n_0$. In brief, if the real number system $\mathbb{R}$ is equipped with its usual metric, then the idea of convergence of sequences in the metric space $\mathbb{R}$ coincides with the ordinary notion of convergence of infinite sequences familiar from elementary calculus. Thus, in particular, if $\{t_n\}$ is a convergent sequence in $\mathbb{R}$ and if for some real number c we have $c \leq t_n [t_n \leq c]$ for all n, then $c \leq \lim_n t_n [\lim_n t_n \leq c]$. Hence, if $c \leq t_n \leq d$ for all n and for some real numbers c and d, then $c \leq \lim_n t_n \leq d$.

Example F. Let $\{x_n = (t_1^{(n)}, \ldots, t_m^{(n)})\}$ be a sequence in $\mathbb{R}^m$ and suppose $a = (a_1, \ldots, a_m)$ belongs to $\mathbb{R}^m$. An easy argument shows that $\lim_n x_n = a$ with respect to the usual metric on $\mathbb{R}^m$ (Ex. A) and also with respect to either of the two other metrics ρ_1 and ρ_∞ of Example B, if and only if $\lim_n t_i^{(n)} = a_i$, $i = 1, \ldots, m$. In brief, a sequence in $\mathbb{R}^m$ converges with respect to any one of these three metrices if and only if it converges *termwise* or *coordinatewise*. Similarly, a sequence $\{x_n = (\xi_1^{(n)}, \ldots, \xi_m^{(n)})\}$ in $\mathbb{C}^m$ converges to a limit $a = (\alpha_1, \ldots, \alpha_m)$ in the usual metric on $\mathbb{C}^m$ if and only if it converges termwise to a, i.e., if and only if $\lim_n \xi_i^{(n)} = \alpha_i$, $i = 1, \ldots, m$, in the metric space $\mathbb{C}$. Moreover, a sequence $\{\xi_n\}$ in $\mathbb{C}$ converges to a limit α in the usual metric on $\mathbb{C}$ if and only if $\lim_n \operatorname{Re} \xi_n = \operatorname{Re} \alpha$ and $\lim_n \operatorname{Im} \xi_n = \operatorname{Im} \alpha$.

Example G. If $\{a_n\}_{n=1}^\infty$ is a sequence of vectors in a normed space $\mathcal{E}$ (Prob. A), then the *infinite series* $\sum_{n=1}^\infty a_n$ is said to *converge to* s, or to have *sum* s (notation: $s = \sum_{n=1}^\infty a_n$), if the corresponding sequence $\{s_n\}_{n=1}^\infty$ of *partial sums*

$$s_n = \sum_{i=1}^n a_i$$

converges to s in $\mathcal{E}$. (If $\{a_n\}_{n=0}^\infty$ is indexed by $\mathbb{N}_0$, we write $\sum_{n=0}^\infty a_n$ for the infinite series, etc.) Just as in elementary analysis, a necessary—but by no means sufficient—condition for the convergence of $\sum_{n=1}^\infty a_n$ is that the sequence $\{a_n\}$ of *terms* should tend to 0.

Example H. Let (X, ρ) be a metric space, let Z be a set, and let $\mathcal{B}(Z; X)$ denote the metric space of all bounded mappings of Z into X equipped with the metric ρ_∞ introduced in Proposition 6.2. A sequence $\{\phi_n\}$ in $\mathcal{B}(Z; X)$ converges to a limit ψ in $\mathcal{B}(Z; X)$ if and only if the following condition is satisfied. Given an arbitrary positive number ε there exists an index n_0 such that $\rho(\phi_n(z), \psi(z)) < \varepsilon$ for all z in Z and for all $n \geq n_0$. In other words, a sequence in $\mathcal{B}(Z; X)$ converges to a limit in that space if and only if it converges to that limit *uniformly on Z*. For this reason the metric ρ_∞ on $\mathcal{B}(Z;X)$ is known as the *metric of uniform convergence* on Z.

Just as nets constitute a natural generalization of infinite sequences (cf. Chapter 1), there is a notion of convergence for nets in a metric space that generalizes the notion of convergence of a sequence.

Definition. Let Λ be a directed set, let $\{x_\lambda\}_{\lambda\in\Lambda}$ be a net in a metric space (X,ρ) indexed by Λ, and let a_0 be a point of X. Then the net $\{x_\lambda\}$ *converges* to a_0, or has *limit* a_0 (notation: $\lim_\lambda x_\lambda = a_0$ or, more simply, $x_\lambda \to a_0$), if for each positive number ε there exists an index λ_0 in Λ such that $\rho(a_0, x_\lambda) < \varepsilon$ whenever $\lambda \geq \lambda_0$. (Just as in the special case of sequences, the limit of a convergent net in a metric space is unique.)

Example I. A nonempty open interval $U = (a,b)$ in $\mathbb{R}$ is directed both upward and downward (as is, in fact, any nonempty simply ordered set). Hence if ϕ is a mapping of U into a metric space X, there are two senses in which ϕ can be convergent. The limit (if it exists) of ϕ as a net indexed by U regarded as an upward directed set is ordinarily denoted by

$$\lim_{t\uparrow b} \phi(t)$$

and is called the limit of ϕ at b *from the left* or *from below*. Dually, the limit (if it exists) of ϕ as a net indexed by U regarded as a downward directed set is ordinarily denoted by

$$\lim_{t\downarrow a} \phi(t)$$

and called the limit of ϕ at a *from the right* or *from above*.

Example J. Let Λ be a directed set and let $\{t_\lambda\}_{\lambda\in\Lambda}$ be a monotone increasing net of real numbers indexed by Λ. If $\{t_\lambda\}$ is bounded above in $\mathbb{R}$, and if $u = \sup_\lambda t_\lambda$, then

$$\lim_\lambda t_\lambda = u.$$

Indeed, if $\varepsilon > 0$, then $u - \varepsilon$ is not an upper bound for $\{t_\lambda\}$, so there exists an index λ_0 such that $t_{\lambda_0} > u - \varepsilon$. Bu then $u - \varepsilon < t_\lambda \leq u < u + \varepsilon$, and therefore $|u - t_\lambda| = u - t_\lambda < \varepsilon$, for every $\lambda \geq \lambda_0$. On the other hand, if $\{t_\lambda\}$ is not bounded above in $\mathbb{R}$, and if a real number T is arbitrarily specified, then there exists an index λ_0 such that $t_{\lambda_0} > T$, whereupon it follows that $t_\lambda > T$ for all $\lambda \geq \lambda_0$. (In this situation we say that $\lim_\lambda t_\lambda = +\infty$. Observe that this notation is here only a formalism, the extended real number $+\infty$ not being a point of the metric space $\mathbb{R}$; see, however, Example 9E.) Dually, a monotone decreasing net $\{t_\lambda\}$ in $\mathbb{R}$ either converges to its infimum $s = \inf_\lambda t_\lambda$ (this occurs when and only when $\{t_\lambda\}$ is bounded below in $\mathbb{R}$), or $\lim_\lambda t_\lambda = -\infty$ (meaning that, for an arbitrarily given real number T there exists an index λ_0 such that $t_\lambda < T$ for all $\lambda \geq \lambda_0$).

Note that a monotone sequence of real numbers constitutes a special case of the situation considered here. Likewise, if $f(t)$ is a monotone increasing real-valued function on an open interval (a, b) in $\mathbb{R}(a < b)$, then $\lim_{t\uparrow b} f(t)$ exists in $\mathbb{R}$ if and only if f is bounded above on (a, b) in $\mathbb{R}$, and if this is the case, then $\lim_{t\uparrow b} f(t) = \sup\{f(t) : a < t < b\}$. Similarly, $\lim_{t\downarrow a} f(t)$ exists in $\mathbb{R}$ if and only if f is bounded below on (a, b) in $\mathbb{R}$, in which case $\lim_{t\downarrow a} f(t) = \inf\{f(t) : a < t < b\}$, and dual remarks hold for a monotone decreasing function on (a, b).

Note, finally, that while these observations and definitions are here formulated for nets in the metric space $\mathbb{R}$, with a few obvious reinterpretations, they apply equally well to nets of extended real numbers, and it is in this larger context that Example J will be applied in the sequel when that is convenient.

Definition. Two metrics ρ and ρ' on the same carrier X are said to be *equivalent* if it is the case that every sequence $\{x_n\}$ in X converges to a limit a with respect to ρ if and only if it also converges to a with respect to ρ'. (Clearly this notion of equivalence is an equivalence relation on the collection of all metrics on a given set X.)

Example K. Let (X, ρ) be a metric space and define ρ' on $X \times X$ by setting

$$\rho'(x, y) = \frac{\rho(x, y)}{1 + \rho(x, y)}, \quad x, y \in X.$$

Then ρ and ρ' are equivalent metrics on X. Indeed, if we assume for the moment that ρ' is a metric, it is clear that if a sequence $\{x_n\}$ converges to a limit a with respect to ρ, then $x_n \to a$ with respect to ρ' as well, since $\rho' \leq \rho$ on $X \times X$. Likewise, if $x_n \to a$ with respect to ρ', then $x_n \to a$ with respect to ρ too, since the function $f(t) = t/(t+1)$ is strictly increasing on the ray $[0, +\infty)$. Hence the only nontrivial chore is to verify the triangle inequality for ρ', and this follows at once from the following:

$$\frac{u+v}{1+u+v} \leq \frac{u+v+uv}{1+u+v+uv} \leq \frac{u}{1+u} + \frac{v}{1+v}, \qquad u, v \geq 0.$$

Thus every metric space (X, ρ) admits an equivalent metric in which X is bounded and diam $X \leq 1$.

The following nonsequential criterion for the equivalence of two metrics is sometimes useful.

Proposition 6.3. *Two metrics ρ and ρ' on the same set X are equivalent if and only if for each point x_0 of X every open ball with center x_0 and positive radius with respect to ρ contains an open ball with center x_0 and positive radius with respect to ρ', and conversely every open ball with center x_0 and positive radius with respect to ρ' contains an open ball with center x_0 and positive radius with respect to ρ.*

PROOF. It is clear that if the stated condition is satisfied, and if $\{x_n\}$ is a sequence in X, then $x_n \to x_0$ with respect to ρ if and only if $x_n \to x_0$ with respect to ρ', and hence that ρ and ρ' are equivalent. Suppose, on the contrary, that the condition is not satisfied. We may assume without loss of generality that there exists a point x_0 in X and a positive radius ε_0 such that the open ball with radius ε_0 and center x_0 with respect to ρ contains no ball with positive radius and center x_0 with respect to ρ'. But then, for each positive integer n, the open ball $D_{1/n}(x_0)$ with respect to ρ' contains at least one point x_n such that $\rho(x_0, x_n) \geq \varepsilon_0$, and the sequence $\{x_n\}$ is therefore convergent to x_0 with respect to ρ' but not with respect to ρ, so ρ and ρ' are not equivalent. □

It is frequently of interest to consider various subsequences of a given sequence in a metric space.

Proposition 6.4. *If X is a metric space, and if $\{x_n\}$ is a sequence in X that converges to some limit a, then $\lim_k x_{n_k} = a$ for every subsequence $\{x_{n_k}\}$ of $\{x_n\}$.*

PROOF. If $\varepsilon > 0$ is given, and if n_0 is selected so that $x_n \in D_\varepsilon(a)$ for all $n \geq n_0$, then $x_{n_k} \in D_\varepsilon(a)$ for all $k \geq n_0$ since (by mathematical induction) $n_k \geq k$ for every index k. (This result has a converse of sorts; see Problem F.) □

Definition. If $\{x_n\}$ is a sequence in a metric space (X, ρ), then a point y of X is a *cluster point* of $\{x_n\}$ if for every $\varepsilon > 0$ and every index n_0 there is an index n such that $n > n_0$ and such that $\rho(y, x_n) < \varepsilon$ (equivalently, if the sequence $\{x_n\}$ belongs to the open ball $D_\varepsilon(y)$ infinitely often for each positive radius ε).

The following result is entirely elementary but serves to tie together several of the concepts introduced thus far.

Proposition 6.5. *Let (X, ρ) be a metric space and let $\{x_n\}$ be a sequence in X. Then a point y of X is a cluster point of $\{x_n\}$ if and only if some subsequence of $\{x_n\}$ converges to y.*

PROOF. If $\{x_{n_k}\}$ is a subsequence of $\{x_n\}$ such that $\lim_k x_{n_k} = y$, and if an index n_0 and a positive number ε are given, then there exists an index k_0 such that $\rho(y, x_{n_k}) < \varepsilon$ for all $k \geq k_0$, and among the indices n_k such that $k \geq k_0$ there are infinitely many that exceed n_0. Thus the condition is sufficient.

Suppose, on the other hand, that y is a cluster point of $\{x_n\}$. Choose first an index n_1 such that $\rho(y, x_{n_1}) < 1$, and suppose that strictly increasing indices $n_1, \ldots, n_k$ have been chosen so that $\rho(y, x_{n_i}) < 1/i$, $i = 1, \ldots, k$. Then there are indices $n > n_k$ such that $\rho(y, x_n) < 1/(k+1)$, and we may take for n_{k+1} the least such index. In this way we obtain by mathematical induction a subsequence $\{x_{n_k}\}$ such that $\rho(y, x_{n_k}) < 1/k$ for every k, so $\lim_k x_{n_k} = y$, and the proposition is proved. □

Considerably more can be said about cluster points of sequences of real numbers. For reasons of future convenience the following discussion is conducted in the more general context of nets of extended real numbers.

Example L. Starting with an arbitrary net $\{t_\lambda\}_{\lambda \in \Lambda}$ of extended real numbers we define a new, similarly indexed net $\{u_\lambda\}_{\lambda \in \Lambda}$ by setting

$$u_\mu = \sup_{\lambda \geq \mu} t_\lambda = \sup T_\mu \ , \ \mu \in \Lambda,$$

where T_μ denotes the *tail* $T_\mu = \{t_\lambda : \lambda \geq \mu\}$. The net $\{u_\mu\}$ is monotone decreasing (since T_μ decreases as μ increases) so we may, and do (Ex. J), define the *upper* or *superior limit* M of the given net $\{t_\lambda\}$ to be the extended real number

$$M = \inf_\mu u_\mu = \lim_\mu u_\mu$$

(notation: $M = \limsup_\lambda t_\lambda$).

Three special cases may be singled out. If $M = -\infty$, then

$$-\infty = \inf_\lambda t_\lambda = \lim_\lambda t_\lambda$$

as well, while $M = +\infty$ if and only if no tail T_μ of $\{t_\lambda\}$ is bounded above in $\mathbb{R}$. On the other hand, if M is *finite*, and if ε is an arbitrary finite positive number, then there is a tail T_{μ_0} with supremum $u_{\mu_0} < M + \varepsilon$, while *every* tail T_μ contains terms exceeding $M - \varepsilon$.

Dually, we define the net $\{s_\mu\}_{\mu \in \Lambda}$, where

$$s_\mu = \inf_{\lambda \geq \mu} t_\lambda = \inf T_\mu \ , \ \mu \in \Lambda,$$

and the *lower* or *inferior limit* $m = \liminf_\lambda t_\lambda$ of $\{t_\lambda\}$ by setting

$$m = \sup_\lambda s_\lambda = \lim_\lambda s_\lambda.$$

Concerning this lower limit it is clear that $m = +\infty$ if and only if $\lim_\lambda t_\lambda = +\infty$, while $m = -\infty$ means simply that no tail T_μ is bounded below in $\mathbb{R}$. Moreover if m is finite and $\varepsilon > 0$, then every tail T_μ of $\{t_\lambda\}$ contains terms less than $m + \varepsilon$, while some tail T_{μ_0} has infimum $s_{\mu_0} > m - \varepsilon$.

From these definitions it is apparent that

$$s_\lambda \leq t_\lambda \leq u_\lambda$$

for every index λ, and hence that $m \leq M$. Moreover, $M = m = \pm\infty$ if and only if $\lim_\lambda t_\lambda = \pm\infty$, while if $\{t_\lambda\}$ is net of finite real numbers such that m and M are equal and finite, then

$$\lim_\lambda t_\lambda = m = M.$$

(Conversely, of course, if $\{t_\lambda\}$ is a net in $\mathbb{R}$ that converges to a limit L in $\mathbb{R}$, then $L = m = M$.)

Finally, let us consider briefly the vitally important special case of an ordinary sequence $\{t_n\}$ in $\mathbb{R}$. Here $M = +\infty$ [$m = -\infty$] means simply that $\{t_n\}$ is unbounded above [below] in $\mathbb{R}$, and $\lim_n t_n = L$ if and only if $m = M = L$. Moreover if M is finite, then M is a cluster point of $\{t_n\}$ and is, indeed, its *largest* cluster point. Dually, if m is finite, then m is the smallest cluster point of $\{t_n\}$. Thus for a sequence $\{t_n\}$ in the metric space $\mathbb{R}$, $-\infty < m \leq M < +\infty$ is characteristic of a bounded sequence with extreme cluster points m and M, while $-\infty < m = M < +\infty$ is characteristic of a convergent sequence (with the *sole* cluster point $m = M$).

Definition. Let X be a metric space and let A be a subset of X. A point a_0 of X is an *adherent point* of A if there exists a sequence $\{x_n\}$ in A that converges to a_0. Similarly, a_0 is a *point of accumulation* or an *accumulation point* of A if there exists a sequence $\{x_n\}$ in $A \backslash \{a_0\}$ that converges to a_0. (Clearly every point of A is an adherent point of A; simple examples show that a point of A need not be an accumulation point of A.)

It is both easy and important to characterize these ideas in nonsequential terms.

Proposition 6.6. *Let A be a subset of a metric space (X, ρ), and let a_0 be a point of X. Then a_0 is an adherent point of A if and only if every open ball with positive radius centered at a_0 meets A or, equivalently, when $A \neq \varnothing$, if and only if $d(a_0, A) = 0$. Moreover, the following conditions are also equivalent:*

(1) *Every open ball $D_\varepsilon(a_0)(\varepsilon > 0)$ contains infinitely many points of A,*
(2) *Every open ball $D_\varepsilon(a_0)(\varepsilon > 0)$ contains a point of A other than a_0,*

(3) *There exists a sequence* $\{x_n\}$ *of pairwise distinct points in* $A\backslash\{a_0\}$ *that converges to* a_0,

(4) *The point* a_0 *is a point of accumulation of* A.

PROOF. Clearly if $A \neq \varnothing$, then every open ball with positive radius and center a_0 meets A if and only if a_0 is at zero distance from A. Likewise, if $\{x_n\}$ is a sequence in A that converges to a_0, and if ε is positive, then $D_\varepsilon(a_0)$ contains a tail of $\{x_n\}$, so the condition stated in the first part of the proposition is necessary. On the other hand, if this condition is satisfied, and if for each positive integer n we choose x_n in $D_{1/n}(a_0) \cap A$, then $\{x_n\}$ is a sequence in A such that $\rho(a_0, x_n) < 1/n$ for all n, and therefore such that $\lim_n x_n = a_0$.

We turn next to the second part of the proposition. It is at once clear that each of (1) and (4) implies (2), and likewise that (3) implies both (1) and (4). Thus the proof will be complete if we show that (2) implies (3). Suppose, accordingly, that (2) holds. Let x_1 be any point of the set $D_1(a_0) \cap A$ other than a_0 itself, and suppose $x_1, \ldots, x_n$ have already been selected in such a way that $\rho(a_0, x_1) > \rho(a_0, x_2) > \ldots > \rho(a_0, x_n) > 0$ and such that $x_i \in D_{1/i}(a_0) \cap A$ for each $i = 1, \ldots, n$. Then, letting $\delta = \rho(a_0, x_n) \wedge 1/(n+1)$, we have but to choose for x_{n+1} an arbitrary point of $D_\delta(a_0) \cap A$ other than a_0 itself in order to complete the inductive construction of a sequence $\{x_n\}$ in A that does all that is required of it. □

Definition. The collection of all adherent points of a subset A of a metric space X is denoted by A^- and is called (for reasons that will be made clear shortly; see Proposition 6.8 below) the *closure* of A. The collection of all points of accumulation of A is denoted by A^*, and is called the *derived set* of A. A subset F of X is *closed* if it contains all of its adherent points, i.e., if $F = F^-$. A subset M of X is *dense* in X if $M^- = X$. More generally, a subset M of a subset A of X is *dense in* A if $M^- \supset A$. A metric space X is *separable* if there exists a countable dense set in X.

Example M. A ray is a closed subset of the metric space $\mathbb{R}$ if and only if it is a closed ray; similarly, a nonempty interval in $\mathbb{R}$ is a closed set in $\mathbb{R}$ if and only if it is a closed interval. (This is, of course, the reason for the choice of the terms "closed ray" and "closed interval" in the first place.) All *closed cells*

$$Z = [a_1, b_1] \times \cdots \times [a_n, b_n] \tag{1}$$

are closed sets in Euclidean space $\mathbb{R}^n$. All closed balls

$$\{x \in X : \rho(x_0, x) \leq r\}$$

with arbitrary center x_0 and radius $r (r \geq 0)$ are closed sets in an arbitrary metric space X (Prob. I). All finite sets are closed in an arbitrary metric space; in particular, singletons are closed sets. The set $\mathbb{Q}$ of rational numbers is dense in $\mathbb{R}$, as is the set T_p of all p-adic fractions (Ths. 2.11 and 2.12). The sets $\mathbb{Q}^n$ and T_p^n are likewise dense in $\mathbb{R}^n$. Thus the metric space $\mathbb{R}^n$ is separable. The set of those points in $\mathbb{Q}^n$ or T_p^n belonging to a closed cell (1) is dense in that cell if the latter is *nondegenerate*, i.e., if $a_i < b_i$, for all $i = 1, \ldots, n$. It follows at once that all cells are also separable.

Proposition 6.7. *If X is a metric space and if A and B are subsets of X such that $A \subset B$, then $A^- \subset B^-$ and $A^* \subset B^*$. Moreover, for any subset A of X we have $A^- = A \cup A^*$.*

PROOF. The first assertion of the proposition is an obvious consequence of the various definitions, as is the fact that A and A^* are both included in A^- for an arbitrary subset A of X. On the other hand, if $a_0 \in A^- \backslash A$, and if $\{x_n\}$ is an arbitrary sequence in A that converges to a_0, then $x_n \neq a_0$ for every index n, so $a_0 \in A^*$. □

Proposition 6.8. *A subset F of a metric space (X, ρ) is closed if and only if it contains all of its points of accumulation. For an arbitrary subset A of X both the closure A^- and the derived set A^* are closed. Moreover, A^- is the smallest closed set in X that contains A.*

PROOF. The first assertion of the proposition is an obvious consequence of Proposition 6.7, and as for the last assertion, it is likewise obvious that if $A \subset F$ where F is closed, then all adherent points of A must belong to $F^- = F$. Hence to complete the proof of the proposition it suffices to show that A^* and A^- are closed. Suppose first that z is an adherent point of A^-, and let ε be a positive number. Then (Prop. 6.6) $D_\varepsilon(z)$ contains a point y of A^-, and if $\eta = \varepsilon - \rho(y, z)$, then $\eta > 0$ and $D_\eta(y)$ contains a point x of A. But then $x \in D_\varepsilon(z)$ by the triangle inequality, which shows $z \in A^-$. Finally, if $z \in (A^*)^-$, the argument goes exactly the same, except y can be chosen in A^*, so that the ball $D_\eta(y)$ contains infinitely many points of A. □

Example N. Let X be a metric space and let $\{E_n\}_{n=1}^\infty$ be a sequence of subsets of X. The collection of all those points x of X with the property that, for every $\varepsilon > 0$, $D_\varepsilon(x)$ meets $\{E_n\}$ eventually is called the *closed limit inferior* of the sequence $\{E_n\}$ (notation: $\operatorname{F}\liminf_n E_n$ or, when possible, simply $\underline{F}$). Dually, the collection of all points x of X with the property that, for every $\varepsilon > 0$, $D_\varepsilon(x)$ meets $\{E_n\}$ infinitely often is the *closed limit superior* of $\{E_n\}$ (notation: $\operatorname{F}\limsup_n E_n$ or, when possible, simply $\overline{F}$). (Using arguments exactly like those in the proof of Proposition 6.8, it is readily seen that the sets $\underline{F}$ and $\overline{F}$ are indeed closed, whether the sets E_n are themselves closed or not; hence the terminology.) Finally, in the event that $\underline{F} = \overline{F}$, this common set is called the *closed limit* of E_n (notation: $\underline{F} = \overline{F} = \mathrm{F}\lim_n E_n$).

Concerning these notions a number of elementary facts are obviously valid and will be stated here with little or no proof. To begin with, these notions of *closed* limits are related to the simpler notions of limit introduced in Chapter 1 for sequences of sets (without any references to a metric; see Example 1U). Thus if $x \in \liminf_n E_n$, then $x \in E_n$ eventually, and it follows at once that $x \in \mathrm{F}\liminf_n E_n$. Similarly, $\limsup_n E_n \subset \mathrm{F}\limsup_n E_n$. (But these limits need not be the same. Consider the sequence $\{C_n\}_{n=1}^{\infty}$ of circles where

$$C_n = \{(x, y) \in \mathbb{R}^2 : (x - n)^2 + y^2 = n^2\}, \quad n \in \mathbb{N}.$$

It is readily verified that both $\lim_n C_n$ and $\mathrm{F}\lim_n C_n$ exist, but $\lim_n C_n = \{(0, 0)\}$ while $\mathrm{F}\lim_n C_n$ consists of the entire y-axis.) We also observe that both the closed limit inferior and the closed limit superior (and therefore the closed limit, as well, if it exists) are the same for the sequence $\{E_n^-\}$ as for the sequence $\{E_n\}$. In the same vein, it should be noted that, as with any reasonable concept of limit, the closed limit superior and the closed limit inferior (and hence the closed limit when it exists) are unaffected by the deletion, adjunction or change of any finite number of terms of the sequence.

Suppose, finally, that $E_n \neq \varnothing$ for all n. Then $x \in \underline{F}$ if and only if there exists a sequence $\{x_n\}$ in X such that $x_n \in E_n$ for all n and such that $\lim_n x_n = x$, or, equivalently, if and only if $\lim_n d(x, E_n) = 0$. Dually, $x \in \overline{F}$ if and only if there exists a sequence $\{x_n\}$ in X such that $x_n \in E_n$ for all n and such that x is a cluster point of $\{x_n\}$, or equivalently, if and only if $\liminf_n d(x, E_n) = 0$. It is also clear that $x \in \overline{F}$ if and only if $x \in \mathrm{F}\liminf_k E_{n_k}$ for some subsequence $\{E_{n_k}\}$ of the given sequence $\{E_n\}$.

Proposition 6.9. *For any metric space X the mapping $A \to A^-$ assigning to each subset A of X its closure A^- is a monotone increasing mapping of 2^X into itself possessing the following properties for all subsets A and B of X:*

(1) $\varnothing^- = \varnothing$,
(2) $A \subset A^-$,
(3) $(A^-)^- = A^-$,
(4) $(A \cup B)^- = A^- \cup B^-$.

PROOF. The only part of the proposition that is not either obvious or already established is (4). Moreover, it is clear that $A^- \cup B^- \subset (A \cup B)^-$ since both A and B are subsets of $A \cup B$. To complete the proof, suppose $a_0 \in (A \cup B)^-$ and let $\{x_n\}$ be a sequence in $A \cup B$ such that $x_n \to a_0$. If $\{x_n\}$ belongs to A infinitely often, then $\{x_n\}$ possesses a subsequence lying in A, and $a_0 \in A^-$ (Prop. 6.4). Otherwise $\{x_n\}$ is eventually not in

A, and is therefore eventually in B, whereupon it follows that $a_0 \in B^-$. Thus $a_0 \in A^- \cup B^-$ in any case, and the proof is complete. □

Concerning the closed sets in a metric space we have the following result.

Theorem 6.10. *In any metric space X the closed sets satisfy the following conditions:*

(1) *X and $\varnothing$ are closed,*
(2) *The union of two (and hence of any finite number of) closed sets is closed,*
(3) *The intersection of an arbitrary nonempty collection of closed sets is closed.*

PROOF. Conditions (1) and (2) are immediate consequences of (1) and (4) of the preceding result, so we need only prove (3). Let $\mathcal{F}$ be a nonempty collection of closed subsets of X, and let $F_0 = \bigcap \mathcal{F}$. Then $F_0 \subset F$ for every F in $\mathcal{F}$, whence it follows that $F_0^- \subset F^- = F$. But then $F_0^- \subset \bigcap \mathcal{F} = F_0$. □

Example O (The Cantor Set). If I is a nondegenerate closed interval $[a, b]$ in $\mathbb{R}$, i.e., if $a < b$, and if $t_1 = (2a + b)/3$ and $t_2 = (a + 2b)/3$, so that $\{a = t_0 < t_1 < t_2 < t_3 = b\}$ is the partition of I into three subintervals of equal length $(b - a)/3$, then the doubleton $\{[a, t_1], [t_2, b]\}$, consisting of the first and third of these subintervals, will be said to be derived from I by *removal of the central third* of I. Likewise, if $\mathcal{S} = \{I_1, \ldots, I_p\}$ is any finite disjoint collection of nondegenerate closed intervals in $\mathbb{R}$, then the corresponding collection $\{I_1, I_2, \ldots, I_{2p-1}, I_{2p}\}$ obtained by replacing each interval I_j by the two subintervals derived from I_j by removal of the central third of I_j will be denoted by $\mathcal{S}^*$. (Note that $\mathcal{S}^*$ is again a finite disjoint system of nondegenerate closed intervals.)

With these preliminaries out of the way we are ready to introduce a construction that is of the greatest importance in all that follows—indeed in all of mathematical analysis. We begin by setting $\mathcal{F}_0 = \{[0, 1]\}$ and $\mathcal{F}_1 = \mathcal{F}_0^* = \{[0, 1/3], [2/3, 1]\}$. Then, supposing $\mathcal{F}_k$ already defined for $k = 0, 1, \ldots, n$, we set, inductively, $\mathcal{F}_{n+1} = \mathcal{F}_n^*$. Clearly then, for each nonnegative integer $n, \mathcal{F}_n$ is a disjoint system of 2^n nondegenerate closed subintervals of the unit interval $[0, 1]$, each having length $1/3^n$. Hence if we define $F_n = \bigcup \mathcal{F}_n, n \in \mathbb{N}_0$, then $\{F_n\}_{n=0}^{\infty}$ is a decreasing sequence of subsets of the unit interval in which each set F_n is the union of 2^n congruent closed subintervals; see Figure 1. According to Theorem 6.10 the sets F_n are all closed, and so therefore is the *Cantor set*

$$C = \bigcap_{n=0}^{\infty} F_n.$$

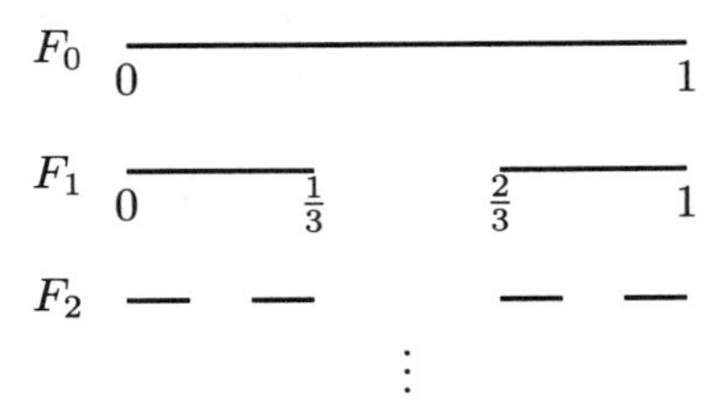

Figure 1

The foregoing construction is of sufficient importance that it is worth scrutinizing a bit more closely. (The notation introduced here will be referred to more than once in the sequel.) To begin with, if $\{\varepsilon_1, \ldots, \varepsilon_n\}$ is an arbitrary element of the set S_n of all finite sequences of length n in which each term ε_i is either zero or one, we shall write $I_{\varepsilon_1,\ldots,\varepsilon_n}$ for the closed interval

$$I_{\varepsilon_1,\ldots,\varepsilon_n} = [0.\eta_1 \ldots \eta_n 00 \ldots, 0.\eta_1 \ldots \eta_n 22 \ldots] \tag{2}$$

in ternary notation, where $\eta_i = 2\varepsilon_i$, $i = 1, \ldots, n$ (Th. 2.12). It is easily seen (by use of mathematical induction) that for each positive integer n the 2^n closed intervals that make up $\mathcal{F}_n$ are precisely the intervals $I_{\varepsilon_1,\ldots,\varepsilon_n}$, $\{\varepsilon_1, \ldots, \varepsilon_n\} \in S_n$. Viewing matters from this vantage point we see that each closed set F_n may be described alternatively as the set of all those real numbers t in [0,1] possessing a ternary expansion

$$t = 0.\eta_1 \ldots \eta_n \ldots \tag{3}$$

in which each of the terms $\eta_1, \ldots, \eta_n$ is either zero or two (i.e., not one). Consequently the Cantor set C may also be described as the set of all those numbers t in [0,1] that possess a *one-free* ternary expansion, that is, an expansion of the form (3) in which all of the terms η_n are either zero or two. (A triadic fraction possesses two ternary expansions and belongs to C if either of these expansions is one-free. Thus $1/3 = 0.100\ldots = 0.022\ldots$ belongs to F_n for every n and therefore belongs to C.)

For any symbol $\{\varepsilon_1, \ldots, \varepsilon_n\}$ in S_n we shall also write $C_{\varepsilon_1,\ldots,\varepsilon_n}$ for the part of C lying in the interval $I_{\varepsilon_1,\ldots,\varepsilon_n}$:

$$C_{\varepsilon_1,\ldots,\varepsilon_n} = C \cap I_{\varepsilon_1,\ldots,\varepsilon_n}.$$

(Thus $C_{\varepsilon_1,\ldots,\varepsilon_n}$ consists of the numbers in $[0,1]$ having a one-free ternary expansion beginning $0.(2\varepsilon_1)\ldots(2\varepsilon_n)$.)

Finally, it will also be convenient to have a standing notation for the open central third of the interval $I_{\varepsilon_1,\ldots,\varepsilon_n}$ (the open interval removed from $I_{\varepsilon_1,\ldots,\varepsilon_n}$ to form $I_{\varepsilon_1,\ldots,\varepsilon_n,0}$ and $I_{\varepsilon_1,\ldots,\varepsilon_n,1}$ in $\mathcal{F}_{n+1}$). We shall denote this open interval by the symbol $U_{\varepsilon_1,\ldots,\varepsilon_n}$ and we observe that, in ternary notation and with $\eta_i = 2\varepsilon_i, i = 1, \ldots, n$,

$$U_{\varepsilon_1,\ldots,\varepsilon_n} = (0.\eta_1 \ldots \eta_n 022\ldots, 0.\eta_1 \ldots \eta_n 200\ldots).$$

(An exceptional role is played in this system of notation by the set $\mathcal{F}_0 = \{[0,1]\}$ and by the central third $(1/3, 2/3)$; this anomaly may be addressed by regarding $[0,1] = [0.00\ldots, 0.22\ldots]$ and $(1/3, 2/3) = (0.022\ldots, 0.200\ldots)$ as corresponding to the empty sequence of zeros and ones.)

Example P. There are a number of important generalizations of the foregoing construction. For one thing, it is possible (and sometimes useful) to replace the ratio 1/3 by other ratios. Let $I = [a,b] (a < b)$ and let $0 < \theta < 1$. Then the points $t_1 = \frac{1}{2}[a+b-\theta(b-a)]$ and $t_2 = \frac{1}{2}[a+b+\theta(b-a)]$ form a partition $\{a = t_0 < t_1 < t_2 < t_3 = b\}$ with the property that the subinterval $[t_1, t_2]$ has length $\theta(b-a)$, while the first and third of the subintervals of this partition are congruent, and we say that the doubleton $\{[a, t_1], [t_2, b]\}$ is derived from I by *removal of the central θth part* of I. Likewise, if $\mathcal{S} = \{I_1, \ldots, I_p\}$ is any finite disjoint collection of nondegenerate closed intervals in $\mathbb{R}$, then the corresponding collection $\{I_1, I_2, \ldots, I_{2p-1}, I_{2p}\}$, obtained by replacing each interval I_j by the two subintervals derived from I_j by removal of the central θth part of I_j will be denoted by $\mathcal{S}^{*(\theta)}$.

Suppose now that $\{\theta_n\}_{n=0}^{\infty}$ is an arbitrarily prescribed sequence of proper ratios (so that each θ_n is subject to the condition $0 < \theta_n < 1$, but to no other restriction). Then, as before, we begin by setting $\mathcal{F}_0 = \{[0,1]\}$ and define, inductively,

$$\mathcal{F}_{n+1} = \mathcal{F}_n^{*(\theta_n)}, \qquad n \in \mathbb{N}_0.$$

Once again, $\mathcal{F}_n$ is a collection of 2^n closed subintervals of [0,1], and the union $F_n = \bigcup \mathcal{F}_n$ is a closed subset of [0,1], as is the set

$$C_{\{\theta_0, \theta_1, \ldots\}} = \bigcap_{n=0}^{\infty} F_n.$$

The set $C_{\{\theta_0,\ldots\}}$ will be called the *generalized Cantor set associated with the sequence* $\{\theta_n\}_{n=0}^{\infty}$. (The Cantor set C of Example O is then the generalized Cantor set associated with the constant sequence $\{1/3, 1/3, \ldots\}$.)

There are also higher dimensional Cantor sets. To indicate how this goes we briefly sketch the construction of a planar Cantor set.

Example Q. Let I and J be nondegenerate closed intervals, and let R be the rectangle $I \times J \subset \mathbb{R}^2$. If for some $\theta, 0 < \theta < 1, \{I_1, I_2\}$ and $\{J_1, J_2\}$ are the pairs of intervals obtained by removing the central θth part of I and J, respectively, then the set $\{I_1 \times J_1, I_1 \times J_2, I_2 \times J_1, I_2 \times J_2\}$ will be said to be derived from R by *removal of the central cross-shaped θth part*, and if $\mathcal{S} = \{R_1, \ldots, R_p\}$ is any finite disjoint collection of nondegenerate rectangles in $\mathbb{R}^2$, then we denote by $\mathcal{S}^{\dagger(\theta)}$ the corresponding collection $\{R_1, \ldots, R_{4p}\}$ of rectangles obtained by replacing each R_j by the four rectangles derived from R_j by removal of the central cross-shaped θth part. Then, just as before, given a sequence $\{\theta_n\}_{n=0}^{\infty}$ of proper ratios, we define a sequence $\{\mathcal{G}_n\}_{n=0}^{\infty}$ inductively, setting $\mathcal{G}_0 = \{[0,1] \times [0,1]\}$ and $\mathcal{G}_{n+1} = \mathcal{G}_n^{\dagger(\theta_n)}$, and consider the sets $G_n = \bigcup \mathcal{G}_n$, $n \in \mathbb{N}_0$. This time $\{G_n\}$ is a nested sequence of closed subsets of the unit square, whence it follows that the intersection $P_{\{\theta_n\}} = \bigcap_{n=0}^{\infty} G_n$, the *planar Cantor set associated with the sequence* $\{\theta_n\}$, is also a closed subset of the unit square. (The reader will have no trouble identifying the set G_n of this construction with the product $F_n \times F_n$, where $\{F_n\}$ is the sequence of subsets of [0,1] introduced above in the construction of $C_{\{\theta_n\}}$—and hence in showing that, in fact,

$$P_{\{\theta_n\}} = C_{\{\theta_n\}} \times C_{\{\theta_n\}}$$

—but the construction set forth here will be of use later. Figure 2 shows the first three of the sets G_n in the important special case $\theta_n \equiv 1/3$.)

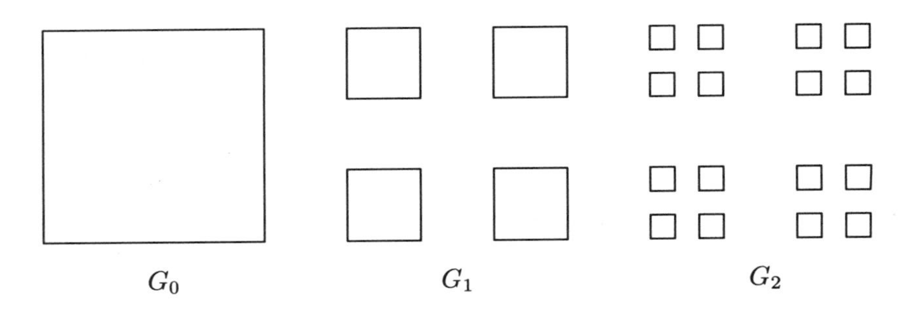

Figure 2

The subsets of a metric space X that are the complements in X of the closed sets are also very important.

Proposition 6.11. *The following properties are equivalent for an arbitrary subset U of a metric space (X, ρ):*

(1) *$X \backslash U$ is closed,*

(2) *For each point x of U there exists a positive radius r such that $D_r(x) \subset U$,*
(3) *U is a union of open balls in X.*

PROOF. If (1) holds and if $x_0 \in U$, then there exists a positive radius r such that $D_r(x_0)$ contains no points of $X \backslash U$ (Prop. 6.6). Thus (2) holds, and it is clear that (2) implies (3). To complete the proof we show that (3) implies (1). Let $\mathcal{D}$ be a collection of open balls in X, and let x_0 be a point of the union $U = \bigcup \mathcal{D}$. Then there exists a center x_1 and a radius r such that $x_0 \in D_r(x_1)$ and $D_r(x_1) \in \mathcal{D}$. Hence $\rho(x_0, x_1) < r$, so $\eta = r - \rho(x_0, x_1) > 0$ and (by the triangle inequality) $D_\eta(x_0) \subset D_r(x_1) \subset U$. Thus x_0 is not an adherent point of $X \backslash U$, and the result follows. □

Definition. A subset U of a metric space X possessing any one (and therefore all) of the properties of the preceding proposition is an *open set* in X, or an *open subset* of X.

Each part of the following result may be obtained simply by taking complements in the corresponding part of Theorem 6.10, and the proof is therefore omitted.

Theorem 6.12. *In any metric space X the open sets satisfy the following conditions:*

(1) *X and $\varnothing$ are open,*
(2) *The intersection of two (and hence of any positive finite number of) open sets is open,*
(3) *The union of an arbitrary collection of open sets is open.*

Example R. An open ball in any metric space X is an open set in X. A ray in $\mathbb{R}$ is an open subset of $\mathbb{R}$ if and only if it is an open ray. Similarly an interval in $\mathbb{R}$ is an open set if and only if it is an open interval. (This is, of course, the reason for the choice of the terms "open ray" and "open interval" in the first place.) All *open cells*

$$Z = (a_1, b_1) \times \cdots \times (a_n, b_n)$$

are open subsets of $\mathbb{R}^n$.

Example S. According to Proposition 6.11 the most general open subset U of $\mathbb{R}$ is the union of some collection $\mathcal{C}$ of nonempty open intervals. This collection $\mathcal{C}$ is not unique in general, and can, indeed, be chosen in infinitely many different ways (unless $U = \varnothing$; the empty set is the union of the empty collection of open intervals). We observe, however, that if $\mathcal{C}_0$ is a nonempty chained subcollection of $\mathcal{C}$ (here and below use is made of the notions introduced in Example 1J), and if $U_0 = \bigcup \mathcal{C}_0$, then $U_0 = (a, b)$ where $a = \inf U_0$ and $b = \sup U_0$. (This follows at once from the fact that the union of two intersecting open intervals is an open interval; the cases $a = -\infty$ and/or $b = +\infty$ are not excluded.) Thus if we partition $\mathcal{C}$ into equivalence classes with respect to the equivalence relation of being chained in $\mathcal{C}$, and if $\mathcal{V}$ denotes the collection of unions of these chained equivalence classes, then $\mathcal{V}$ is a disjoint collection of nonempty open intervals and open rays with union U.

Thus, summarizing, we see that the most general open set U in $\mathbb{R}$ can be expressed as the union of a pairwise disjoint collection $\mathcal{V}$ of nonempty open sets, each of which is either an open interval or an open ray (or $\mathbb{R}$), and *this* representation of U *is* unique. Indeed, if $\widetilde{\mathcal{V}}$ is another such collection of open intervals such that $U = \bigcup \widetilde{\mathcal{V}}$, and if $W_0 \in \widetilde{\mathcal{V}}$, let V_0 be an element of $\mathcal{V}$ such that $W_0 \cap V_0 \neq \varnothing$. (Such a set V_0 exists since $\varnothing \neq W_0 \subset \bigcup \mathcal{V}$.) Suppose c is an endpoint of V_0 that belongs to W_0. Then some other set V in $\mathcal{V}$ must contain c, and must therefore meet V_0, which is impossible. Hence no endpoint of V_0 belongs to W_0, and a consideration of cases reveals that $W_0 \subset V_0$. But then $V_0 = W_0$ by symmetry. This shows that every set in $\widetilde{\mathcal{V}}$ is also in $\mathcal{V}$, whence, invoking symmetry once again, we conclude that $\widetilde{\mathcal{V}} = \mathcal{V}$. We observe also that the collection $\mathcal{V}$ is necessarily countable. Indeed, as noted in Theorem 2.11, each V in $\mathcal{V}$ contains rational numbers, and if we fix one such number r_V for each V in $\mathcal{V}$, then the mapping $V \to r_V$ is a one-to-one mapping of $\mathcal{V}$ into the countable set $\mathbb{Q}$. (In this connection see also Problem R.)

The intervals V belonging to the collection $\mathcal{V}$ will be called *constituent intervals* of U. Moreover, if F is an arbitrary closed set in $\mathbb{R}$, then the constituent intervals of $\mathbb{R} \backslash F$ will be referred to as the (*open*) *intervals contiguous* to F. (If F is bounded above and/or below, one or two of these intervals will be *contiguous* (*open*) *rays*.)

Proposition 6.13. *A subset M of a metric space X is dense in X if and only if every nonempty open subset of X contains points of M.*

PROOF. It is clear that the stated condition is sufficient, so it is enough to prove its necessity. Suppose, accordingly, that U is a nonempty open subset of X such that $U \cap M = \varnothing$. Then the closed set $F = X \backslash U$ contains M and therefore M^-, so $M^- \neq X$. □

In connection with the notions of open and closed sets the following terminology is also frequently useful.

Definition. If A is a subset of a metric space X, then a covering of A consisting of subsets of X is said to be *open* if every set in the covering is open in X. Likewise, a *closed* covering of A is a covering of A composed exclusively of closed subsets of X.

Those properties of a metric space that depend only on the notions of open set, closed set, etc., and do not require the actual values of the metric for their definitions, are called *topological* properties. As it turns out, these are exactly the properties that are shared by metric spaces equipped with equivalent metrics.

Proposition 6.14. *Let ρ and ρ' be metrics on a set X. The following conditions are equivalent:*

(1) *ρ and ρ' are equivalent,*
(2) *Every subset of X has the same closure with respect to both ρ and ρ',*
(3) *Precisely the same subsets of X are closed in (X, ρ) and (X, ρ'),*
(4) *Precisely the same subsets of X are open in (X, ρ) and (X, ρ').*

PROOF. It is obvious that (3) and (4) are equivalent (since open and closed sets are complements of one another), and almost equally obvious that (2) and (3) are equivalent (recall Proposition 6.8). Moreover, a point a_0 belongs to the closure of a set A in X if and only if there exists a sequence in A that converges to a_0, which shows that (1) implies (2). Hence it suffices to verify that (2) implies (1). Suppose then that (1) is false, and suppose also, without loss of generality, that a sequence $\{x_n\}$ is given in X that converges to a limit a_0 with respect to ρ but not with respect to ρ'. Then there exists a positive radius ε_0 so small that $\{x_n\}$ is not eventually in the open ball $U = \{x \in X : \rho'(x, a_0) < \varepsilon_0\}$, and consequently there exists a subsequence $\{x_{n_k}\}$ of $\{x_n\}$ contained in $X \backslash U$. Let A denote the set of points in the subsequence $\{x_{n_k}\}$. Then, of course, $A \subset X \backslash U$, and since $X \backslash U$ is closed with respect to ρ', the closure of A with respect to ρ' is also contained in $X \backslash U$. In particular, the closure of A with respect to ρ' does not contain the point a_0. But a_0 is a point of the closure of A with respect to ρ since $\{x_{n_k}\}$ is a subsequence of a sequence converging to a_0, and is therefore itself convergent to a_0, with respect to ρ (Prop. 6.4). □

Since subspaces of a metric space X are metric spaces in their own right, the notions of closed set and open set are meaningful in any subspace of X, and it is frequently important to be able to identify those subsets of a subspace of X that are either open or closed.

Proposition 6.15. *Let B be a subset of a subspace A of a metric space X. Then B is open relative to A if and only if $B = A \cap U$, where U is an open set in X. Similarly, B is closed relative to A if and only if $B = A \cap F$, where F is a closed set in X. Moreover, if $B^{-(A)}$ denotes the closure of B relative to A, then $B^{-(A)} = B^- \cap A$.*

PROOF. If for each point a of A and positive number r we write $D_r(a)^{(A)}$ for the open ball with center a and radius r in the subspace A, then it is obvious that $D_r(a)^{(A)}$ is simply $D_r(a) \cap A$. Hence any union of open balls in the subspace A is just the intersection with A of the corresponding union of open balls in X. Since open and closed sets are the complements of one another, this proves both of the first two assertions of the proposition, and the third is an immediate consequence of the second. □

Corollary 6.16. *Let X be a metric space. The open subsets of an open subspace of X are open sets in X. Dually, the closed subsets of a closed subspace of X are closed sets in X.*

Example T. If a and b are irrational real numbers such that $a < b$, then the set $\{r \in \mathbb{Q} : a < r < b\}$ is both open and closed in the metric space $\mathbb{Q}$.

Throughout our treatment of the theory of metric spaces a very special role has been played by a particular collection of open sets, viz., the open balls. In various particular settings other special collections of open sets play important special roles. In this connection the following concept is of interest.

Definition. A collection $\mathcal{B}$ of open subsets of a metric space X is a *base* (*of open sets*), or a *topological base*, for X if every open set in X is the union of some subcollection of $\mathcal{B}$.

The following criterion is very elementary but still useful.

Proposition 6.17. *Let X be a metric space and let $\mathcal{V}$ be a collection of open subsets of X. Then $\mathcal{V}$ is a topological base for X if and only if for every open set U in X and every point x of U there exists a set V in $\mathcal{V}$ such that $x \in V \subset U$.*

PROOF. For each open set U in X let us write $\mathcal{V}_U$ for the collection $\{V \in \mathcal{V} : V \subset U\}$. If $\mathcal{V}$ is a base of open sets for X, then by definition U is the union of the collection $\mathcal{V}_U$, so, if $x \in U$, there must be some set V in $\mathcal{V}_U$ such that $x \in V$, and the criterion is satisfied. On the other hand, if the criterion is satisfied, and if U is any open set in X, then it is clear that $U = \bigcup \mathcal{V}_U$, which shows that $\mathcal{V}$ is a base of open sets for X. □

Note. It is clear that in applications of Proposition 6.17 it always suffices to verify the criterion for those sets U belonging to some already known base of open sets.

According to the above definition a collection $\mathcal{B}$ of open subsets of a metric space X is a topological base for X if it is, so to speak, large enough. Thus if $\mathcal{B}$ is a topological base for X and if $\mathcal{U}$ is some other collection of open sets in X, then $\mathcal{B} \cup \mathcal{U}$ is also a base for X. Accordingly, it is sometimes of interest to seek topological bases that are "small" in that they consist exclusively of sets that are special in one way or another. In this connection the following concept is important.

Definition. A metric space X is said to satisfy the *second axiom of countability* if there exists a countable base of open subsets of X.

Readers curious to know what became of the "first" axiom of countability are counseled to be patient. This matter will be cleared up in Chapter 9.

Example U. An open cell

$$W = (a_1, b_1) \times \cdots \times (a_n, b_n)$$

in $\mathbb{R}^n$ is an open *cube* if the edge lengths of W are all equal, i.e., if $b_1 - a_1 = \cdots = b_n - a_n$. Likewise W will be said to be *rational* if all of the endpoints $a_1, b_1, \ldots, a_n, b_n$ are rational numbers. We shall show that the collection of all rational open cubes constitutes a base of open sets for $\mathbb{R}^n$ by verifying the criterion of Proposition 6.17. (Since the cardinal number of this base is $\aleph_0^{2n} = \aleph_0$, this will show that $\mathbb{R}^n$ satisfies the second axiom of countability.)

Indeed, let U be open in $\mathbb{R}^n$ and let $x = (s_1, \ldots, s_n)$ be a point of U. We first choose $\varepsilon > 0$ so that $D_\varepsilon(x) \subset U$, and then a rational number q such that $0 < q < \varepsilon/(2\sqrt{n})$. Next we select rational numbers r_i so that $|r_i - s_i| < q$, $i = 1, \ldots, n$, and set

$$W = (r_1 - q, r_1 + q) \times \cdots \times (r_n - q, r_n + q).$$

Then W is a rational open cube containing x, and a brief calculation discloses that $W \subset D_\varepsilon(x)$, and hence that $x \in W \subset U$.

Example V. The system $\mathcal{C}$ of all sets of the form $C_{\varepsilon_1, \ldots, \varepsilon_n}$, $n \in \mathbb{N}$, is a topological base for the Cantor set C (see Example O). Indeed, $C_{\varepsilon_1, \ldots, \varepsilon_n}$ is open (relative to C!) since an open interval slightly larger than $I_{\varepsilon_1, \ldots, \varepsilon_n}$ also meets C in $C_{\varepsilon_1, \ldots, \varepsilon_n}$, so it remains only to verify the criterion of Proposition 6.17 for sets relatively open in C. But if $t \in C \cap U$, where U is an open set in $\mathbb{R}$, we may first choose $\varepsilon > 0$ so that $(t - \varepsilon, t + \varepsilon) \subset U$, and then a positive integer n large enough so that $1/3^n < \varepsilon$. There is then a unique interval $I_{\varepsilon_1, \ldots, \varepsilon_n}$ in $\mathcal{F}_n$ that contains t, and it is clear that $t \in C_{\varepsilon_1, \ldots, \varepsilon_n} \subset U$. Thus the metric space C also satisfies the second axiom of countability.

Example W. Let $\{X_1, \dots, X_n\}$ be a finite sequence of metric spaces, and let the product $\Pi = X_1 \times \cdots \times X_n$ be equipped with an arbitrary product metric (Prob. H). If U_i is an open set in X_i, $i = 1, \dots, n$, then the product $V = U_1 \times \cdots \times U_n$ is an open subset of Π. (If $\{(x_1^{(m)}, \dots, x_n^{(m)})\}_{m=1}^{\infty}$ is an arbitrary convergent sequence of points in $\Pi \backslash V$, and if $(a_1, \dots, a_n) = \lim_m (x_1^{(m)}, \dots, x_n^{(m)})$, then for at least one index i the sequence $\{x_i^{(m)}\}_{m=1}^{\infty}$ must lie in $X_i \backslash U_i$ infinitely often, whence it follows that $a_i \in X_i \backslash U_i$, and hence that $(a_1, \dots, a_n) \in \Pi \backslash V$. Thus $\Pi \backslash V$ is closed, and V is open.) As a matter of fact, if $\mathcal{B}_i$ is any base of open sets for X_i, $i = 1, \dots, n$, and if we set $\mathcal{B} = \{U_1 \times \cdots \times U_n : U_i \in \mathcal{B}_i,\ i = 1, \dots, n\}$, then $\mathcal{B}$ is a base of open sets for Π. To see that this is so, suppose the criterion of Proposition 6.17 is not satisfied. Then there exist an open set W in Π and a point $(a_1, \dots, a_n)$ in W such that there is no set V in $\mathcal{B}$ such that $(a_1, \dots, a_n) \in V \subset W$. For every positive integer m and for each index i there exists a set U_i in $\mathcal{B}_i$ such that $a_i \in U_i \subset D_{1/m}(a_i)$ (since $\mathcal{B}_i$ is a base of open sets for X_i). But then $(a_1, \dots, a_n) \in U_1 \times \cdots \times U_n \subset D_{1/m}(a_1) \times \cdots \times D_{1/m}(a_n)$, and since, as we are supposing, no set in $\mathcal{B}$ both contains $(a_1, \dots, a_n)$ and is contained in W, it follows that there exists a point $(x_1^{(m)}, \dots, x_n^{(m)})$ in $\Pi \backslash W$ that also belongs to $D_{1/m}(a_1) \times \cdots \times D_{1/m}(a_n)$. The sequence $\{(x_1^{(m)}, \dots, x_n^{(m)})\}_{m=1}^{\infty}$ thus constructed obviously converges in Π to the limit $(a_1, \dots, a_n)$, which is impossible since W is an open set. Thus the criterion of Proposition 6.17 must be satisfied, and $\mathcal{B}$ is a base of open sets for Π. (This serves to show that the product of a finite number of metric spaces satisfying the second axiom of countability also satisfies the second axiom of countability provided the product is equipped with a product metric.)

As it turns out, a metric space satisfies the second axiom of countability if and only if it possesses a very common property that we have already had occasion to mention.

Theorem 6.18. *A metric space (X, ρ) satisfies the second axiom of countability if and only if it is separable.*

PROOF. Suppose first that X satisfies the second axiom of countability, let $\mathcal{B}_0$ be a countable topological base for X, and for each set V in $\mathcal{B}_0$ let x_V be a point of V. (If the empty set should belong to $\mathcal{B}_0$, we simply delete it.) Then the countable set of points $\{x_V\}_{V \in \mathcal{B}_0}$ is clearly dense in X (Prop 6.13), so X is separable. On the other hand, if X is separable, let M be a countable subset of X such that $M^- = X$, and let E be a countable set of positive numbers with the property that $\inf E = 0$. We complete the proof by showing that the (countable) collection of open balls $\mathcal{B}_0 = \{D_r(x) : x \in M, r \in E\}$ is a base of open sets for X. Indeed, let U be open in X and let x be a point of U. We first choose $\varepsilon > 0$ so that

$D_\varepsilon(x) \subset U$ and then select a number r from E such that $r < \varepsilon/2$. Let y be an element of M such that $\rho(x,y) < r$ and consider the ball $D_r(y)$ —an element of $\mathcal{B}_0$ by definition. Clearly $x \in D_r(y)$, while if $z \in D_r(y)$, then $\rho(x,z) < 2r < \varepsilon$, so $z \in D_\varepsilon(x)$. Thus $D_r(y) \subset D_\varepsilon(x) \subset U$, and $\mathcal{B}_0$ is a base of open sets for X by Proposition 6.17. □

Corollary 6.19. *Every subspace of a separable metric space is itself separable.*

PROOF. Let X be a separable metric space and let A be a subspace of X. If $\mathcal{B}$ is a countable base for X, and if $\mathcal{B}_A = \{V \cap A : V \in \mathcal{B}\}$, then it is clear that $\mathcal{B}_A$ is countable and it is also easy to see that $\mathcal{B}_A$ is a base for A. Indeed, the sets of $\mathcal{B}_A$ are open relative to A by Proposition 6.15, while if B is any relatively open subset of A, then there exists an open set U in X such that $B = U \cap A$. But if $\mathcal{B}_0$ is a subcollection of $\mathcal{B}$ such that $U = \bigcup \mathcal{B}_0$, then $B = \bigcup_{V \in \mathcal{B}_0}(V \cap A)$. Thus $\mathcal{B}_A$ is a countable base of open sets for A, so A is separable. □

We conclude this chapter with a brief discussion of a concept that generalizes the notion of a metric in a modest but frequently useful way.

Definition. If X is an arbitrary set, then a real-valued function σ on $X \times X$ is a *pseudometric* on X if the following conditions are satisfied for all elements x, y, and z of X:

(1) $\sigma(x,y) \geq 0$; $\sigma(x,x) = 0$,
(2) $\sigma(x,y) = \sigma(y,x)$,
(3) $\sigma(x,z) \leq \sigma(x,y) + \sigma(y,z)$.

The value of the pseudometric σ at a pair (x,y) of points of X is the *pseudodistance* between x and y. A pair (X,σ) consisting of a set X equipped with a pseudometric σ is a *pseudometric space.*

As is apparent from this definition, the sole difference between a metric and a pseudometric is that it is possible for a pseudometric to vanish at a pair (x,y) even though $x \neq y$. The inequality (3) is known as the *triangle inequality* for pseudometric spaces.

Example X. Let X be a metric space, let $\mathcal{B}_0$ denote the collection of all bounded nonempty subsets of X, and let τ be the nonnegative function defined on $\mathcal{B}_0 \times \mathcal{B}_0$ in Example D. If we define

$$\sigma(A,B) = \tau(A,B) \vee \tau(B,A), \quad A, B \in \mathcal{B}_0,$$

then σ is a nonnegative function on $\mathcal{B}_0 \times \mathcal{B}_0$ such that $\sigma(A,A) = 0$ for each A in $\mathcal{B}_0$, and it is obvious that σ is symmetric, that is, satisfies condition

(2) in the above definition. Moreover, if A, B and C are arbitrary elements of $\mathcal{B}_0$, then

$$\tau(A,C) \leq \tau(A,B) + \tau(B,C) \leq \sigma(A,B) + \sigma(B,C)$$

(see Example D once again). Likewise, of course,

$$\tau(C,A) \leq \tau(C,B) + \tau(B,A) \leq \sigma(A,B) + \sigma(B,C).$$

Thus $\sigma(A,C) \leq \sigma(A,B) + \sigma(B,C)$, so σ satisfies the triangle inequality too, and is therefore a pseudometric on $\mathcal{B}_0$.

Suppose now that $\sigma(A,B) = 0$ for some pair of sets in $\mathcal{B}_0$. Then $\tau(A,B) = \tau(B,A) = 0$ and therefore $A \subset B^-$ and $B \subset A^-$ (see Problem J). In other words, $A^- = B^-$, and it is easily seen that $A^- = B^-$ implies, in turn, that $\sigma(A,B) = 0$. Hence if we denote by $\mathcal{H}$ the subset of $\mathcal{B}_0$ consisting of all the *closed* bounded nonempty subsets of X, then the restriction $\sigma_0 = \sigma \mid (\mathcal{H} \times \mathcal{H})$ is actually a metric on $\mathcal{H}$. This metric σ_0 is known as the *Hausdorff metric* on $\mathcal{H}$. If $\{E_n\}$ is a sequence in $\mathcal{H}$ and if A_0 is an element of $\mathcal{H}$ such that $\sigma_0(A_0, E_n) \to 0$, we shall call A_0 the (*Hausdorff*) *metric limit* of $\{E_n\}$.

For the pseudometric σ defined in the foregoing example it turned out that there was an obvious and natural way to restrict σ so as to obtain a true metric. Things do not work out so agreeably in general, but there is always a natural way of obtaining a metric from a given pseudometric.

Proposition 6.20. *Let σ be a pseudometric on a set X and for each pair of elements x, y of X let us write $x \sim y$ to signify that $\sigma(x,y) = 0$. Then $\sim$ is an equivalence relation on X, and if $[x]$ and $[y]$ denote the equivalence classes of an arbitrary pair x and y of elements of X, then setting*

$$\rho([x],[y]) = \sigma(x,y) \tag{4}$$

defines a metric on the quotient space $X/\sim$. (The metric ρ is called the metric *associated with σ, and the metric space $(X/\sim, \rho)$ is the metric* space *associated with the pseudometric space (X,σ).)*

PROOF. The fact that $\sim$ is an equivalence relation may be taken as obvious. If x' is an arbitrary element of the equivalence class $[x]$ and y' an arbitrary element of $[y]$, then

$$\sigma(x',y') \leq \sigma(x'x) + \sigma(x,y) + \sigma(y,y') = \sigma(x,y),$$

and, similarly, $\sigma(x,y) \leq \sigma(x',y')$. Thus (4) does in fact define a nonnegative function ρ on $(X/\sim) \times (X/\sim)$, and it remains only to show that ρ

is a metric. To this end we first note that ρ is symmetric along with σ. Furthermore, if x, y and z are arbitrary elements of X, then

$$\rho([x],[z]) = \sigma(x,z) \leq \sigma(x,y) + \sigma(y,z) = \rho([x],[y]) + \rho([y],[z]),$$

so ρ also satisfies the triangle inequality. Finally, if $\rho([x],[y]) = 0$, then $\sigma(x,y) = 0$, so $x \sim y$ and therefore $[x] = [y]$. □

PROBLEMS

A. Let $\mathcal{E}$ be a vector space. A *norm* on $\mathcal{E}$ is a nonnegative real-valued function $\| \ \|$ on $\mathcal{E}$ satisfying the following conditions for all vectors x and y in $\mathcal{E}$ and all scalars α:

(1) $\|x\| > 0$ whenever $x \neq 0$,

(2) $\|\alpha x\| = |\alpha|\|x\|$,

(3) $\|x + y\| \leq \|x\| + \|y\|$.

A vector space equipped with such a norm is a *normed space*; the inequality (3) is the *triangle inequality* for the norm $\| \ \|$. Show that if $\mathcal{E}$ is a normed space equipped with the norm $\| \ \|$, then setting

$$\rho(x,y) = \|x - y\|, \quad x, y \in \mathcal{E},$$

defines a metric on $\mathcal{E}$ (called the metric *defined by* $\| \ \|$), and that this metric is *invariant* in the sense that $\rho(x,y) = \rho(x+a, y+a)$ for all vectors x, y and a in $\mathcal{E}$. Show also that the metrics on $\mathbb{R}^n$ and $\mathbb{C}^n$ introduced in Example B are defined by suitably chosen norms $\| \ \|_1$ and $\| \ \|_\infty$.

B. The (*Euclidean*) *inner product* (x, y) of two vectors $x = (\xi_1, \ldots, \xi_n)$ and $y = (\eta_1, \ldots, \eta_n)$ in $\mathbb{C}^n$ is, by definition,

$$(x,y) = \sum_{i=1}^{n} \xi_i \overline{\eta}_i.$$

The same formula also defines the (*Euclidean*) *inner product* on $\mathbb{R}^n$, but there, of course, the complex conjugate bars are unnecessary. Verify that the inner product on $\mathbb{R}^n$ is a positive definite symmetric bilinear functional, while the inner product on $\mathbb{C}^n$ is a positive definite sesquilinear functional (cf. Problem 3Q).

(i) Show that the inner products on both $\mathbb{R}^n$ and $\mathbb{C}^n$ satisfy the inequality

$$|(x,y)| \leq (x,x)^{1/2}(y,y)^{1/2}, \quad x, y \in \mathbb{R}^n[\mathbb{C}^n] \tag{5}$$

(called the *Cauchy-Schwarz inequality*), and use this inequality to prove that

$$\|x\|_2 = (x,x)^{1/2}, \qquad x \in \mathbb{R}^n[\mathbb{C}^n],$$

defines a norm $\| \|_2$ on $\mathbb{R}^n[\mathbb{C}^n]$ (called the *usual* or *Euclidean* norm). Verify that this norm defines the usual metric on $\mathbb{R}^n[\mathbb{C}^n]$, and hence that these "usual metrics" are indeed metrics. (Hint: It suffices to deal with two points $x = (u_1, \ldots, u_n)$ and $y = (v_1, \ldots, v_n)$ with nonnegative coordinates. The verification of the Cauchy-Schwarz inequality may be achieved in a number of ways. The most elementary, but also the most laborious, approach is simply to expand the difference $(x,x)(y,y) - (x,y)^2$ in terms of coordinates and show directly, by carefully grouping terms, that this difference is a sum of squares. An alternate and much less onerous approach is based on elementary algebra (Prob. 3Q). Yet another approach consists in observing the elementary inequality

$$u_i v_i \le (u_i^2 + v_i^2)/2$$

for each index $i = 1, \ldots, n$, and adding these inequalities in the normalized case $(x,x) = (y,y) = 1$.)

(ii) Prove that the Cauchy-Schwarz inequality (5) reduces to equality when and only when the vectors x and y are linearly dependent, and conclude that the triangle inequality for $\| \|_2$ likewise reduces to the equality $\|x + y\|_2 = \|x\|_2 + \|y\|_2$ when and only when one of the vectors is a nonnegative multiple of the other.

C. For an arbitrary n-tuple $a = (\alpha_1, \ldots, \alpha_n)$ in $\mathbb{C}^n$ and an arbitrary positive real number p we write

$$\|a\|_p = \left(\sum_{i=1}^{n} |\alpha_i|^p \right)^{1/p}$$

(Note that this notation agrees with that introduced in Problem B.)

(i) If $p > 1$, then there exists a unique positive number q satisfying the condition $1/p + 1/q = 1$. (The number q is called the *Hölder conjugate* of p, and p and q are said to be *Hölder conjugates*.) Show that if p and q are Hölder conjugates and x and y are any two vectors in $\mathbb{C}^n$, then

$$|(x,y)| \le \|x\|_p \|y\|_q. \tag{6}$$

(This inequality, which is clearly a generalization of the Cauchy-Schwarz inequality of Problem B, is known as the *Hölder inequality*.) Use the Hölder inequality to prove that $\| \|_p$ is a norm on $\mathbb{C}^n$ (and $\mathbb{R}^n$) whenever $p > 1$. (The triangle inequality for the norm $\| \|_p, p > 1$, is called the *Minkowski inequality*. The metric on $\mathbb{C}^n$ defined by $\| \|_p$ will be denoted by ρ_p. This notation is in accord with that in Example B.) (Hint: Use the methods of calculus to verify first that

$$uv \le \frac{u^p}{p} + \frac{v^q}{q}$$

for any two nonnegative numbers u and v, and use this elementary inequality to derive the Hölder inequality in the case $\|x\|_p = \|y\|_q = 1$. Finally, to verify the Minkowski inequality, write

$$\sum_{i=1}^{n} |\xi_i + \eta_i|^p \leq \sum_{i=1}^{n} |\xi_i||\xi_i + \eta_i|^{p-1} + \sum_{i=1}^{n} |\eta_i||\xi_i + \eta_i|^{p-1}, \tag{7}$$

and apply the Hölder inequality to each of the sums in the right member of (7).)

(ii) Show that $\| \ \|_p$ does *not* satisfy the triangle inequality and is therefore not a norm on $\mathbb{C}^n$ for $0 < p < 1$ (unless, of course, $n = 1$). (Hint: Consider $x = (1, 0)$ and $y = (0, 1)$ for the case $n = 2$.)

(iii) Prove that, for each element a of $\mathbb{C}^n$, $\|a\|_p$ converges monotonely downward (as a function of p) to $\|a\|_\infty$ as p tends to infinity. (Hint: Use the fact that $\lim_{p\to+\infty} c^{1/p} = 1$ for every positive number c to show that $\lim_{p\to+\infty} \|a\|_p = 1$ whenever $\|a\|_\infty = 1$.)

D. For each real number p such that $p \geq 1$ let (ℓ_p) denote the set of those infinite sequences $\{\xi_n\}_{n=0}^\infty$ of complex numbers such that

$$\sum_{n=0}^{\infty} |\xi_n|^p < +\infty,$$

and for each element $x = \{\xi_n\}$ of (ℓ_p) define $\|x\|_p = \left(\sum_{n=0}^\infty |\xi_n|^p\right)^{1/p}$. Show that (ℓ_p) is a complex linear space and that $\| \ \|_p$ is a norm on (ℓ_p). Show also that setting $\|x\|_\infty = \sup_{n\in\mathbb{N}_0} |\xi_n|$ defines a norm on the linear space (ℓ_∞) of all bounded complex sequences $x = \{\xi_n\}_{n=0}^\infty$.

(i) Verify that

$$(\ell_1) \subsetneqq (\ell_p) \subsetneqq (\ell_{p'}) \subsetneqq (\ell_\infty)$$

whenever $1 < p < p' < +\infty$, and also that if x belongs to (ℓ_{p_0}) for some finite $p_0 \geq 1$, then $\lim_{p\to+\infty} \|x\|_p = \|x\|_\infty$. (Hint: The infinite versions of the *Hölder* and *Minkowski inequalities* needed to verify that $\| \ \|_p$ satisfies the triangle inequality may be obtained by passing to the limit in the finite versions of those inequalities obtained in the preceding problem; it is probably more instructive, however, to use the techniques developed in solving that problem to derive these inequalities directly.)

(ii) We write e_m for the special sequence $e_m = \{\delta_{mn}\}_{n=0}^\infty$ where δ_{mn} denotes the Kronecker delta (see Example 3F). The sequence $\{e_n\}_{n=0}^\infty$ belongs to (ℓ_p) for all $1 \leq p \leq +\infty$. For each p determine which elements $x = \{\xi_n\}_{n=0}^\infty$ of (ℓ_p) have the property that

$$x = \sum_{n=0}^{\infty} \xi_n e_n.$$

E. Let X be a set and let $\mathcal{B}_{\mathbb{R}}(X)[\mathcal{B}_{\mathbb{C}}(X)]$ denote the linear space of all bounded real-valued [complex-valued] functions on X. The metric of uniform convergence on $\mathcal{B}_{\mathbb{R}}(X)[\mathcal{B}_{\mathbb{C}}(X)]$ (Example H; see also Example 3G) is defined by a uniquely determined norm on $\mathcal{B}_{\mathbb{R}}(X)[\mathcal{B}_{\mathbb{C}}(X)]$. (This norm is known, affectionately if somewhat inelegantly, as the *sup norm*.) More generally, if $\mathcal{E}$ is an arbitrary linear space equipped with the invariant metric defined by some norm, then the metric of uniform convergence on $\mathcal{B}(X;\mathcal{E})$ is also defined by a norm.

F. Let X be a metric space and let $\{x_n\}_{n=1}^{\infty}$ be a sequence in X.

(i) Show that if $\{x_n\}$ converges to a limit a_0, then every *permutation* of $\{x_n\}$, that is, every sequence of the form $\{x_{\sigma(n)}\}$, where σ is a permutation of $\mathbb{N}$ (Prob. 1C), likewise converges to a_0.

(ii) Show that $\{x_n\}$ converges to a limit a_0 if and only if every subsequence of $\{x_n\}$ possesses a subsequence that converges to a_0.

G. Let $\{X_n\}_{n=0}^{\infty}$ be a sequence of nonempty sets, and let $\Pi = \prod_{n=0}^{\infty} X_n$ be their Cartesian product. If $x = \{x_n\}_{n=0}^{\infty}$ and $y = \{y_n\}_{n=0}^{\infty}$ are any two distinct elements of Π, then there exists a smallest index—call it $m(x,y)$—such that $x_{m(x,y)} \neq y_{m(x,y)}$. Show that if $z = \{z_n\}_{n=0}^{\infty}$ is some third element of Π, and if $m = m(x,y) \wedge m(y,z)$, then $x_n = z_n$ for all $n = 1, \ldots, m-1$, and hence that $m(x,z) \geq m$. Use this fact to show that if we define

$$\rho(x,y) = \begin{cases} 0, & x = y, \\ 1/(1+m(x,y)), & x \neq y, \end{cases} \qquad x, y \in \Pi,$$

then $\rho(x,z) \leq \rho(x,y) \vee \rho(y,z)$ for any three elements x, y and z of Π, and conclude that ρ is a metric on Π. (This metric is usually called the *Baire metric*.) Explain what it means for a sequence $\{x^{(m)}\}_{m=1}^{\infty}$ of elements of Π to converge to a limit a in the Baire metric.

H. (i) Suppose given a finite sequence $\{(X_1,\rho_1), \ldots, (X_n,\rho_n)\}$ of metric spaces, and let $x = (x_1, \ldots, x_n)$ and $y = (y_1, \ldots, y_n)$ denote arbitrary elements of the product $\Pi = X_1 \times \cdots \times X_n$. Show that for each real number p such that $p \geq 1$, the function

$$\rho_p(x,y) = [\rho_1(x_1,y_1)^p + \cdots + \rho_n(x_n,y_n)^p]^{1/p}$$

is a metric on Π. Show also that

$$\rho_\infty(x,y) = \rho_1(x_1,y_1) \vee \cdots \vee \rho_n(x_n,y_n)$$

is another metric on Π and that all of these metrics are equivalent in that a sequence $\{x^{(m)} = (x_1^{(m)}, \ldots, x_n^{(m)})\}_{m=1}^{\infty}$ in Π converges to a point $a = (a_1, \ldots, a_n)$ if and only if it converges termwise to a, i.e., if and only if $\lim_m x_i^{(m)} = a_i, i = 1, \ldots, n$. (A metric on a product such as

Π having this last property is called a *product metric*; thus each of the metrics ρ_p is a product metric.)

(ii) Suppose given an infinite sequence $\{(X_n, \rho_n)\}_{n=0}^{\infty}$ of metric spaces, and let $\{k_n\}_{n=0}^{\infty}$ be an arbitrary sequence of strictly positive numbers such that $\sum_{n=0}^{\infty} k_n < +\infty$. Verify that if, for an arbitrary pair $x = \{x_n\}_{n=0}^{\infty}$ and $y = \{y_n\}_{n=0}^{\infty}$ of elements of the product $\Pi = \prod_{n=0}^{\infty} X_n$, we set

$$\rho(x, y) = \sum_{n=0}^{\infty} k_n \frac{\rho_n(x_n, y_n)}{1 + \rho_n(x_n, y_n)},$$

then ρ is a metric on Π with the property that a sequence $\{x_m = \{x_n^{(m)}\}_{n=0}^{\infty}\}_{m=1}^{\infty}$ converges in Π to a limit $a = \{a_n\}_{n=0}^{\infty}$ with respect to the metric ρ if and only if it converges termwise to a, i.e., if and only if $\lim_m x_n^{(m)} = a_n$, $n \in \mathbb{N}_0$. Conclude that any two such metrics on Π are equivalent. (Here too it will be convenient to refer to any such metric as a *product metric* on Π.) (Hint: Recall Example K.) Conclude also, in particular, that

$$\rho(x, y) = \sum_{n=0}^{\infty} \frac{1}{2^n} \frac{|\xi_n - \eta_n|}{1 + |\xi_n - \eta_n|}, \quad x = \{\xi_n\}_{n=0}^{\infty}, y = \{\eta_n\}_{n=0}^{\infty},$$

defines a metric on the space (s) of all sequences $\{\alpha_n\}_{n=0}^{\infty}$ of complex numbers. (This is the *metric of pointwise convergence* on (s).)

(iii) Let S denote the space $\mathbb{N}^{\mathbb{N}_0} = \mathbb{N} \times \mathbb{N} \times \cdots$ of all sequences $s = \{k_n\}_{n=0}^{\infty}$ of positive integers. Show that the relative metric on S that it receives as a subspace of (s) is a product metric on S if each copy of $\mathbb{N}$ is equipped with the discrete metric (Ex. C), and that this metric is also equivalent to the Baire metric on S (Prob. G). (Hint: Examine what it means for a sequence to converge in S in all three cases.)

I. Let (X, ρ) be a metric space. Show that the metric ρ satisfies the inequality

$$|\rho(x_1, y_1) - \rho(x_2, y_2)| \leq \rho(x_1, x_2) + \rho(y_1, y_2)$$

for all quadruples x_1, x_2, y_1 and y_2 of points of X. Conclude from this that if $\{x_n\}$ and $\{y_n\}$ are sequences of points in X converging to limits a_0 and b_0, respectively, then $\lim_n \rho(x_n, y_n) = \rho(a_0, b_0)$. Show that diam $A^- =$ diam A for every subset A of X.

J. Let (X, ρ) be a metric space and let A be a nonempty subset of X.

(i) Show that $d(x, A) \leq \rho(x, y) + d(y, A)$ for each pair x, y of points of X, and use this fact to prove that the function $f_A(x) = d(x, A), x \in X$, satisfies the inequality

$$|f_A(x) - f_A(y)| \leq \rho(x, y), \qquad x, y \in X.$$

(ii) Show that if $\{x_n\}_{n=1}^{\infty}$ is a convergent sequence in X, then the sequence $\{f_A(x_n)\}_{n=1}^{\infty}$ converges in $\mathbb{R}$ to $f_A(\lim_n x_n)$, and conclude that if c is an arbitrary nonnegative number, the sets $\{x \in X : f_A(x) \leq c\}$ and $\{x \in X : f_A(x) \geq c\}$ are both closed. Show, in particular, that $\{x \in X : f_A(x) = 0\}$ coincides with A^-.

(iii) Let B and C be bounded nonempty subsets of X, and let τ be the function defined in Example D. Prove that $\tau(B, C) = 0$ if and only if C is dense in B, i.e., if and only if $B \subset C^-$.

K. (i) Which of the spaces (ℓ_p) of Problem D are separable and which are not?

(ii) Determine when the product Π of a sequence of sets is separable when equipped with the Baire metric (Prob. G).

(iii) Determine when the space $\mathcal{B}(Z; X)$ of all bounded mappings of a set Z into a metric space X is separable in the metric of uniform convergence (Ex. H).

L. Prove that the cardinal number of a separable metric space cannot exceed $\aleph$, the power of the continuum (Chap. 4). (Hint: There are $\aleph$ sequences in a countably infinite set.)

M. Two (similarly indexed) sequences $\{x_n\}$ and $\{y_n\}$ in a metric space (X, ρ) are said to be *equiconvergent* if $\lim_n \rho(x_n, y_n) = 0$. Show that a sequence $\{x_n\}$ converges to a limit a in X if and only if it is equiconvergent with the constant sequence $\{a, a, \ldots\}$. Show too that if either of two equiconvergent sequences $\{x_n\}$ and $\{y_n\}$ is convergent, then both are, and $\lim_n x_n = \lim_n y_n$. Show, finally, that if A is an arbitrary subset of X and $\{x_n\}$ is any sequence in A^-, then there exists an equiconvergent sequence $\{y_n\}$ in A.

N. Show that every monotone increasing sequence $\{E_n\}$ of subsets of a metric space X possesses a closed limit (Ex. N) by finding that limit. Similarly, find $\mathrm{F}\lim_n E_n$ for an arbitrary decreasing sequence $\{E_n\}$ of subsets of X.

O. A subset D of a metric space X is *dense in itself* if every point of D is an accumulation point of D, i.e., if $D \subset D^*$.

(i) If a set A is *not* dense in itself, then there exists a point x_0 in A and a positive radius ε such that $D_\varepsilon(x_0) \cap A = \{x_0\}$. Such a point is an *isolated point* of A. (Thus A is dense in itself if and only if it does not possess any isolated points.) A subset A of a metric space X consisting entirely of isolated points is a *discrete* subset of X. Give an example of a countable discrete subset M of a metric space X with the property that M^* has the power of the continuum and show that this result cannot be improved upon in that the derived set of a countable subset of an arbitrary metric space has cardinal number at most $\aleph$.

(ii) A subset P of a metric space X that is both closed and dense in itself, i.e., that is such that $P^* = P$, is called *perfect*. Show that the Cantor set C (Ex. O) is a perfect subset of $\mathbb{R}$. Show also that the generalized Cantor sets of Examples P and Q are all perfect.

P. A subset A of a metric space X is known as a G_δ in X if there exists a sequence $\{U_n\}_{n=1}^{\infty}$ of open sets in X such that

$$A = \bigcap_{n=1}^{\infty} U_n.$$

Dually, a subset B of X is an F_σ in X if there exists a sequence $\{F_n\}_{n=1}^{\infty}$ of closed sets in X such that

$$B = \bigcup_{n=1}^{\infty} F_n.$$

(In this notation the letter "G" stands for "*Gebiet*," a word meaning "open set" in German, while "δ" suggests countable intersection because the German for "intersection" is "*Durchschnitt*"; similarly, the letter "F" stands for "*fermé*", meaning "closed" in French, while "σ" suggests countable union, because the French for "union" used to be "*somme*".)

(i) Verify that these notions are truly dual in the sense that the complement in X of an arbitrary G_δ in X is an F_σ in X, and vice versa. Show too that the intersection of an arbitrary nonempty countable collection of G_δs in X is a G_δ in X, and, dually, that the union of an arbitrary countable collection of F_δs in X is an F_σ in X.

(ii) Show that every closed subset of a metric space X is a G_δ in X, and dually, that every open set in X is an F_σ in X. (Hint: If F is a nonempty closed subset of X, and if $f(x) = d(x, F)$, $x \in X$, then $F = \{x \in X : f(x) = 0\}$, while each set of the form $\{x \in X : f(x) < c\}$ is open; see Problem J.)

(iii) Let A be a subspace of a metric space X, and suppose A is a G_δ in X. Prove that a G_δ in A is also a G_δ in X. State and prove the analogous fact concerning F_σ subsets of an F_σ. (Hint: Show first that a relatively open subset of A is a G_δ in X.)

Q. Let X be a metric space and let A be a subset of X. A point x is an *interior point* of A if there exists a positive number ε such that $D_\varepsilon(x) \subset A$. The set of all interior points of A is called the *interior* of A and is denoted by A°.

(i) Show that for an arbitrary subset A of X the interior A° is open and is, in fact, the largest open subset of A. Show that a subset U of X is itself open if and only if $U = U^\circ$, and that interiors are dual to closures in the sense that $(X\backslash A)^- = X\backslash A^\circ$ (and $(X\backslash A)^\circ = X\backslash A^-$) for every subset A of X.

(ii) Formulate and prove the dual of Proposition 6.9.

(iii) For any subset A of X the difference $A^-\backslash A^\circ$ is the *boundary* of A (notation: ∂A). Verify that ∂A is a closed set that may equally well be described as the intersection $A^- \cap (X\backslash A)^-$, and thus consists of the set of all those points x in X with the property that every ball $D_\varepsilon(x)(\varepsilon > 0)$ meets both A and $X\backslash A$. Verify also that a subset A of X is closed if and only if $\partial A \subset A$, while A is open if and only if $A \cap \partial A = \varnothing$.

(iv) Verify that for arbitrary subsets A and B of a metric space X the boundaries of $A \cup B, A \cap B, A\backslash B$ and $A\nabla B$ are all contained in the union $(\partial A) \cup (\partial B)$.

R. For any metric space X there is a smallest cardinal number c with the property that there exists a topological base $\mathcal{B}$ for X such that card $\mathcal{B} = c$ (why?), and this cardinal number is called the *weight* of X. (Thus X satisfies the second axiom of countability if and only if the weight of X is $\leq \aleph_0$.) Generalize Theorem 6.18 by showing that the weight of an arbitrary metric space X may also be described as the smallest cardinal number c with the property that there exists a set M in X with card $M = c$ and $M^- = X$. Show too that if the weight of X is c, then the cardinal number of a disjoint collection of nonempty open subsets of X cannot exceed c. (Hint: Recall Proposition 4.6. It is best to treat spaces of finite weight as a special case.)

S. Let $\mathcal{B}_0$ be a base of open sets for a metric space X and let $\mathcal{U}$ be an arbitrary collection of open sets in X. Prove that $\mathcal{U}$ contains a subcollection $\mathcal{U}_0$ such that $\bigcup \mathcal{U}_0 = \bigcup \mathcal{U}$ and card $\mathcal{U}_0 \leq$ card $\mathcal{B}_0$. (Hint: The collection $\mathcal{V} = \bigcup_{U \in \mathcal{U}} \{V \in \mathcal{B}_0 : V \subset U\}$ is contained in $\mathcal{B}_0$ and every set in $\mathcal{V}$ is contained in some set belonging to $\mathcal{U}$.)

(i) Use the foregoing result to establish that if X has weight c, and if $\mathcal{U}$ is an arbitrary collection of open subsets of X, then there exists a subcollection $\mathcal{U}_0$ with $\bigcup \mathcal{U}_0 = \bigcup \mathcal{U}$ and card $\mathcal{U}_0 \leq c$.

(ii) Verify also that if X has weight c and if $\mathcal{B}$ is some other base of open subsets of X, then $\mathcal{B}$ contains a base of open sets $\mathcal{B}_1$ for X such that card $\mathcal{B}_1 = c$. (Thus, in particular, if X is separable and if $\mathcal{B}$ is an arbitrary base of open sets for X, then $\mathcal{B}$ contains a countable base for X.) (Hint: Recall Problems 4I and 4S. Here again it is best to treat spaces of finite weight as a special case.)

T. Let X be a metric space, let α be an ordinal number, and suppose given a strictly increasing indexed family $\{U_\xi\}_{\xi<\alpha}$ of open subsets of X. (Thus the index set is the ordinal number segment $W(\alpha)$ and $U_\xi \subsetneqq U_\eta$ whenever

$\xi < \eta < \alpha$.) Show that the cardinal number of α cannot exceed the weight of X. (In particular, then, if X is separable, α must be countable.) (Hint: If $\mathcal{B}$ is any base of open subsets of X and $\xi + 1 \in W(\alpha)$, then $U_{\xi+1}$ must contain at least one element V_ξ of $\mathcal{B}$ that U_ξ does not.)

U. The idea of the derived set leads naturally to the notion of higher order derivatives. Indeed, we define the *first derivative* $A^{(1)}$ of a subset A of a metric space X to be A^*, the *second derivative* $A^{(2)}$ to be $A^{**} = A^{(1)*}$, and so forth. (For technical reasons it is desirable to set $A^{(0)}$ equal to A^- instead of A itself.) For some purposes it is important to press on with this inductive definition into the realm of the transfinite. Suppose, accordingly, that α is an ordinal number, that η is an ordinal number belonging to the ordinal number segment $W(\alpha)$, and that derivatives $A^{(\xi)}$ of every order $\xi, \xi < \eta$, have been defined for a subset A of a metric space X. We then define the *derivative* $A^{(\eta)}$ *of order* η as follows: If η is an ordinal number of type I (Prob. 5G) and if, say, $\eta = \xi_0 + 1$, then we set $A^{(\eta)} = A^{(\xi_0)*}$; if, on the other hand, η is a limit number, we set

$$A^{(\eta)} = \bigcap_{\xi < \eta} A^{(\xi)}.$$

In this way the derivative of each order η in $W(\alpha)$ of an arbitrary subset A of X is defined by transfinite induction (Th. 5.12). (We leave to the reader the routine verification that if α is replaced in this definition by some other ordinal number α', then the resulting notion of higher order derivatives is unchanged on the ordinal number segment $W(\alpha \wedge \alpha')$.)

(i) For an arbitrary subset A of a metric space X the indexed family $\{A^{(\xi)}\}_{\xi<\alpha}$ thus obtained is a monotone decreasing family of closed sets. Show that if the ordinal number α is taken large enough, then there necessarily exists a unique smallest ordinal number η_0 in $W(\alpha)$ (with η_0 independent of α) such that $A^{(\eta_0+1)} = A^{(\eta_0)}$, and that $A^{(\xi)}$ then coincides with $A^{(\eta_0)}$ for all ξ such that $\eta_0 \leq \xi < \alpha$. This smallest ordinal number η_0 such that $A^{(\xi)}$ is constant on $W(\alpha) \backslash W(\eta_0)$ is the *derivation order* of A. (Hint: Take for α the initial number ω_c of some cardinal number c such that c is greater than the weight of X.)

(ii) Prove that if X is a separable metric space, then the derivation order of every subset of X is countable. (Hint: Use Problem T.)

(iii) Show, conversely, by transfinite induction that if a is an arbitrary real number and η is an arbitrary countable ordinal number, then there exists a set A of real numbers suc. that $A^{(\eta)} = \{a\}$. (Hint: This is one of those situations in which it is easier for technical reasons to prove more than is required. Let a and b be real numbers such that $a < b$, and let $\{s_n\}$ be a strictly decreasing sequence in the open interval (a, b) such that $\lim_n s_n = a$. Show first that if A_n is an arbitrary nonempty subset of the interval $[s_{n+1}, s_n)$, $n \in \mathbb{N}$, then

$$\left(\bigcup_{n=1}^{\infty} A_n\right)^* = \{a\} \cup \bigcup_{n=1}^{\infty} A_n^*,$$

and use this fact to prove that for an arbitrary countable η there exists a countable closed subset F of $[a, b)$ such that $F^{(\eta)} = \{a\}$.)

(iv) Show, finally, that for an arbitrary countable ordinal number η there exists a closed subset F of $\mathbb{R}$ having η as its derivation order.

V. A point x in a metric space X is a *condensation point* of a set $A \subset X$ if the intersection $D_\varepsilon(x) \cap A$ is *uncountable* for every open ball $D_\varepsilon(x)$ with $\varepsilon > 0$. Show that the set $A^\dagger$ of all condensation points of an arbitrary set A in a metric space X is a closed set contained in the derived set A^*. Give examples of sets A for which $A^\dagger \neq A^*$ and for which $A^\dagger = A^*$. Show that an arbitrary closed subset F of a separable metric space can be expressed uniquely as the disjoint union of a countable set and a perfect set P having the property that every point of P is a condensation point of P. (The set P is called the *Bendixson kernel* of F.) Verify that $P \subset F^{(\xi)}$ for all derivatives of F, and hence that $P \subset F^{(\eta)}$ where η denotes the derivation order of F (Prob. U). (Hint: Set $P = F^\dagger$.) Show also that the Cantor set is its own Bendixson kernel.

W. (i) Show that if A is a countable set, then there exists a sequence $\{x_n\}$ in A such that every point a of A is taken on infinitely often by $\{x_n\}$. Conclude that if A is an arbitrary countable subset of a metric space X, then there exists in X an infinite sequence $\{x_n\}$ with the property that every point of A is a cluster point of $\{x_n\}$.

(ii) Prove that the set of all cluster points of an arbitrary sequence $\{x_n\}$ in a metric space X is a closed subset of X. Show also, in the converse direction, that if X is a separable metric space, then there exists a single infinite sequence $\{x_n\}$ in X with the property that if F is an arbitrary closed subset of X, then $\{x_n\}$ possesses a subsequence $\{x_{n_k}\}$ having F for the set of all of its cluster points.

X. Let $\mathcal{B}_0$ denote the collection of all bounded nonempty subsets of a metric space X, let $\{E_n\}_{n=1}^\infty$ be a sequence in $\mathcal{B}_0$, and let $\overline{F}$ and $\underline{F}$ denote the closed limit superior and the closed limit inferior of $\{E_n\}$, respectively (see Example N).

(i) If τ is the function defined on $\mathcal{B}_0 \times \mathcal{B}_0$ in Example D, and if $\lim_n \tau(A, E_n) = 0$ for some element A of $\mathcal{B}_0$, then $A \subset \underline{F}$. Show, conversely, that if $\underline{F} \in \mathcal{B}_0$ and if it is not the case that $\lim_n \tau(\underline{F}, E_n) = 0$, then there exists a sequence $\{x_k\}$ in $\underline{F}$ such that $\{x_k\}$ has no cluster point in X. (Hint: If $\tau(\underline{F}, E_n) \not\to 0$, then there exist a strictly increasing sequence $\{n_k\}_{k=1}^\infty$ of positive integers and a sequence $\{x_k\}_{k=1}^\infty$ in $\underline{F}$ such that the sequence $\{d(x_k, E_{n_k})\}$ is bounded away from zero; use the fact that $\underline{F}$ is closed.)

(ii) Show, dually, that if $\lim_n \tau(E_n, A) = 0$ for some element A of $\mathcal{B}_0$, then $\overline{F} \subset A^-$, and also that if $\overline{F} \in \mathcal{B}_0$ and $\tau(E_n, \overline{F}) \not\to 0$, then there exist a strictly increasing sequence $\{n_k\}$ of positive integers and a sequence

$\{x_k\}$ of points of X such that $x_k \in E_{n_k}$ for every index k and such that the sequence $\{x_k\}$ has no cluster point in X. (Hint: If $\tau(E_n, \overline{F}) \not\to 0$, then there exists a strictly increasing sequence $\{n_k\}$ of positive integers such that for each index k a point x_k of E_{n_k} can be selected in such a way that the sequence $\{d(x_k, \overline{F})\}$ is bounded away from zero.)

(iii) Conclude that if σ_0 denotes the Hausdorff metric on the space $\mathcal{H}$ consisting of the closed sets in $\mathcal{B}_0$ (so that $\sigma_0(A, B) = \tau(A, B) \vee \tau(B, A)$; see Example X), if the sets E_n all belong to $\mathcal{H}$, and if the metric limit $\lim_n E_n$ exists in $\mathcal{H}$, then

$$\lim_n E_n = \mathrm{F}\lim_n E_n.$$

(That is, whenever the metric limit of a sequence in $\mathcal{H}$ exists, the closed limit of that sequence also exists and coincides with the metric limit.) Conclude, in the other direction, that if $F = \overline{F} = \underline{F} \in \mathcal{H}$, and if $\sigma_0(F, E_n) \not\to 0$, then there exists a sequence in X that possesses no cluster point.

(iv) Show also that if the sequence $\{E_n\}$ lies in $\mathcal{H}$, and if an element A_0 of $\mathcal{H}$ is a metric cluster point of $\{E_n\}$ (so that there exists a subsequence $\{E_{n_k}\}$ of $\{E_n\}$ such that A_0 is the metric limit of $\{E_{n_k}\}$), then

$$\underline{F} \subset A_0 \subset \overline{F}.$$

Conclude that if the closed limit $F = \mathrm{F}\lim_n E_n$ exists, then F is the only possible metric cluster point of $\{E_n\}$.

(v) Give an example of a sequence $\{E_n\}$ of closed bounded nonempty subsets of $\mathbb{R}^2$ such that $F = \mathrm{F}\lim_n E_n$ exists and is bounded and nonempty, and such that the sequence $\{\sigma_0(F, E_n)\}$ tends to $+\infty$.

Y. (i) Let (Y, ρ) be a metric space and let ϕ be an arbitrary mapping of a set X into Y. Then setting

$$\sigma(x, x') = \rho(\phi(x), \phi(x')), \qquad x, x' \in X,$$

defines a pseudometric on X, as does

$$\sigma'(x, x') = \frac{\rho(\phi(x), \phi(x'))}{1 + \rho(\phi(x), \phi(x'))}, \qquad x, x' \in X.$$

Show that the metrics associated with σ and σ' are equivalent, and also that σ and σ' are themselves metrics on X if and only if ϕ is one-to-one.

(ii) Let $\{\sigma_n\}_{n=1}^N$ be a finite sequence of pseudometrics on a set X. Verify that

$$\widetilde{\sigma}_1(x, y) = \sum_{n=1}^{N} \sigma_n(x, y)$$

and

$$\widetilde{\sigma}_\infty(x, y) = \sigma_1(x, y) \vee \cdots \vee \sigma_N(x, y)$$

are also pseudometrics on X. Indeed,

$$\widetilde{\sigma}_p(x, y) = [\sigma_1(x, y)^p + \cdots + \sigma_N(x, y)^p]^{1/p}$$

is a pseudometric on X for each $p, 1 \leq p < +\infty$. Under what conditions are the pseudometrics $\widetilde{\sigma}_p$ actually metrics on X?

(iii) Let $\{\sigma_n\}_{n=1}^{\infty}$ be an infinite sequence of pseudometrics on X. Show that

$$\widetilde{\sigma}(x, y) = \sum_{n=1}^{\infty} \frac{1}{2^n} \frac{\sigma_n(x, y)}{1 + \sigma_n(x, y)}$$

defines a pseudometric $\widetilde{\sigma}$ on X, and give conditions on the sequence $\{\sigma_n\}$ in order that $\widetilde{\sigma}$ be a metric.

7 Continuity and limits

A mapping between metric spaces is *continuous* if it preserves closeness. This idea is made precise in the following definition.

Definition. Let (X, ρ) and (Y, σ) be metric spaces, let ϕ be a mapping of X into Y, and let x_0 be a point of X. Then ϕ is *continuous at* x_0 (and x_0 is a *point of continuity* of ϕ) if for each positive number ε there exists a positive number δ such that $\rho(x, x_0) < \delta$ implies $\sigma(\phi(x), \ \phi(x_0)) < \varepsilon$, or, equivalently, such that $\phi(D_\delta(x_0)) \subset D_\varepsilon(\phi(x_0))$. Conversely, if ϕ is not continuous at x_0, then ϕ is *discontinuous at* x_0 (and x_0 is a *point of discontinuity* of ϕ). Finally, ϕ is *continuous* (*on* X) if it is continuous at every point of X.

Continuity is a topological property, as the following two elementary propositions clearly show.

Proposition 7.1. *Let ϕ be a mapping of a metric space X into a metric space Y, and let x_0 be a point of X. Then ϕ is continuous at x_0 if and only if the sequence $\{\phi(x_n)\}$ converges in Y to $\phi(x_0)$ whenever $\{x_n\}$ is a sequence in X that converges to x_0. Moreover, if ϕ is continuous at x_0, and if $\{x_\lambda\}_{\lambda \in \Lambda}$ is an arbitrary net in X that converges to x_0, then the net $\{\phi(x_\lambda)\}$ converges to $\phi(x_0)$.*

Proof. Suppose first that ϕ is continuous at x_0 and that $\{x_\lambda\}$ is a net tending to x_0 in X. If $\varepsilon > 0$, then there exist a positive number δ such that $\phi(D_\delta(x_0)) \subset D_\varepsilon((\phi(x_0))$ and an index λ_0 such that $x_\lambda \in D_\delta(x_0)$ for all $\lambda \geq \lambda_0$. But then $\phi(x_\lambda) \in D_\varepsilon(\phi(x_0))$ for $\lambda \geq \lambda_0$, which shows that $\phi(x_\lambda) \to \phi(x_0)$.

Suppose next that ϕ is discontinuous at x_0. Then there exists a positive number ε_0 so small that $D_{\varepsilon_0}(\phi(x_0))$ contains no set of the form $\phi(D_\delta(x_0))$ for any positive δ. Hence for each positive integer n there exists at least one point x_n of $D_{1/n}(x_0)$ such that $\phi(x_n) \notin D_{\varepsilon_0}(\phi(x_0))$. But then the sequence $\{x_n\}$ converges to x_0 in X while the sequence $\{\phi(x_n)\}$ clearly fails to converge to $\phi(x_0)$ in Y. □

Proposition 7.2. *Let ρ and ρ' be equivalent metrics on a set X, and let σ and σ' be equivalent metrics on a set Y. Then a mapping $\phi : X \to Y$ is continuous at a point x_0 of X as a mapping of (X, ρ) into (Y, σ) if and only if it is continuous at x_0 as a mapping of (X, ρ') into (Y, σ'). Hence ϕ is continuous as a mapping of (X, ρ) into (Y, σ) if and only if it is continuous as a mapping of (X, ρ') into (Y, σ').*

PROOF. In view of Proposition 7.1, both parts of this proposition are immediate consequences of the definition of equivalence. □

Example A. Let $\phi_1, \ldots, \phi_k$ be mappings of a metric space X into metric spaces $Y_1, \ldots, Y_k$, respectively, and let $\Pi = Y_1 \times \ldots \times Y_k$ be equipped with a product metric (see Problem 6H). Then the mapping Φ of X into Π defined by setting

$$\Phi(x) = (\phi_1(x), \ldots, \phi_k(x)), \quad x \in X,$$

is continuous at a point x_0 of X if and only if all of the mappings $\phi_i, i = 1, \ldots, k$, are continuous there. Indeed, if $\{x_n\}$ is a sequence in X converging to x_0, then the sequence $\{\Phi(x_n)\}$ converges to $\Phi(x_0)$ when and only when each sequence $\{\phi_i(x_n)\}$ converges to $\phi_i(x_0), i = 1, \ldots, k$, so the desired result follows at once from Proposition 7.1.

Definition. A mapping ϕ of a metric space (X, ρ) into a metric space (Y, σ) is said to be *Lipschitzian*, or to satisfy a *Lipschitz condition*, with *Lipschitz constant* M, if

$$\sigma(\phi(x), \phi(x')) \leq M\rho(x, x'), \quad x, x' \in X,$$

for some positive number M. (In the event that the Lipschitz constant M can be taken to be one or less, the mapping ϕ is also sometimes said to be *contractive*.)

Clearly a Lipschitzian mapping is continuous (set $\delta = \varepsilon/M$), so examples of Lipschitzian mappings are automatically examples of continuous mappings as well.

Example B. If A is a nonempty subset of a metric space (X, ρ), then the function $f_A(x) = d(x, A), x \in X$, is contractive (Prob. 6J). Likewise, the metric ρ itself is contractive as a mapping of $X \times X$ into $\mathbb{R}$ if $X \times X$ is equipped with the metric ρ_1 (Prob. 6I). In particular, the mapping $\zeta \to |\zeta|$ is contractive on $\mathbb{C}$, as is the mapping $x \to \|x\|$ on any normed space $\mathcal{E}$. Obviously all constant mappings and all isometries are contractive. A somewhat more interesting example goes as follows. Let $X_1, \ldots, X_n$ be metric spaces, and let Π denote the product $\Pi = X_1 \times \ldots \times X_n$ equipped with any one of the metrics $\rho_p, 1 \leq p \leq +\infty$ (see Problem 6H). Then the projection π_i of Π onto X_i is contractive for each $i = 1, \ldots, n$.

Example C. The operations of addition and subtraction are Lipschitzian when viewed as complex-valued functions on $\mathbb{C}^2$ (and therefore also, of course, when viewed as real-valued functions on $\mathbb{R}^2$). Indeed,

$$(\xi_1 \pm \eta_1) - (\xi_2 \pm \eta_2) = (\xi_1 - \xi_2) \pm (\eta_1 - \eta_2)\,,$$

so

$$|(\xi_1 \pm \eta_1) - (\xi_2 \pm \eta_2)| \leq |\xi_1 - \xi_2| + |\eta_1 - \eta_2|$$

for all complex numbers $\xi_1, \xi_2, \eta_1, \eta_2$. Thus addition and subtraction satisfy a Lipschitz condition with Lipschitz constant $M = 1$ if $\mathbb{C}^2$ is equipped with the metric ρ_1, and they also satisfy a Lipschitz condition (with $M = \sqrt{2}$) if $\mathbb{C}^2$ is equipped with its usual metric. (Similarly, for any normed space $\mathcal{E}$, the operations of vector addition and subtraction are Lipschitzian on $\mathcal{E} \times \mathcal{E}$.)

Example D. Let $\mathcal{E}$ and $\mathcal{F}$ be normed spaces, let $T : \mathcal{E} \to \mathcal{F}$ be a linear transformation of $\mathcal{E}$ into $\mathcal{F}$, and suppose there is some point x_0 of $\mathcal{E}$ at which T is continuous. Then there exists a positive number δ such that $\|x - x_0\| < \delta$ (with x in $\mathcal{E}$) implies that $\|Tx - Tx_0\| = \|T(x - x_0)\| < 1$. But then for any δ' such that $0 < \delta' < \delta$ it is the case that $\|z\| \leq \delta'$ implies that $\|T(x_0 + z) - Tx_0\| = \|Tz\| \leq 1$. Let us fix one such number δ' and set $M = 1/\delta'$. If for an arbitrary nonzero vector z in $\mathcal{E}$ we set $z' = z/\|z\|$, then $\delta'\|Tz'\| = \|T\delta' z'\| \leq 1$, so $\|Tz'\| \leq M$. Hence the number M satisfies the following condition:

$$\|Tz\| \leq M\|z\| \quad , \quad z \in \mathcal{E}. \tag{1}$$

But then, of course, $\|Tx - Tx'\| = \|T(x - x')\| \leq M\|x - x'\|$ for any two vectors x and x' in $\mathcal{E}$, so T is Lipschitzian with Lipschitz constant M. Thus we see that the following four conditions are equivalent for a linear transformation T between normed spaces $\mathcal{E}$ and $\mathcal{F}$:

(a) T is continuous at some one vector in $\mathcal{E}$,
(b) There is a number M satisfying (1) for T,

(c) T is Lipschitzian,
(d) T is continuous on $\mathcal{E}$.

A linear transformation T of a normed space $\mathcal{E}$ into a normed space $\mathcal{F}$ is said to be *bounded* if it possesses any one, and hence all four, of these properties. (It should be emphasized that this terminology, which is at variance with our standard use of the term "bounded" in the context of a general metric space, applies only to linear transformations.)

Suppose now that T is a bounded linear transformation of $\mathcal{E}$ into $\mathcal{F}$, and consider the (nonempty) set $\mathcal{M}$ of all those real numbers M such that (1) holds for M and T. It is obvious that $\mathcal{M}$ is a ray to the right in $\mathbb{R}$. Somewhat less obviously, the ray $\mathcal{M}$ is closed. (Indeed, if $M_0 = \inf \mathcal{M}$, and if for some vector x_0 in $\mathcal{E}$ we have $\|Tx_0\| > M_0\|x_0\|$, then $x_0 \neq 0$ and $\|Tx_0\|/\|x_0\| > M_0$, so there exists a number M in $\mathcal{M}$ such that $M < \|Tx_0\|/\|x_0\|$. But then $\|Tx_0\| > M\|x_0\|$, which is impossible.) This least element of $\mathcal{M}$ is called the *norm* of T and is denoted by $\|T\|$. (It is sometimes useful to know that the norm of a bounded linear transformation is also given by the supremum $\sup\{\|Tx\| : \|x\| \leq 1\}$. Indeed, if we set $M_1 = \sup\{\|Tx\| : \|x\| \leq 1\}$, then it is obvious that $M_1 \leq \|T\|$; on the other hand, if $x \neq 0$, then

$$\|T(x/\|x\|)\| = \|Tx\|/\|x\| \leq M_1,$$

whence it follows that $M_1 \in \mathcal{M}$, and hence that $\|T\| \leq M_1$.)

Example E. Every linear transformation T of $\mathbb{R}^n$ into $\mathbb{R}^m$ is bounded. Indeed, if $A = (a_{ij})$ is the $m \times n$ matrix defining T (so that the image $(t_1, \ldots, t_m)$ under T of a point $(s_1, \ldots, s_n)$ is given by

$$\begin{pmatrix} t_1 \\ \vdots \\ t_m \end{pmatrix} = A \begin{pmatrix} s_n \\ \vdots \\ s_n \end{pmatrix};$$

see Example 3Q), then $\rho_2(T(s_1, \ldots, s_n), T(s_1', \ldots, s_n'))$ is given by

$$\left[\sum_{i=1}^{m} \left(\sum_{j=1}^{n} a_{ij}(s_j - s_j')\right)^2\right]^{\frac{1}{2}}$$

Hence, if we write $x = (s_1, \ldots, s_n)$ and $x' = (s_1', \ldots, s_n')$, we have

$$\begin{aligned} \rho_2(Tx, Tx') &\leq \left[\sum_{i=1}^{m} \left(\sum_{j=1}^{n} a_{ij}^2\right)\left(\sum_{j=1}^{n} (s_j - s_j')^2\right)\right]^{\frac{1}{2}} \\ &= \rho_2(x, x') \left[\sum_{i=1}^{m} \sum_{j=1}^{n} a_{ij}^2\right]^{\frac{1}{2}} \end{aligned}$$

by the Cauchy-Schwarz inequality (Prob. 6B). Thus when $\mathbb{R}^n$ and $\mathbb{R}^m$ are equipped with their usual Euclidean metrics, $\|T\|$ is dominated by the constant

$$N(A) = \left[\sum_{i=1}^{m}\sum_{j=1}^{n} a_{ij}^2\right]^{\frac{1}{2}},$$

called the *Hilbert–Schmidt norm* of A.

It is frequently necessary to deal with mappings that are defined only on some subset of a metric space. The following definition is nothing but a formalization of various conventions agreed upon earlier and is included here only to avoid misunderstandings and to fix terminology.

Definition. Let X and Y be metric spaces, and let ϕ be a mapping of some subset A of X into Y. Then ϕ is said to be *continuous* at a point x_0 of A *relative to* A if the restriction $\phi|A$ is continuous at x_0. If ϕ is continuous at a point x_0 of A relative to A, then x_0 is a *point of continuity* of ϕ *relative to* A, and if ϕ is not continuous at a point x_0 of A relative to A, then ϕ is *discontinuous* at x_0 *relative to* A, and x_0 is a *point of discontinuity* of ϕ *relative to* A. Similarly, ϕ is *continuous on* A (*relative to* A) if the mapping $\phi|A : A \to Y$ is continuous on the subspace A.

The following proposition merely states, in the context of a relative metric, the definition of continuity and the criteria for continuity set forth in Proposition 7.1, and no proof is needed or given.

Proposition 7.3. *Let X and Y be metric spaces, and let ϕ be a mapping of a subset A of X into Y. Then ϕ is continuous at a point x_0 of A relative to A if and only if for each positive number ε there exists a positive number δ such that $\phi(D_\delta(x_0) \cap A) \subset D_\varepsilon(\phi(x_0))$. Equivalently, ϕ is continuous at x_0 relative to A if and only if $\{\phi(x_n)\}$ converges to $\phi(x_0)$ whenever $\{x_n\}$ is a sequence in A that converges to x_0.*

Example F. The characteristic function $\chi_{\mathbb{Q}}$ of the rational numbers (Prob. 1R), regarded as a mapping of $\mathbb{R}$ into itself, is discontinuous at every point of $\mathbb{R}$. Nonetheless, this function is continuous on $\mathbb{Q}$ relative to $\mathbb{Q}$ (being constantly equal to one there) and is likewise continuous on the set $\mathbb{R}\backslash\mathbb{Q}$ of irrational numbers relative to $\mathbb{R}\backslash\mathbb{Q}$ (being constantly equal to zero there).

The following result provides a pair of important criteria for the continuity of a mapping between metric spaces. (For other such criteria see Problem B.)

Theorem 7.4. *Let X and Y be metric spaces and let ϕ be a mapping of X into Y. Then the following are equivalent:*
(1) ϕ is continuous,
(2) For every open set U in Y the inverse image $\phi^{-1}(U)$ is open in X,
(3) For every closed set F in Y the inverse image $\phi^{-1}(F)$ is closed in X.

PROOF. Since $\phi^{-1}(Y \backslash B) = X \backslash \phi^{-1}(B)$ for every subset B of Y, and since closed and open sets are the complements of one another, it is clear that (2) and (3) are equivalent. Moreover, if (2) is satisfied, if $x_0 \in X$, and if ε is a positive number, then $\phi^{-1}(D_\varepsilon(\phi(x_0)))$ is an open set in X containing x_0, whence it follows that there exists a positive radius δ such that $D_\delta(x_0) \subset \phi^{-1}(D_\varepsilon(\phi(x_0)))$. But then $\phi(D_\delta(x_0)) \subset D_\varepsilon(\phi(x_0))$, which shows that ϕ is continuous at x_0, and since x_0 is an arbitrary point of X, it follows that ϕ is continuous. Hence to complete the proof it suffices to show that (1) implies (2).

Suppose, accordingly, that ϕ is continuous. Let U be an open set in Y, and let $x_0 \in \phi^{-1}(U)$. Then $\phi(x_0) \in U$ so there exists a positive radius ε such that $D_\varepsilon(\phi(x_0)) \subset U$. But then there also exists a positive radius δ such that $\phi(D_\delta(x_0)) \subset D_\varepsilon(\phi(x_0)) \subset U$, and hence such that $D_\delta(x_0) \subset \phi^{-1}(U)$. Thus, $\phi^{-1}(U)$ is open (Prop. 6.11), and the theorem is proved. □

Example G. Let X be a metric space, let f be a continuous real-valued function on X, and for each real number t, set $U_t = \{x \in X : f(x) < t\}$. Then $\{U_t\}_{t \in \mathbb{R}}$ is a nested family of open sets in X satisfying the following three conditions:

(1) $\bigcup_{t \in \mathbb{R}} U_t = X$ and $\bigcap_{t \in \mathbb{R}} U_t = \varnothing$,
(2) If $s < t$, then $U_s^- \subset U_t$,
(3) For each t in $\mathbb{R}$, $U_t = \bigcup_{s<t} U_s$.

(To verify (2) note that U_s is a subset of the closed set $\{x \in X : f(x) \leq s\}$, and that this set is, in turn, contained in U_t.)

Suppose, conversely, that M is some dense subset of $\mathbb{R}$ and that a family $\{V_t\}_{t \in M}$ of open subsets of X is given satisfying the following conditions:

(1′) $\bigcup_{t \in M} V_t = X$ and $\bigcap_{t \in M} V_t = \varnothing$,
(2′) If $s < t$ $(s, t \in M)$, then $V_s^- \subset V_t$.

For each point x of X we write $R_x = \{t \in M : x \in V_t\}$. Because of (2′) it is clear that if $t \in R_x$, then $M \cap [t, +\infty) \subset R_x$. But then by (1′), R_x is nonempty and bounded below in $\mathbb{R}$ for every x in X. Hence we may, and do, define a real-valued function f on X by setting

$$f(x) = \inf R_x$$

for each x in X. (This infimum is taken in $\mathbb{R}$, of course, and need not belong to M.) If now $x_0 \in X$ and ε is a positive number, then the interval

$(f(x_0)-\varepsilon, f(x_0))$ contains points t and t' of M such that $t < t'$, and it is clear that $t' \notin R_{x_0}$. Thus $x_0 \notin V_{t'}$, and consequently by (2′), $x_0 \notin V_t^-$. On the other hand, the interval $(f(x_0), f(x_0)+\varepsilon)$ certainly contains an element t'' of R_{x_0}, so $x_0 \in V_{t''}$. Thus $W = V_{t''} \backslash V_t^-$ is an open subset of X containing x_0, and if x is any point of W, then R_x contains t'' but not t. Hence $t \le f(x) \le t''$, and therefore $|f(x) - f(x_0)| < \varepsilon$ for all x in W, whence it follows that f is continuous at x_0. Since x_0 is an arbitrary point of X, this shows that f is a continuous function on X. Moreover it is readily verified that, for each element t of M,

$$\{x \in X : f(x) < t\} \subset V_t \subset \{x \in X : f(x) \le t\},$$

and also that $V_t = \{x \in X : f(x) < t\}$ for every t in M if and only if the given family $\{V_t\}_{t \in M}$ satisfies the additional condition

(3′) for each t in M, $V_t = \bigcup_{\substack{s<t \\ s \in M}} V_s$.

(It is also easily seen that if (3′) holds, then the continuous real-valued function f is uniquely determined by the fact that $V_t = \{x \in X : f(x) < t\}$ for every element t of M.) This construction is an important source of continuous functions, and we shall return to it later.

Several important corollaries of Theorem 7.4, along with some related examples, follow. The first of these corollaries is a straightforward consequence of the theorem and the definition of relative continuity, and no proof is given.

Corollary 7.5. *Let X and Y be metric spaces, and let ϕ be a mapping of a subset A of X into Y. Then ϕ is continuous on A if and only if $\phi^{-1}(U)$ is relatively open in A whenever U is an open subset of Y.*

Corollary 7.6. *A continuous mapping on a metric space X is completely determined by its action on any dense subset of X. That is, if ϕ and ψ are two continuous mappings of X into a metric space Y, and if ϕ and ψ agree on some dense subset of X, then $\phi = \psi$.*

PROOF. Let ϕ and ψ be continuous mappings of (X, ρ) into (Y, σ), and suppose $\phi(x_0) \ne \psi(x_0)$ for some point x_0 of X. Let $d = \sigma(\phi(x_0), \psi(x_0))$. Then $U = \phi^{-1}(D_{d/2}(\phi(x_0)))$ and $V = \psi^{-1}(D_{d/2}(\psi(x_0)))$ are nonempty open subsets of X, as is $U \cap V$ (since all three sets contain x_0), and it is clear that $\phi(x) \ne \psi(x)$ for each x in $U \cap V$. Thus, the set $\{x \in X : \phi(x) = \psi(x)\}$ is not dense in X, contrary to hypothesis. □

Example H. Each continuous mapping f of $\mathbb{R}$ into $\mathbb{R}$ is uniquely determined by the restriction $f|\mathbb{Q}$. Hence $f \to f|\mathbb{Q}$ is a one-to-one mapping of the set of all continuous real-valued functions on $\mathbb{R}$ into $\mathbb{R}^{\mathbb{Q}}$. Since card $\mathbb{R}^{\mathbb{Q}} = \aleph^{\aleph_0} = \aleph$ (Prob. 4F) it is clear that there are at most $\aleph$ continuous real-valued functions on $\mathbb{R}$, and since the cardinal number of the set of constant functions is $\aleph$, this shows that the cardinal number of the set of all continuous real-valued functions on $\mathbb{R}$ is exactly $\aleph$. The same line of reasoning establishes the same conclusion for the set of continuous mappings of any separable metric space into a metric space Y with card $Y = \aleph$.

Corollary 7.7. *The level sets (Ex. 1H) of an arbitrary continuous mapping of a metric space X into a metric space Y are closed.*

PROOF. Singletons are closed sets in an arbitrary metric space. Hence, if ϕ is a continuous mapping on X, a set of the form $\{x \in X : \phi(x) = \phi(x_0)\} = \phi^{-1}(\{\phi(x_0)\})$ is closed too. □

Definition. A mapping ϕ of a metric space X into a metric space Y is said to be *open* [*closed*] if $\phi(A)$ is open [closed] in Y whenever A is open [closed] in X.

The following result is nothing more than a paraphrase of Theorem 7.4 in the special case of a one-to-one mapping, and no proof is required.

Corollary 7.8. *A one-to-one mapping ϕ of a metric space X into a metric space Y is continuous if and only if the inverse mapping ϕ^{-1} (regarded as a mapping of the range of ϕ onto X) is open or, equivalently, closed.*

Definition. A one-to-one mapping ϕ of a metric space X onto a metric space Y is a *homeomorphism* if both ϕ and ϕ^{-1} are continuous, and if such a homeomorphism exists, then X and Y are *homeomorphic*. (According to Corollary 7.8, there are a number of equivalent formulations of what is required of a one-to-one correspondence between two metric spaces in order that it be a homeomorphism. Thus, a one-to-one mapping ϕ of X onto Y is a homeomorphism if and only if ϕ is both continuous and open. Likewise, ϕ is a homeomorphism if and only if the mapping of the power class 2^X onto 2^Y induced by ϕ (Prob. 1F) carries the collection of all open subsets of X onto the collection of all open subsets of Y.) A homeomorphism may also be described as a mapping that preserves all topological properties (see Proposition 6.14).

Example I. Every isometry is a homeomorphism. Somewhat more interestingly, let ρ and ρ' be two metrics on the same carrier X. Then ρ and ρ' are equivalent if and only if the identity mapping $\iota : (X, \rho) \to (X, \rho')$ is a homeomorphism.

Example J. For an arbitrary open set U in $\mathbb{R}$, the set $\mathcal{U}$ of the constituent intervals of U (Ex. 6S) forms an ordered set in a natural way. Indeed, if I and I' are two distinct elements of $\mathcal{U}$ then either every element of I is less than each element of I', in which case we write $I < I'$ or every element of I' is less than each element of I, in which case we write $I' < I$, and it is clear that this definition turns $\mathcal{U}$ into a simply ordered set. In this way we associate with each closed set F in $\mathbb{R}$ a countable linearly ordered set $\mathcal{U}_F$ of open intervals, namely, the set of intervals contiguous to F ordered as just indicated. This correspondence is most interesting when F has empty interior (Prob. 6Q), for if $F^\circ = \varnothing$ and $t \in F$, then it is easily seen that t is the supremum of the set $U_t = \{s \in \mathbb{R}\backslash F : s < t\}$, and also that U_t is the union of a nonempty initial segment $\mathcal{A}$ of the ordered set $\mathcal{U}_F$ such that $\mathcal{A} \neq \mathcal{U}_F$. On the other hand, if $\mathcal{A}$ is an arbitrary nonempty initial segment of $\mathcal{U}_F$ other than $\mathcal{U}_F$ itself, then $U = \bigcup \mathcal{A}$ is a nonempty open subset of $\mathbb{R}$ that is bounded above in $\mathbb{R}$ (by any element of any interval in $\mathcal{U}_F$ not belonging to $\mathcal{A}$). It is clear that $t = \sup U \in F$, and also that (in the notation just introduced) $U = U_t$. Thus a closed subset F of $\mathbb{R}$ such that $F^\circ = \varnothing$ is, in a natural way, in one-to-one correspondence with the collection of all initial segments of the associated simply ordered set $\mathcal{U}_F$ (different from $\varnothing$ and $\mathcal{U}_F$ itself), and it is clear that this correspondence is, in fact, an order isomorphism.

Suppose next that K is a bounded and perfect subset of $\mathbb{R}$ having empty interior (Prob. 6O). Then the simply ordered set $\mathcal{U}_K$ is not only countable but also possesses a greatest and a least element and, more importantly, the property that between any two distinct elements of $\mathcal{U}_K$ there is a third element different from both. Indeed, if I and I' are elements of $\mathcal{U}_K$ such that $I < I'$, and if $c = \sup I$ and $d = \inf I'$ then $c, d \in K$ with $c \leq d$. But $c = d$ is impossible since K is perfect by hypothesis and can therefore have no isolated points. Hence $c < d$ and $[c, d] \not\subset K$ (since $K^\circ = \varnothing$). Hence some element J of $\mathcal{U}_K$ is contained in $[c, d]$, and we have $I < J < I'$.

Suppose, finally, that L is some other bounded perfect set in $\mathbb{R}$ such that $L^\circ = \varnothing$. Then, according to Example 4G, the simply ordered sets $\mathcal{U}_K$ and $\mathcal{U}_L$ are order isomorphic. Moreover, if ϕ is any one order isomorphism of $\mathcal{U}_K$ onto $\mathcal{U}_L$, then ϕ automatically maps the collection of all proper initial segments of $\mathcal{U}_K$ order isomorphically onto the collection of all proper initial segments of $\mathcal{U}_L$. Thus, composing order isomorphisms, we obtain an order isomorphism ψ of K onto L. But also, more to the point for present purposes, the mapping ψ is automatically a homeomorphism as well. Indeed, let s_0 be a point of K, and let $t_0 = \psi(s_0)$. If ε is an arbitrary positive number, then there exists an interval J in $\mathcal{U}_L$ such that $t_0 - \varepsilon < \sup J \leq t_0$ (since the interval $(t_0 - \varepsilon, t_0)$ must contain a point of $\mathbb{R}\backslash L$), and there also exists an interval J' in $\mathcal{U}_L$ such that $t_0 \leq \inf J' < t_0 + \varepsilon$ (since the interval $(t_0, t_0 + \varepsilon)$ must also contain a point of $\mathbb{R}\backslash L$). Let $\phi(I) = J$ and $\phi(I') = J'$. Then $\sup I \leq s_0$ (since I must belong to the initial segment in $\mathcal{U}_K$ whose union has supremum s_0), while $\inf I' \geq s_0$

(since I' does not belong to this initial segment). But now, if $a = \inf I$ and $b = \sup I'$, then $a < s_0 < b$, and if s is any point of K such that $a < s < b$, then I belongs to the initial segment of $\mathcal{U}_K$ determined by s, while I' does not, whereupon it follows that $\sup J \leq \psi(s) \leq \inf J'$, and therefore that $t_0 - \varepsilon < \psi(s) < t_0 + \varepsilon$. Thus ψ is continuous, and since the roles of K and L are interchangeable in this argument, it follows by symmetry that ψ is a homeomorphism. In particular, any two of the linear Cantor sets of Example 6P are homeomorphic (in many ways) in order preserving fashion. (For this reason it is customary to refer to an arbitrary bounded and perfect subset of $\mathbb{R}$ having empty interior as a *generalized Cantor set* in $\mathbb{R}$.)

It is an important fact that continuity is a *local* property. This idea is made precise in the following result.

Proposition 7.9. *Let X be a metric space, let ϕ be a mapping of X into a metric space Y, and let U be an open set in X. Then ϕ is continuous at a point x_0 of U relative to U if and only if ϕ is continuous at x_0 (relative to X).*

PROOF. For every sufficiently small positive radius δ we have $D_\delta(x_0) \subset U$, and therefore $D_\delta(x_0) \cap U = D_\delta(x_0)$. □

Corollary 7.10. *If two mappings ϕ and ψ of a metric space X into a metric space Y coincide on some open subset U of X, then ϕ and ψ are continuous (and discontinuous) at precisely the same points of U.*

Corollary 7.11. *A mapping ϕ of a metric space X into a metric space Y is continuous if and only if there exists an open covering of X consisting of sets U such that ϕ is continuous on U relative to U.*

Example K. If $\mathcal{U}$ is an open covering of a metric space X, and if $\{\phi_U\}_{U \in \mathcal{U}}$ is an arbitrary coherent collection of mappings (Prob. 1M) with the property that each ϕ_U is a continuous mapping of U into a metric space Y, then it results at once from Corollary 7.11 that the supremum of the family $\{\phi_U\}$ is likewise a continuous mapping of X into Y. Moreover it is obvious that the assumption that the covering sets are open cannot be omitted in this construction (cf. Example F), but it is possible to replace that assumption by other conditions. Thus, suppose $\mathcal{F}$ is a *finite closed* covering of X, let $\{\phi_F\}_{F \in \mathcal{F}}$ be a coherent collection of mappings with the property that each ϕ_F is a continuous mapping of F into Y, and let ϕ denote the supremum of $\{\phi_F\}$. If $x_0 \in X$ and if $F_1, \ldots, F_k$ is an enumeration of all of those sets in $\mathcal{F}$ that contain x_0, then for a given positive number ε there exist positive numbers $\delta_1, \ldots, \delta_k$ such that

$$\phi\left(D_{\delta_i}(x_0) \cap F_i\right) \left(= \phi_{F_i}\left(D_{\delta_i}(x_0) \cap F_i\right)\right) \subset D_\varepsilon(\phi(x_0)), \; i = 1, \ldots, k.$$

Moreover, the union of all those sets in $\mathcal{F}$ that do not contain x_0 is itself a closed set not containing x_0, so there exists a positive number η such that $D_\eta(x_0)$ meets only the sets $F_1, \ldots, F_k$, and if $\delta = \eta \wedge \delta_1 \wedge \ldots \wedge \delta_k$, then it is clear that $\phi(D_\delta(x_0)) \subset D_\varepsilon(\phi(x_0))$, and hence that ϕ is continuous.

It is possible to improve upon this last construction significantly. Suppose $\mathcal{F}$ is a closed covering of X that is *locally finite*, meaning that each point x of X is contained in some open set that meets only finitely many of the covering sets. Then, once again, if $\{\phi_F\}_{F\in\mathcal{F}}$ is a coherent collection of mappings where, as before, each ϕ_F is a continuous mapping of F into Y, then the supremum ϕ of the family $\{\phi_F\}$ is a continuous mapping. Indeed, if $x_0 \in X$ and if U is an open set containing x_0 that meets only finitely many covering sets, then $\phi|U$ is continuous on U by the foregoing observation, and the continuity of ϕ follows immediately by Corollary 7.11.

Definition. A mapping ϕ of a metric space X into a metric space Y is *locally Lipschitzian* if for each point x of X there is an open set U in X containing x such that $\phi|U$ is Lipschitzian.

Since Lipschitzian mappings are necessarily continuous, it follows at once from Corollary 7.11 that locally Lipschitzian mappings are also continuous.

Example L. To show that the multiplication mapping $m(\xi,\eta) = \xi\eta$ is continuous as a complex-valued function on $\mathbb{C}^2$ it suffices to prove that m is locally Lipschitzian with respect to any one of the equivalent metrics of Problem 6C. But now

$$\xi\eta - \xi'\eta' = \xi(\eta - \eta') + (\xi - \xi')\eta',$$

so

$$|m(\xi,\eta) - m(\xi',\eta')| \leq |\xi|\ |\eta - \eta'| + |\eta'|\ |\xi - \xi'|,$$

whence it follows that m is Lipschitzian with respect to the metric ρ_1 (with Lipschitz constant K) on the open set $\{(\xi,\eta) \in \mathbb{C}^2 : |\xi|, |\eta| < K\}$, and those open sets clearly cover $\mathbb{C}^2$. Similarly, the division mapping $d(\xi,\eta) = \xi/\eta$ is continuous on the open subset of $\mathbb{C}^2$ on which it is defined, viz., the set $\{(\xi,\eta) \in \mathbb{C}^2 : \eta \neq 0\}$. For exactly similar reasons, if $\mathcal{E}$ is a complex [real] normed space, the mapping $(\lambda, x) \to \lambda x$ is continuous on $\mathbb{C} \times \mathcal{E}[\mathbb{R} \times \mathcal{E}]$.

The following result, while very important, is an almost trivial consequence of the definition of continuity.

Proposition 7.12. *Let X, Y, and Z be metric spaces, let ϕ be a mapping of X into Y, and let ψ be a mapping of Y into Z, so that $\psi \circ \phi : X \to Z$ is a mapping of X into Z. If ϕ is continuous at a point x_0 of X and ψ is continuous at the point $y_0 = \phi(x_0)$, then $\psi \circ \phi$ is also continuous* *at x_0. In particular, if ϕ and ψ are both continuous, then $\psi \circ \phi$ is also continuous.*

PROOF. It suffices to prove the first assertion of the proposition. Let ϕ be continuous at x_0, let ψ be continuous at $y_0 = \phi(x_0)$, and let ε be a positive number. Then there exists $\eta > 0$ such that $\psi(D_\eta(y_0)) \subset D_\varepsilon(\psi(y_0))$, and there also exists a positive number δ such that $\phi(D_\delta(x_0)) \subset D_\eta(y_0)$. But then, of course, $(\psi \circ \phi)(D_\delta(x_0)) \subset D_\varepsilon((\psi \circ \phi)(x_0))$. □

Example M. Putting Proposition 7.12 together with Examples A, C, and L, we see that if $f_1, \ldots, f_k$ are any continuous complex-valued functions on a metric space X, and if $p(\lambda_1, \ldots, \lambda_k)$ is an arbitrary complex polynomial, then the complex-valued function $x \to p(f_1(x), \ldots, f_k(x))$ is continuous. Similarly, if $r(\lambda_1, \ldots, \lambda_k)$ is an arbitrary complex rational function (i.e., if $r(\lambda_1, \ldots, \lambda_k) = p(\lambda_1, \ldots, \lambda_k)/q(\lambda_1, \ldots, \lambda_k)$, where p and q are polynomials), then the complex-valued function $x \to r(f_1(x), \ldots, f_k(x))$ is continuous on the (open) subset of X on which it is defined. Similarly, the function $f(t) = t/(1 + |t|)$ is a continuous mapping of $\mathbb{R}$ onto $(-1, +1)$, and its inverse $f^{-1}(t) = t/(1 - |t|), |t| < 1$, is also continuous. Thus f is a homeomorphism of $\mathbb{R}$ onto $(-1, +1)$ (cf. Example 2B). In the same vein, the mapping $x \to x/\|x\|$ is continuous on the set $\mathcal{E}\backslash(0)$ in any normed space $\mathcal{E}$.

Example N. Let x_0 and x_1 be points of Euclidean space $\mathbb{R}^n$ and let a and b be real numbers such that $a < b$. The affine mapping ϕ defined by setting

$$\phi(t) = \frac{t-a}{b-a} x_1 + \frac{b-t}{b-a} x_0$$

maps $\mathbb{R}$ continuously into $\mathbb{R}^n$, and the restriction $\phi|[a, b]$ likewise maps the interval $[a, b]$ continuously onto the line segment $\ell(x_0, x_1)$ joining x_0 to x_1 (Prob. 3S). This mapping is the *linear parametrization* on $[a, b]$ of $\ell(x_0, x_1)$.

Suppose now that $\mathcal{P} = \{a = t_0 < \ldots < t_N = b\}$ is a partition of $[a, b]$, and let ψ_0 be an arbitrary mapping of the finite set $\{t_0, \ldots, t_N\}$ into $\mathbb{R}^n$. Then there exists a unique mapping ψ of $[a, b]$ into $\mathbb{R}^n$ with the property that for each $i = 1, \ldots, N, \psi|[t_{i-1}, t_i]$ is the linear parametrization on the subinterval $[t_{i-1}, t_i]$ of the line segment $\ell(\psi_0(t_{i-1}), \psi_0(t_i))$. The mapping ψ, which is continuous by virtue of Example K, is said to be *piecewise linear* with respect to $\mathcal{P}$ and to be the *piecewise linear extension* of ψ_0. This construction goes through without change if $\mathbb{R}^n$ is replaced by an arbitrary normed space.

It is an elementary but profoundly important fact that the limit of a uniformly convergent sequence of continuous mappings is continuous.

Proposition 7.13. *Let (X, ρ) and (Y, σ) be metric spaces and let $\{\phi_n\}_{n=1}^{\infty}$ be a sequence of bounded mappings of X into Y converging uniformly to a limit ϕ. If infinitely many of the mappings ϕ_n are continuous at a point x_0 of X, then ϕ is also continuous at x_0.*

PROOF. Suppose ϕ_n is continuous at x_0 for infinitely many values of n, and let ε be a positive number. If the positive integer N is chosen large enough so that $\rho_\infty(\phi, \phi_n) < \varepsilon/3$ for all $n \geq N$, and if n_0 is a positive integer such that $n_0 \geq N$ and such that ϕ_{n_0} is continuous at x_0, then there exists a positive number δ such that $\rho(x_0, x) < \delta$ implies $\sigma(\phi_{n_0}(x_0), \phi_{n_0}(x)) < \varepsilon/3$. But then by the triangle inequality we have

$$\begin{aligned}\sigma(\phi(x_0), \phi(x)) &\leq \sigma(\phi(x_0), \phi_{n_0}(x_0)) + \sigma(\phi_{n_0}(x_0), \phi_{n_0}(x)) \\ &\quad + \sigma(\phi_{n_0}(x), \phi(x)) < \varepsilon/3 + \varepsilon/3 + \varepsilon/3 = \varepsilon\end{aligned}$$

whenever $\rho(x_0, x) < \delta$. Thus ϕ is continuous at x_0. □

Corollary 7.14. *The limit of a uniformly convergent sequence of bounded continuous mappings of one metric space X into another metric space Y is itself continuous. In other words, the set of continuous mappings in the metric space $\mathcal{B}(X; Y)$ is closed in the metric of uniform convergence.*

We have already introduced a number of concepts related to the basic notion of continuity of mappings between metric spaces. Here is yet another one.

Definition. Let (X, ρ) and (Y, σ) be metric spaces and let $\phi : X \to Y$ be a mapping of X into Y. Then ϕ is *uniformly continuous* on X if for each positive number ε there exists a positive number δ such that $\rho(x, x') < \delta$ implies $\sigma(\phi(x), \phi(x')) < \varepsilon$ for all x, x' in X.

A mapping such as ϕ is continuous on X if for *each individual point* x_0 of X and for each positive number ε there exists a positive number δ such that $\rho(x, x_0) < \delta$ implies $\sigma(\phi(x), \phi(x_0)) < \varepsilon$. The distinguishing feature of *uniform* continuity is the requirement that, given ε, it must be possible to choose δ so as to make this implication valid for all points x_0 simultaneously, or, as it is said, *uniformly in* x_0. Thus it is clear that every uniformly continuous mapping is continuous, but the converse is false.

Example O. All Lipschitzian mappings are automatically uniformly continuous. (Indeed, if ϕ satisfies a Lipschitz condition with Lipschitz constant M, then for given positive ε we have but to set $\delta = \varepsilon/M$ in the definition of uniform continuity.) Thus the projections of any finite product $X_1 \times \ldots \times X_n$ of metric spaces onto the various factors are uniformly continuous with respect to any of the standard product metrics ρ_p (Prob. 6H).

Example P. The function $f(t) = t^2$, regarded as a mapping of $\mathbb{R}$ into itself, is neither uniformly continuous nor Lipschitzian, but it is locally Lipschitzian. Thus, being locally Lipschitzian does not imply uniform continuity. The function $g(t) = \sqrt{t}$, regarded as a mapping of the closed ray $\mathbb{R}_+ = [0, +\infty)$ onto itself, is uniformly continuous (indeed, it is readily seen that if $s, t \geq 0$ and if $|s - t| < \varepsilon^2$, then $|\sqrt{s} - \sqrt{t}| < \varepsilon$) but not locally Lipschitzian (since g is not Lipschitzian in any open set containing the point $t = 0$). Thus the uniform continuity of a mapping does not imply that it is either Lipschitzian or locally Lipschitzian.

Next, for each positive integer n, set

$$g_n(t) = \begin{cases} nt, & 0 \leq t \leq 1/n^2, \\ \sqrt{t}, & t > 1/n^2. \end{cases}$$

Each g_n is a continuous monotone increasing mapping of $\mathbb{R}_+$ onto itself, and the sequence $\{g_n\}_{n=1}^\infty$ is readily seen to converge monotonely upward and uniformly to the limit g. But while each function g_n is Lipschitzian on $\mathbb{R}_+$ (with Lipschitz constant n), the limit g is not even locally Lipschitzian there. (A further refinement of these observations will be found in Problem L.) On the other hand, an obvious modification of the proof of Proposition 7.13 establishes the following result: *The limit of a uniformly convergent sequence of uniformly continuous mappings is uniformly continuous.*

Example Q (The Cantor-Lebesgue Function). We define a real-valued function h_0 on the Cantor set C as follows: If $t \in C$ and $t = 0.\eta_1\eta_2\ldots\eta_n\ldots$ is the unique one-free ternary expansion of t, then

$$h_0(t) = 0.\frac{\eta_1}{2}\,\frac{\eta_2}{2}\;\cdots\;, \tag{2}$$

where the expansion in the right-hand member of (2) is taken to be a *binary* expansion. (See Theorem 2.12; for details concerning the Cantor set and its construction, cf. Example 6O.) It is obvious that the mapping h_0 is monotone increasing, and it is also easily seen that h_0 maps C *onto* the unit interval $[0, 1]$. (Indeed, if $0.\varepsilon_1\varepsilon_2\;\ldots$ is a binary expansion of an arbitrary element s of $[0, 1]$, and if we set $\eta_n = 2\varepsilon_n, n \in \mathbb{N}$, then $0.\eta_1\eta_2\ldots$ is the ternary expansion of an element t of C such that $h_0(t) = s$; alternatively, it may be observed that h_0 is just the composition $\psi \circ \phi^{-1}$ where ϕ and ψ are as in Example 4I.) Moreover, h_0 is uniformly continuous. Indeed, if ε is a given positive number and if n is chosen large enough so that $1/2^n < \varepsilon$, then $|s - t| < 1/3^n$ with s and t in C implies that the one-free ternary expansions of s and t agree through the first n terms, and hence that the binary expansions of $h_0(s)$ and $h_0(t)$ also agree through the first n terms.

But then $|h_0(s) - h_0(t)| \leq 1/2^n < \varepsilon$. (However, the uniformly continuous mapping h_0 is not locally Lipschitzian on C as may be seen from the fact that if we set $s = 0$ and $t = 1/3^n$, then $h_0(t) - h_0(s) = 1/2^n = (3/2)^n(t-s)$. In this connection, see Problem M.)

Suppose now that U is one of the bounded open intervals contiguous to C. Then, as has been noted (cf. Example 6O once again), there exists a sequence $\{\eta_1, \ldots, \eta_k\}$ of zeros and twos (possibly vacuous) such that

$$U = (0.\eta_1 \ldots \eta_k 022 \ldots, 0.\eta_1 \ldots \eta_k 200 \ldots)$$

(in ternary notation), and it is at once clear that

$$h_0\left(0.\eta_1 \ldots \eta_k 022 \ldots\right) = h_0\left(0.\eta_1 \ldots \eta_k 200 \ldots\right) = 0.\frac{\eta_1}{2} \ldots \frac{\eta_k}{2} 100 \ldots .$$

Thus h_0 assumes the same value at the endpoints of every bounded interval contiguous to C, from which it follows at once that there exists a unique monotone increasing function h on $[0,1]$ that agrees with h_0 on C, and that the function h is a continuous mapping of $[0,1]$ onto itself. This remarkable function, which is, of course, constant on each of the bounded open intervals contiguous to C, is known as the *Cantor-Lebesgue function*. (In connection with this construction and the following one, see also Problem J.)

Example R (A Peano Curve). In a similar fashion we can define a continuous mapping ϕ_0 of the Cantor set C onto the unit square $Q = [0,1] \times [0,1]$ by setting

$$\phi_0(0.\eta_1 \ldots \eta_n \ldots) = \left(0.\frac{\eta_1}{2}\frac{\eta_3}{2} \ldots \frac{\eta_{2n-1}}{2} \ldots, 0.\frac{\eta_2}{2}\frac{\eta_4}{2} \ldots \frac{\eta_{2n}}{2} \ldots\right), \qquad (3)$$

where, as in the preceding example, the expansion in the left-hand member of (3) is the one-free ternary expansion of the general element of C, while the two expansions in the right-hand member of (3) are binary expansions. (The proof that ϕ_0 maps C onto Q in a uniformly continuous fashion is substantially the same as that in Example Q and is omitted.) Here, as before, the mapping ϕ_0 may be extended to a continuous mapping ϕ of the entire unit interval $[0,1]$ onto Q by defining ϕ to be linear on each of the bounded open intervals contiguous to C. If this is done, the interval $[1/3, 2/3]$ is mapped linearly by ϕ onto the line segment $\ell((1/2,1),(1/2,0))$, the intervals U_0^- and U_1^- onto the line segments shown in Figure 3, and so on. (We shall return to this construction in Chapter 8.)

It is customary to call an arbitrary continuous mapping of a real interval $[a,b]$ into a metric space X a *curve* in X (with *parameter interval* $[a,b]$). Thus the mapping ϕ just constructed is a planar curve, i.e., a curve in $\mathbb{R}^2$, with the startling property that the range of ϕ contains a nonempty open subset of $\mathbb{R}^2$. Such curves were first discovered by G. Peano [21] and are known generically as *Peano (space-filling) curves*. The interested reader will have no difficulty in defining similar Peano curves that fill nonempty open subsets of Euclidean space $\mathbb{R}^n$ of arbitrary dimension (cf. also Problem 8I).

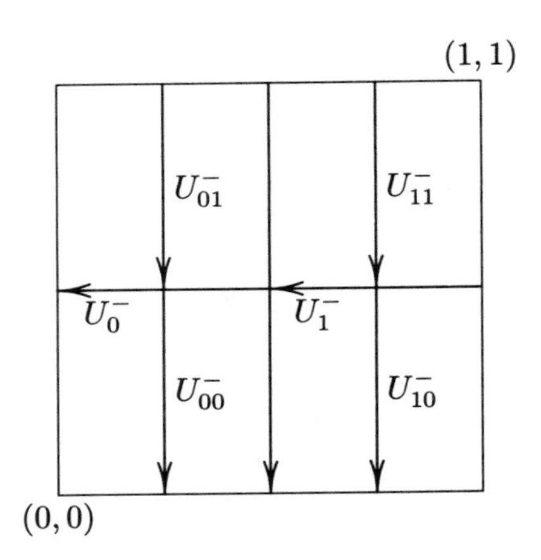

Figure 3

Before turning to the subject of limits, we conclude our initial discussion of continuity with the introduction of a notion that is frequently useful in the study of real-valued functions. For technical reasons it is desirable to define the concept for extended real-valued functions.

Definition. An extended real-valued function f on a metric space (X, ρ) is *upper semicontinuous* at a point x_0 of X if for each extended real number u such that $f(x_0) < u$ there exists a positive number δ such that $\rho(x, x_0) < \delta$ implies $f(x) < u$ (or, in other words, such that $f(x) < u$ for all x in $D_\delta(x_0)$). Dually, f is *lower semicontinuous* at x_0 if for each extended real number s such that $f(x_0) > s$ there exists a positive number δ such that $\rho(x, x_0) < \delta$ implies $f(x) > s$. Finally, f is *upper [lower] semicontinuous on* X if it is upper [lower] semicontinuous at every point of X.

Since no metric has been defined on the extended real number system, it would be inaccurate to say that these definitions constitute generalizations of the earlier notion of continuity (but see Example 9K). The exact state of affairs is set forth in the following elementary proposition, the proof of which is omitted.

Proposition 7.15. *If f is an extended real-valued function defined on a metric space (X, ρ), and if x_0 is a point of X at which f is finite-valued ($-\infty < f(x_0) < +\infty$), then f is upper [lower] semicontinuous at x_0 if and only if for each given positive number ε there exists $\delta > 0$ such that $\rho(x, x_0) < \delta$ implies $f(x) < f(x_0) + \varepsilon$ $[f(x) > f(x_0) - \varepsilon]$. Thus, a finite real-valued function f on X is continuous on X [at a point x_0 of X] if and only if it is both upper and lower semicontinuous on X [at x_0]. An extended real-valued function f on X is automatically upper semicontinuous at any point at which it assumes the value $+\infty$; dually, f is automatically lower semicontinuous at any point at which it assumes the value $-\infty$.*

The following results are reminiscent of Proposition 7.1 and Theorem 7.4.

Proposition 7.16. *Let f be an extended real-valued function defined on a metric space X. Then f is upper semicontinuous at a point x_0 of X if and only if*

$$\limsup_n f(x_n) \leq f(x_0)$$

for every sequence $\{x_n\}_{n=1}^{\infty}$ in X such that $x_n \to x_0$ (cf. Example 6L). Dually, f is lower semicontinuous at x_0 if and only if

$$\liminf_n f(x_n) \geq f(x_0)$$

for every sequence $\{x_n\}_{n=1}^{\infty}$ in X such that $x_n \to x_0$.

PROOF. We treat only the former assertion; the latter result is proved similarly. Furthermore, it clearly suffices to deal with the case $f(x_0) < +\infty$. Suppose that f is upper semicontinuous at x_0 and let $\{x_n\}$ be a sequence in X converging to x_0. If ε is a positive number, and if $\delta > 0$ is chosen so that $f(x) < f(x_0) + \varepsilon$ for all x in $D_\delta(x_0)$, then it is clear that $f(x_n) < f(x_0) + \varepsilon$ eventually, and hence that $\limsup_n f(x_n) \leq f(x_0) + \varepsilon$. Since ε is arbitrary, this proves that $\limsup_n f(x_n) \leq f(x_0)$, as was to be shown.

Suppose, on the other hand, that f is not upper semicontinuous at x_0. Then there exists a positive number ε_0 so small that for every positive integer n the ball $D_{1/n}(x_0)$ contains points x such that $f(x) \geq f(x_0) + \varepsilon_0$. But if, for each n, x_n is such a point of $D_{1/n}(x_0)$, then the sequence $\{x_n\}$ converges to x_0 while the sequence $\{f(x_n)\}$ is bounded below by $f(x_0) + \varepsilon_0$, whence it follows at once that $\limsup_n f(x_n) \geq f(x_0) + \varepsilon_0 > f(x_0)$. □

Proposition 7.17. *An extended real-valued function f on a metric space X is upper semicontinuous on X if and only if $\{x \in X : f(x) < u\}$ is open in X for every finite real number u. Dually, f is lower semicontinuous if* *and only if the set* $\{x \in X : f(x) > s\}$ *is open in* X *for every finite real number* s.

PROOF. Once again it suffices to prove the former assertion (cf. Problem N). Suppose first that f is upper semicontinuous, let u be a (finite) real number, and set $U = \{x \in X : f(x) < u\}$. If $x_0 \in U$, then by definition there exists $\delta > 0$ such that $D_\delta(x_0) \subset U$, so U is open, and the stated condition is satisfied. Suppose, on the other hand, that the stated condition is satisfied. If $f(x_0) < u$, where u is an extended real number, then either u is finite, in which case it is clear that $f(x) < u$ holds on some open ball $D_\delta(x_0)$, $\delta > 0$, or else $u = +\infty$, in which case we have but to select a finite real number u' such that $f(x_0) < u'$ in order to see that f is finite-valued on an open ball $D_\delta(x_0), \delta > 0$. □

The foregoing result has the following two interesting consequences (cf. also Problem Q).

Corollary 7.18. *Let* f *be an extended real-valued function on a metric space* X. *If* f *is upper semicontinuous, then for each extended real number* t *the set* $\{x \in X : f(x) \leq t\}$ *is a* G_δ *in* X. *Dually, if* f *is lower semicontinuous, then the set* $\{x \in X : f(x) \geq t\}$ *is a* G_δ *for each extended real number* t.

PROOF. It suffices as before to prove the former assertion, so we assume f to be upper semicontinuous. The set $\{x \in X : f(x) \leq -\infty\}$ is the intersection of the sequence of open sets $\{x \in X : f(x) < -n\}, n \in \mathbb{N}$, while $\{x \in X : f(x) \leq +\infty\} = X$. On the other hand, if t is finite, then $\{x \in X : f(x) \leq t\}$ is the intersection of the sequence of open sets $\{x \in X : f(x) < t + (1/n)\}, n \in \mathbb{N}$. □

Corollary 7.19. *The infimum (taken pointwise in* $R^\natural$*) of an arbitrary collection of upper semicontinuous functions on a metric space* X *is again upper semicontinuous on* X. *Dually, the supremum of an arbitrary collection of lower semicontinuous functions is lower semicontinuous. In particular, the pointwise limit of a monotone decreasing [increasing] sequence of continuous real-valued functions on* X *is upper [lower] semicontinuous.*

PROOF. It suffices to prove the first assertion of the corollary. Let each function f in a collection $\mathcal{F}$ be upper semicontinuous, and let $g(x) = \inf\{f(x) : f \in \mathcal{F}\}$, $x \in X$. If u denotes a finite real number, then $g(x) < u$ if and only if $f(x) < u$ for some f in $\mathcal{F}$. In other words, the set $\{x \in X : g(x) < u\}$ coincides with the union

$$\bigcup_{f \in \mathcal{F}} \{x \in X : f(x) < u\},$$

and is therefore an open set. □

It is a remarkable fact that the last assertion of Corollary 7.19 has a valid converse (provided one sticks to finite-valued functions).

Proposition 7.20 (Baire [1]). *If h is an arbitrary finite-valued upper semicontinuous function on a metric space X, then there exists a monotone decreasing sequence $\{f_n\}$ of continuous real-valued functions on X that converges pointwise to h. Dually, every finite-valued lower semicontinuous real-valued function k on X is the pointwise limit of a monotone increasing sequence $\{g_n\}$ of continuous real-valued functions.*

PROOF. As before, it suffices to prove either one of the two dual assertions, and this time it seems slightly more convenient to deal with the latter. Suppose, then, that k is a finite-valued lower semicontinuous function on X, and let us suppose also, to begin with, that k is bounded and, in fact, takes its values in the open unit interval $(0, 1)$. For each positive integer N, and for each $i = 1, \ldots, N$, set $U_{i,N} = \{x \in X : k(x) > i/N\}$. It is obvious that the sets $U_{1,N}, \ldots, U_{N,N}$ are nested ($U_{1,N} \supset \ldots \supset U_{N,N}$), and also that $U_{N,N} = \varnothing$ for each N. Moreover, since k is lower semicontinuous by hypothesis, the sets $U_{i,N}$ are all open. Hence, for each $i = 1, \ldots, N-1$, there exists a monotone increasing sequence $\{g_m^{i,N}\}_{m=1}^{\infty}$ of continuous real-valued functions on X converging pointwise to the function $k_{i,N} = \chi_{U_{i,N}}$ (see Problem F). Next let us define

$$k_N = \frac{1}{N}(k_{1,N} + \ldots + k_{N-1,N})$$

and also

$$g_m^N = \frac{1}{N}\left(g_m^{1,N} + \ldots + g_m^{N-1,N}\right)$$

for all positive integers m and N. The functions g_m^N are all continuous on X and the sequence $\{g_m^N\}_{m=1}^{\infty}$ clearly tends upward to k_N. Also if x is in $U_{i,N} \backslash U_{i+1,N}$ for some $i = 1, \ldots, N-1$, then $i/N < k(x) \leq (i+1)/N$, while $k_N(x) = i/N$. Likewise, if $x \notin U_{1,N}$, then $k(x) \leq 1/N$ while $k_N(x) = 0$. Thus at every point x of X we have

$$k_N(x) < k(x) \leq k_N(x) + \frac{1}{N},$$

and therefore $\lim_m g_m^N(x) = k_N(x) \geq k(x) - (1/N)$. Finally, we define

$$g_m = g_m^1 \vee \ldots \vee g_m^m,$$

for each positive integer m. Then g_m is real-valued and continuous (since each of the functions $g_m^1, \ldots, g_m^m$ is), and $g_m \le k$ (since this is also true of each of the functions $g_m^1, \ldots, g_m^m$). Moreover the sequence $\{g_m\}$ is clearly monotone increasing, and its pointwise limit (which is the same as its pointwise supremum) is at least as great as that of the sequence $\{g_m^N\}$ for any given index N. Thus $\lim_m g_m(x) = k(x)$ at every point x of X.

Suppose next that k is merely bounded. It is easy to find positive numbers A and B such that the function $\widetilde{k} = Ak + B$ takes its values in $(0,1)$. Since $\widetilde{k}$ is lower semicontinuous along with k (Prob. N), there exists a monotone increasing sequence $\{\widetilde{g}_m\}$ of continuous functions tending pointwise to $\widetilde{k}$, and if we set $g_m = (1/A)(\widetilde{g}_m - B)$, the functions g_m are continuous, and the sequence $\{g_m\}$ tends pointwise upward to k.

Suppose, finally, that k is an arbitrary finite-valued lower semicontinuous function on X. We have recourse once again to the standard homeomorphism $\widehat{\phi}(t) = t/(1+|t|)$ of $\mathbb{R}$ onto $(-1,+1)$ (Ex. 2B). The composition

$$\widehat{k} = \widehat{\phi} \circ k = \frac{k}{1+|k|}$$

is lower semicontinuous along with k (Prob. O), and is also bounded. Hence there exists a monotone increasing sequence $\{\widehat{g}_m\}$ of continuous real-valued functions tending pointwise to $\widehat{k}$. Moreover, we may and do assume each function $\widehat{g}_m$ to be bounded below by -1 (since we may simply replace $\widehat{g}_m$ by $\widehat{g}_m \vee -1$; cf. Problem H). What is needed, however, is a sequence—say $\{\widehat{g}'_m\}$—that tends upward to $\widehat{k}$ and is such that each $\widehat{g}'_m$ takes its values in $(-1,+1)$, and while it is clear that no $\widehat{g}_m$ takes on the value $+1$ (since $\widehat{k}$ does not, and $\widehat{g}_m \le \widehat{k}$), there is no reason to suppose that the functions $\widehat{g}_m$ do not assume the value -1 at various points of X. To take care of this technicality we define

$$\widehat{g}'_m = \frac{1}{2}\widehat{g}_m + \frac{1}{4}\widehat{g}_{m+1} + \ldots = \sum_{n=1}^{\infty} 2^{-n}\widehat{g}_{m+n-1}.$$

As the sum of a uniformly convergent series of continuous functions, the function $\widehat{g}'_m$ is continuous for each index m (Cor. 7.14). Moreover, it is readily seen that

$$\widehat{g}_m \le \widehat{g}'_m \le \widehat{g}'_{m+1} \le \widehat{k}$$

for each index m (since $\sum_{n=1}^{\infty} 2^{-n} = 1$). Hence the sequence $\{\widehat{g}'_m\}$ tends pointwise upward to $\widehat{k}$. But also (and this is what we were after) the equality $\widehat{g}'_m(x) = \widehat{g}_m(x)$ occurs at a point x when and only when $\widehat{g}_m(x) = \widehat{g}_{m+1}(x) = \ldots = \widehat{g}_{m+n}(x) \ldots$, which implies that $\widehat{g}_m(x) = \widehat{k}(x) > -1$.

Thus each function $\widehat{g}'_m$ takes all of its values in the open interval $(-1, +1)$, and we have but to define

$$g_m = \widehat{\phi}^{-1} \circ \widehat{g}'_m = \frac{\widehat{g}'_m}{1 - |\widehat{g}'_m|}, \quad m \in \mathbb{N},$$

to obtain a sequence of continuous real-valued functions $\{g_m\}$ tending pointwise upward to the originally given function k. □

The following result is noteworthy, not only for its content but also for its delicate proof.

Proposition 7.21 (Hahn Interpolation Theorem [9]). *Let h and k denote, respectively, a finite-valued upper semicontinuous function and a finite-valued lower semicontinuous function on the same metric space X, and suppose $h(x) \leq k(x)$ at each point x of X. Then there exists a continuous function c on X such that $h \leq c \leq k$.*

PROOF. By Proposition 7.20 there exists a monotone decreasing sequence $\{f_n\}_{n=1}^{\infty}$ of continuous real-valued functions on X tending pointwise to h, and likewise a monotone increasing sequence $\{g_n\}_{n=1}^{\infty}$ of continuous functions tending pointwise to k. Using these two sequences, we define a third sequence $\{p_n\}_{n=1}^{\infty}$ as follows:

$$p_{2k-1} = f_k - g_k, \; p_{2k} = f_k - g_{k+1}, \; k \in \mathbb{N}.$$

From this definition it is clear that $\{p_n\}$ is a monotone decreasing sequence of continuous functions converging pointwise to the difference $h - k \leq 0$. Somewhat less obvious, but still readily verified, is the fact that the infinite series

$$g_1 + \sum_{n=1}^{\infty} (-1)^{n+1} p_n \tag{4}$$

is telescoping. Indeed, if s_n denotes the nth partial sum of (4), then $\{s_n\}$ is the sequence $\{g_1, f_1, g_2, f_2, \ldots\}$ (so that for each index k, $s_{2k-1} = g_k$ while $s_{2k} = f_k$).

Next we set $q_n = p_n^+$, $n \in \mathbb{N}$, and modify the series (4) by replacing each p_n in it by the corresponding nonnegative q_n, thus constructing the series

$$g_1 + \sum_{n=1}^{\infty} (-1)^{n+1} q_n. \tag{5}$$

Inasmuch as the sequence $\{q_n\}$ is monotone decreasing along with $\{p_n\}$, and tends pointwise to $(h - k)^+ = 0$, the series (5) is pointwise convergent by the alternating series test. Hence we may, and do, define a function c on X by setting

$$c(x) = g_1(x) + \sum_{n=1}^{\infty}(-1)^{n+1}q_n(x), \qquad x \in X.$$

Concerning the function c we note first that at each point x of X the sequence $\{q_n(x)\}$ is obtained from the monotone decreasing sequence $\{p_n(x)\}$ by replacing all terms by zeros beginning with the first negative $p_n(x)$ (if there is one). Suppose the sequence $\{p_n(x)\}$ is eventually negative and that the first negative $p_n(x)$ occurs when n is odd—say for $n = 2k - 1$. Then $c(x)$ is equal to $s_{2k-1}(x) = g_k(x)$, and we have $f_k(x) < g_k(x)$. Thus,

$$h(x) \leq f_k(x) < c(x) = g_k(x) \leq k(x).$$

On the other hand, if the sequence $\{p_n(x)\}$ is eventually negative, and if the first negative term $p_n(x)$ occurs when n is even—say for $n = 2k$—then $c(x) = s_{2k}(x) = f_k(x)$, and we have $f_k(x) < g_{k+1}(x)$. Thus, once again,

$$h(x) \leq f_k(x) = c(x) < g_{k+1}(x) \leq k(x).$$

Finally, if x is a point of X at which the sequence $\{p_n(x)\}$ is always nonnegative, then $q_n(x) = p_n(x)$ for all n, the series (5) coincides with (4) at x, and we have

$$c(x) = h(x) = k(x).$$

Thus in all cases we find $h(x) \leq c(x) \leq k(x)$; the function c is everywhere between the functions h and k. But also—and this is the striking feature of the entire construction—as the pointwise limit of the alternating series (5), the function c is both the pointwise limit of the increasing sequence of odd partial sums and of the decreasing sequence of even partial sums of (5). Thus c is a finite-valued function that is both lower and upper semicontinuous (Cor. 7.19) and is therefore continuous (Prop. 7.15). □

Note. It is essential in this proposition that the larger of the two functions be lower semicontinuous and the smaller upper semicontinuous, as may readily be seen by consideration of the characteristic functions of the open and closed unit intervals.

Example S (Tietze Extension Theorem). Let F be a nonempty closed subset of a metric space X. If a real-valued function h_0 is defined and upper semicontinuous on F, and if h_0 is bounded below on F by a, then the extension

$$h(x) = \begin{cases} h_0(x), & x \in F, \\ a, & x \in X \backslash F, \end{cases}$$

is readily seen to be upper semicontinuous on all of X. Dually, if k_0 is defined, lower semicontinuous, and bounded above on F by b, then

$$k(x) = \begin{cases} k_0(x), & x \in F, \\ b, & x \in X \backslash F, \end{cases}$$

defines a lower semicontinuous function on X.

Suppose now that f_0 is a real-valued function that is defined, continuous and bounded on F, and set $a = \inf_{z \in F} f_0(z)$ and $b = \sup_{z \in F} f_0(z)$. Then the functions h and k defined by

$$h(x) = \begin{cases} f_0(x), & x \in F, \\ a, & x \in X \backslash F, \end{cases} \quad \text{and} \quad k(x) = \begin{cases} f_0(x), & x \in F, \\ b, & x \in X \backslash F, \end{cases}$$

are, respectively, upper and lower semicontinuous on X, with $h \leq k$. Hence, by the Hahn interpolation theorem, there exists a continuous real-valued function f on X such that $h \leq f \leq k$. But then $a \leq f(x) \leq b$ on X and $f = f_0$ on F. Thus *a bounded continuous real-valued function f_0 defined on a nonempty closed subset of a metric space X admits a continuous extension f to all of X having the same upper and lower bounds as f_0.*

We conclude this chapter with a discussion of the idea of the limit at a point of a mapping of one metric space into another.

Definition. Let A be a subset of a metric space X, let ϕ be a mapping of A into a metric space Y, and let $a_0 \in A^-$. Then a point y_0 of Y is the *limit of $\phi(x)$ as x approaches* (or *tends to*) a_0 *through* A (notation: $\lim_{\substack{x \to a_0 \\ x \in A}} \phi(x)$ or, when $A = X$, simply $\lim_{x \to a_0} \phi(x)$) if for every positive number ε there exists a positive number δ such that $\phi(D_\delta(a_0) \cap A)$ is contained in $D_\varepsilon(y_0)$.

Note. When $X = \mathbb{R}$ and ϕ is defined on a nonempty open interval $U = (a, b)$, the notion of the limit of ϕ as t tends to a or b through U clearly agrees with the earlier $\lim_{t \downarrow a} \phi(t)$ or $\lim_{t \uparrow b} \phi(t)$ introduced in Example 6I.

Proposition 7.22. *If ϕ is a mapping of a subset A of a metric space (X, ρ) into a metric space (Y, σ), and if $a_0 \in A^-$, then either there is no point y_0 of Y such that $y_0 = \lim_{\substack{x \to a_0 \\ x \in A}} \phi(x)$ (in which case the limit of $\phi(x)$ as x tends to a_0 through A* fails to exist) *or there is exactly one such limit.*

Proof. Suppose $y_0 = \lim_{\substack{x \to a_0 \\ x \in A}} \phi(x)$ and $y_1 = \lim_{\substack{x \to a_0 \\ x \in A}} \phi(x)$ where $y_0 \neq y_1$. If ε is sufficiently small ($\varepsilon < \sigma(y_0, y_1)/2$), the balls $D_\varepsilon(y_0)$ and $D_\varepsilon(y_1)$ are disjoint. On the other hand, by hypothesis, there exist positive numbers δ_0 and δ_1 such that $\phi(D_{\delta_0}(a_0) \cap A) \subset D_\varepsilon(y_0)$ and $\phi(D_{\delta_1}(a_0) \cap A) \subset D_\varepsilon(y_1)$.

But then if $\delta = \delta_0 \wedge \delta_1$, we must have $D_\delta(a_0) \cap A = \varnothing$, which is impossible since $a_0 \in A^-$. □

There is a sequential criterion for the existence of limits. The following argument may be compared to the proof of Proposition 7.1.

Proposition 7.23. *Let X and Y be metric spaces, let ϕ be a mapping of a subset A of X into Y, and let a_0 be a point of X belonging to A^-. Then $y_0 = \lim_{\substack{x \to a_0 \\ x \in A}} \phi(x)$ if and only if $\phi(x_n) \to y_0$ for every sequence $\{x_n\}$ in A such that $x_n \to a_0$. Moreover, if $y_0 = \lim_{\substack{x \to a_0 \\ x \in A}} \phi(x)$, and if $\{x_\lambda\}_{\lambda \in \Lambda}$ is an arbitrary net in A that converges to a_0, then $\lim_\lambda \phi(x_\lambda) = y_0$.*

PROOF. Suppose first that $y_0 = \lim_{\substack{x \to a_0 \\ x \in A}} \phi(x)$. Let $\{x_\lambda\}$ be a net in A converging to a_0, and let ε be a positive number. By definition there exist a positive number δ such that $\phi(D_\delta(a_0) \cap A) \subset D_\varepsilon(y_0)$ and an index λ_0 such that $x_\lambda \in D_\delta(a_0)$ for all $\lambda \geq \lambda_0$. But then $\phi(x_\lambda) \in D_\varepsilon(y_0)$ for all $\lambda \geq \lambda_0$, and since ε is arbitrary, this shows that $\lim_\lambda \phi(x_\lambda) = y_0$. Thus the stated criterion is necessary. To see that the corresponding sequential criterion is sufficient, suppose it is not the case that $\lim_{\substack{x \to a_0 \\ x \in A}} \phi(x) = y_0$. Then there exists a positive number ε_0 so small that $D_{\varepsilon_0}(y_0)$ does not contain any set of the form $\phi(D_\delta(a_0) \cap A), \delta > 0$. Hence, in particular, for each positive integer n there exists a point x_n of $D_{1/n}(a_0) \cap A$ such that $\phi(x_n) \notin D_{\varepsilon_0}(y_0)$. But then $\{x_n\}$ is a sequence in A that converges to a_0, while $\{\phi(x_n)\}$ clearly does not converge to y_0. □

In dealing with limits, and in other contexts as well, the following notion is sometimes extremely useful.

Definition. A nonempty collection $\mathcal{B}$ of nonempty subsets of a set X is a *filter base* in X if for every nonempty finite subcollection $\{B_1, \ldots, B_n\}$ of $\mathcal{B}$ there is a set B in $\mathcal{B}$ such that $B \subset B_1 \cap \ldots \cap B_n$. A filter base $\mathcal{B}$ in a metric space X is said to be *convergent* to a point a_0 of X, or to have *limit* a_0 (notation: $a_0 = \lim \mathcal{B}$), if for every positive number ε the ball $D_\varepsilon(a_0)$ contains a set B belonging to $\mathcal{B}$. Likewise, a point a of X is an *adherent point* of $\mathcal{B}$ or, more generally, of an arbitrary nonempty collection $\mathcal{C}$ of subsets of X, if for every positive number ε the ball $D_\varepsilon(a)$ meets every set C of $\mathcal{C}$, or, equivalently, if $a \in \bigcap_{C \in \mathcal{C}} C^-$. (Just as in the case of nets, the limit of a filter base in a metric space is unique if it exists. Adherent points of a filter base, on the other hand, may exist in abundance.)

Example T. Any nonempty nested collection of nonempty subsets of a set X is a filter base. Hence if a_0 is a point of a metric space X, and if M

is an arbitrary nonempty set of positive real numbers, then the collection $\mathcal{D}_M = \{D_r(a_0) : r \in M\}$ is a filter base in X, and this filter base converges to a_0 whenever $\inf M = 0$. Moreover, if M is a set of positive real numbers such that $\inf M = 0$, and if A is a subset of X, then the *trace* on A of the filter base $\mathcal{D}_M$, that is, $\{D_r(a_0) \cap A : r \in M\}$, is a filter base in A (and in X) when and only when $a_0 \in A^-$.

The following elementary result is a more or less immediate consequence of the foregoing example. (There is an analogous criterion for continuity phrased in terms of filter bases; see Problem V.)

Proposition 7.24. *Let ϕ be a mapping of a subset A of a metric space X into a metric space Y, and let a_0 be a point of A^-. Then a point y_0 of Y is the limit of $\phi(x)$ as x approaches a_0 through A if and only if $y_0 = \lim \phi(\mathcal{D})$ where $\mathcal{D} = \{D_r(a_0) \cap A : r > 0\}$ (see Problem S(v)). Moreover if, for any one set M of positive real numbers such that $\inf M = 0$, we have $y_0 = \lim \phi(\mathcal{D}_M)$, where $\mathcal{D}_M = \{D_r(a_0) \cap A : r \in M\}$, then $y_0 = \lim_{\substack{x \to a_0 \\ x \in A}} \phi(x)$. Finally, if $y_0 = \lim_{\substack{x \to a_0 \\ x \in A}} \phi(x)$, and if $\mathcal{B}$ is an arbitrary filter base in A such that $a_0 = \lim \mathcal{B}$, then $y_0 = \lim \phi(\mathcal{B})$.*

The relations between continuity and limits are obvious on the basis of either the definition or any of the foregoing criteria.

Proposition 7.25. *Let ϕ be a mapping of a subset A of a metric space X into a metric space Y, and let a_0 be a point of A^-. If $a_0 \notin A$, then the limit of $\phi(x)$ as x tends to a_0 through A exists if and only if there exists a (necessarily unique) point y_0 of Y with the property that, if ϕ is extended to a mapping $\widehat{\phi}$ on $A \cup \{a_0\}$ by defining $\widehat{\phi}(a_0) = y_0$, the extended mapping $\widehat{\phi}$ is continuous at a_0 relative to $A \cup \{a_0\}$. Moreover, if such a point y_0 exists, then $y_0 = \lim_{\substack{x \to a_0 \\ x \in A}} \phi(x)$. On the other hand, if $a_0 \in A$, then the limit of $\phi(x)$ as x tends to a_0 through A exists if and only if ϕ is continuous at a_0 (relative to A), and in this case the limit must coincide with $\phi(a_0)$.*

Note. As this last proposition clearly shows, when the point a_0 belongs to the domain of definition of the mapping, our notion of the limit of the mapping at a_0 differs from the one customarily introduced in calculus courses. This latter notion of limit, which corresponds in our notation to

$$\lim_{\substack{x \to a_0 \\ x \in A \setminus \{a_0\}}} \phi(x),$$

is the subject of Example U.

Example U. If x_0 is a point of Euclidean space $\mathbb{R}^n$, then a subset N of $\mathbb{R}^n$ is called a *punctured neighborhood* of x_0 if there exists a positive radius r such that $N \supset D_r(x_0)\backslash\{x_0\}$. Let g be a complex-valued function defined and never equal to zero on some punctured neighborhood of x_0, and let f be an arbitrary complex-valued function defined on a punctured neighborhood of x_0. Then one says that f *vanishes at* x_0 *to the same order as* g (or that f is "*big oh*" of g) (notation: $f(x) = O(g(x))$), if there exists some (sufficiently small) punctured neighborhood of x_0 on which the function $f(x)/g(x)$ is bounded. Likewise, one says that f *vanishes at* x_0 *to a higher order than* g (or that f is "*little oh*" of g) (notation: $f(x) = o(g(x))$), if

$$\lim_{\substack{x \to x_0 \\ x \in N}} f(x)/g(x) = 0,$$

where N denotes any punctured neighborhood of x_0 on which $f(x)/g(x)$ is defined. (It is obvious that such punctured neighborhoods exist, and that the defined concept is independent of which such punctured neighborhood is used.)

Of particular interest in mathematical analysis is the case in which $g(x)$ equals some power of the polar distance $r(x) = \rho_2(x, x_0)$ from x to x_0. If f is a real-valued function defined in a punctured neighborhood of x_0 and if a denotes a positive number, then f *vanishes at* x_0 *to order* a if $f = O(r^a)$, while f vanishes at x_0 to a *higher order than* a if $f = o(r^a)$.

Example V. Let U be an open set in $\mathbb{R}^n$, let ϕ be a mapping of U into $\mathbb{R}^m$, and let x_0 be a point of U. In the language of advanced calculus the mapping ϕ is *differentiable* at x_0 if there exists a linear transformation T of $\mathbb{R}^n$ into $\mathbb{R}^m$ such that

$$\|\phi(x) - \phi(x_0) - T(x - x_0)\|_2 = o(\|x - x_0\|_2). \tag{6}$$

From this definition it is clear, on the one hand, that if ϕ is differentiable at x_0, then it is automatically continuous there. (Recall that T is continuous; see Example E.) On the other hand, it is not hard to see that if ϕ is differentiable at x_0, and $f_1, \ldots, f_m$ denote the *coordinate functions* of ϕ, so that $\phi(x) = (f_1(x), \ldots, f_m(x)), x \in U$, then the matrix $A = (a_{ij})$ defining the linear transformation T in (6) (see Example 3Q) is necessarily given by

$$a_{ij} = \left. \frac{\partial f_i}{\partial x_j} \right|_{x = x_0}$$

for each pair (i, j) of indices, $i = 1, \ldots, m$; $j = 1, \ldots, n$. (In particular, all of these partial derivatives must exist at any point at which ϕ is differentiable.) The linear transformation T, which is thus seen to be uniquely determined by (6) when it exists, is called the *differential* of ϕ (notation: $d\phi$).

While the limit of a mapping at a point is defined in such a way as to provide an extension of the given mapping that is continuous at just that one point, it turns out that the formation of limits automatically gives rise to a continuous mapping on the set where those limits exist.

Proposition 7.26. *Let X and Y be metric spaces, let ϕ be a mapping of a subset A of X into Y, let $\widetilde{A}$ denote the subset of A^- consisting of all those points a at which the limit of $\phi(x)$ exists as x tends to a through A, and for each point a of $\widetilde{A}$ set*

$$\widetilde{\phi}(a) = \lim_{\substack{x \to a \\ x \in A}} \phi(x).$$

Then $\widetilde{\phi}$ is a continuous mapping of $\widetilde{A}$ into Y.

PROOF. Let a_0 be a fixed point of $\widetilde{A}$, let $y_0 = \widetilde{\phi}(a_0)$, let ε be a positive number, and let δ be a positive number such that $\phi(D_\delta(a_0) \cap A) \subset D_{\varepsilon/2}(y_0)$. (Such a δ exists by the definition of $\widetilde{\phi}$.) We shall show that $\widetilde{\phi}(D_\delta(a_0) \cap \widetilde{A})$ is contained in $D_\varepsilon(y_0)$, thus proving that $\widetilde{\phi}$ is continuous at a_0 relative to $\widetilde{A}$, and since a_0 is an arbitrary point of $\widetilde{A}$, this will complete the proof. To this end let a belong to $D_\delta(a_0) \cap \widetilde{A}$, and let η be a positive radius small enough so that $D_\eta(a) \subset D_\delta(a_0)$. Then $\phi(D_\eta(a) \cap A) \subset D_{\varepsilon/2}(y_0)$, and since $\widetilde{\phi}(a) \in [\phi(D_\eta(a) \cap A)]^-$, it follows that $\widetilde{\phi}(a)$ belongs to $[D_{\varepsilon/2}(y_0)]^-$. Thus $\widetilde{\phi}(D_\delta(a_0) \cap \widetilde{A}) \subset D_\varepsilon(y_0)$. □

Note. The sets A and $\widetilde{A}$ of the preceding proposition may intersect or not (see Example X below), but the mappings ϕ and $\widetilde{\phi}$ are coherent in any case by virtue of Proposition 7.25 (cf. Problem 1M). Thus, $\phi \cup \widetilde{\phi}$ is an extension of ϕ to $A \cup \widetilde{A}$, but this mapping is continuous only at the points of $\widetilde{A}$. Special interest attaches to the case $A \subset \widetilde{A}$, which is, of course, the case in which ϕ is continuous on A to begin with.

Definition. Let X and Y be metric spaces, and let ϕ be a continuous mapping of a subset A of X into Y, so that the set $\widetilde{A}$ of Proposition 7.26 contains A. Then the mapping $\widetilde{\phi}$ is a continuous extension of ϕ, and we say that $\widetilde{\phi}$ results from *extending ϕ by continuity*.

Example W. Let C denote the Cantor set (Ex. 6O), and let f_0 be the real-valued function on $[0,1]\backslash C$ that is constantly equal to $0.\varepsilon_1 \ldots \varepsilon_k 100\ldots$ (binary expansion) on the contiguous interval $U_{\varepsilon_1,\ldots,\varepsilon_k}$. Then f_0 agrees with the Cantor-Lebesgue function h of Example Q on a dense subset of $[0,1]$, so h results from extending the function f_0 by continuity. Similarly, the space-filling curve of Example R may be recaptured as the extension by continuity of an explicitly defined mapping of $[0,1]\backslash C$ into $\mathbb{R}^2$.

For real-valued functions the notion of limit can be split into two dual notions, just as the concept of continuity is split into the two dual concepts of semicontinuity. Here again it is convenient to deal with extended real-valued functions.

Definition. Let A be a subset of a metric space X, and let f be an extended real-valued function defined on A. For each point a_0 of the closure A^- we define the *limit superior* (or *upper limit*) of $f(x)$ as x tends to a_0 through A (notation: $\limsup_{\substack{x \to a_0 \\ x \in A}} f(x)$ or, when $A = X$, $\limsup_{x \to a_0} f(x)$) as follows: For each positive radius ε we first set $M(f; a_0, \varepsilon) = \sup\{f(x) : x \in D_\varepsilon(a_0) \cap A\}$, and then define

$$\limsup_{\substack{x \to a_0 \\ x \in A}} f(x) = \inf_{\varepsilon > 0} M(f; a_0, \varepsilon) = \lim_{\varepsilon \downarrow 0} M(f; a_0, \varepsilon).$$

(This latter equation is correct since $M(f; a_0, \varepsilon)$ is a monotone increasing function of ε, as is readily seen; cf. Example 6J.) Dually, we define $m(f; a_0, \varepsilon) = \inf\{f(x) : x \in D_\varepsilon(a_0) \cap A\}$ for each $\varepsilon > 0$ and then define the *limit inferior* (or *lower limit*) of $f(x)$ as x tends to a_0 through A (notation: $\liminf_{\substack{x \to a_0 \\ x \in A}} f(x)$ or, when $A = X, \liminf_{x \to a_0} f(x)$) by setting

$$\liminf_{\substack{x \to a_0 \\ x \in A}} f(x) = \sup_{\varepsilon > 0} m(f; a_0, \varepsilon) = \lim_{\varepsilon \downarrow 0} m(f; a_0, \varepsilon).$$

(The latter equation is valid this time because $m(f; a_0, \varepsilon)$ is a monotone decreasing function of ε.)

From these definitions it is apparent that for any extended real-valued function f on a subset A of a metric space X we have

$$\liminf_{\substack{x \to a_0 \\ x \in A}} f(x) \leq \limsup_{\substack{x \to a_0 \\ x \in A}} f(x)$$

at every point a_0 of A^-, and

$$\liminf_{\substack{x \to x_0 \\ x \in A}} f(x) \leq f(x_0) \leq \limsup_{\substack{x \to x_0 \\ x \in A}} f(x)$$

at every point x_0 of A. Moreover it is clear that if f is finite-valued on A, then the limit of $f(x)$ as x tends to a point a_0 of A^- exists at a point a_0 of A^- when and only when $\liminf_{\substack{x \to a_0 \\ x \in A}} f(x)$ and $\limsup_{\substack{x \to a_0 \\ x \in A}} f(x)$ are equal and finite, in which case we have

$$\lim_{\substack{x \to a_0 \\ x \in A}} f(x) = \liminf_{\substack{x \to a_0 \\ x \in A}} f(x) = \limsup_{\substack{x \to a_0 \\ x \in A}} f(x).$$

Other properties of these concepts are set forth in the following proposition.

Proposition 7.27. *Let A be a subset of a metric space X, let f be an extended real-valued function defined on A, and for each point a_0 of A^- set*

$$m(a_0) = \liminf_{\substack{x \to a_0 \\ x \in A}} f(x), \quad M(a_0) = \limsup_{\substack{x \to a_0 \\ x \in A}} f(x).$$

Then $m \leq f \leq M$ on A, while $m \leq M$ holds everywhere on A^-. Moreover M is an upper semicontinuous function on A^- with the property that f is upper semicontinuous at a point x_0 of A (relative to A) if and only if $f(x_0) = M(x_0)$, and also with the property that if h is an arbitrary upper semicontinuous function on A^- such that $f \leq h$ on A, then $M \leq h$ as well. Dually, m is a lower semicontinuous function on A^- with the property that f is lower semicontinuous at a point x_0 of A if and only if $f(x_0) = m(x_0)$, and also with the property that if k is an arbitrary lower semicontinuous function on A^- such that $k \leq f$ on A, then $k \leq m$ as well. (Because of these extremal properties, M and m are sometimes called the upper *and* lower envelopes *of f, respectively; they are also called the* upper *and* lower functions *of f.)*

PROOF. As always, it is enough to prove the half of the proposition bearing on upper semicontinuity. Suppose, to begin with, that f is upper semicontinuous (relative to A) at some point x_0 of A. If $f(x_0) = +\infty$, then it is certain that $M(x_0) = f(x_0)$. Otherwise, let u be a finite real number such that $f(x_0) < u$. Then there exists a positive number δ such that $f(x) < u$ for every x in $D_\delta(x_0) \cap A$, whence it is clear that $M(x_0) \leq u$. Since u is an arbitrary real number exceeding $f(x_0)$, this shows that $M(x_0) \leq f(x_0)$, and hence that $M(x_0) = f(x_0)$.

Next we show that M is upper semicontinuous on A^-. To this end let u be a finite real number, and suppose $M(a_0) < u$ for some a_0 in A^-. Then there is a positive number δ such that

$$M(f; a_0, \delta) = \sup\{f(x) : x \in D_\delta(a_0) \cap A\} < u.$$

But then for any point a of $D_\delta(a_0) \cap A^-$ it is readily seen that

$$M(a) \leq M(f; a_0, \delta) < u$$

as well. Thus the set $\{a \in A^- : M(a) < u\}$ is open relative to A^- (Prop. 6.15) and the result follows by Proposition 7.17.

Suppose next that at some point x_0 of A we have $f(x_0) = M(x_0) < +\infty$. (If $f(x_0) = +\infty$, then f is automatically upper semicontinuous at x_0.) If u is a finite real number such that $f(x_0) < u$, then by what has just been seen there is an open set U in X containing x_0 such that $M(x) < u$ for all x in $A^- \cap U$. But then, of course, $f(x) < u$ for all x in $A \cap U$ since $f \leq M$ on A. Hence f is upper semicontinuous at x_0.

Finally, let h be an extended real-valued function defined and upper semicontinuous on A^- (relative to A^-) such that $h \geq f$ on A. Then, of course,

$$\limsup_{\substack{x \to a_0 \\ x \in A}} f(x) \leq \limsup_{\substack{x \to a_0 \\ x \in A}} h(x) \leq \limsup_{\substack{x \to a_0 \\ x \in A^-}} h(x)$$

at each point a_0 of A^-, and since h is upper semicontinuous by hypothesis, this inequality reduces to the inequality $M(a_0) \leq h(a_0)$. □

Example X. Let $X = (0, +\infty)$, and for each rational number r in X set $g_0(r) = 1/n$ where $r = m/n$ in *lowest terms* (meaning that m and n are relatively prime positive integers). Then g_0 is a real-valued function defined on the set $\mathbb{Q}_+$ of all positive rational numbers and taking its values in the set $\{1/n : n \in \mathbb{N}\}$ of reciprocals of positive integers. Moreover, if n is an arbitrary positive integer, and if $F_n = \{r \in \mathbb{Q}_+ : g_0(r) \geq 1/n\}$, then F_n coincides with the union over $k = 1, \ldots, n$ of the arithmetic progressions $A_k = \{j/k : j \in \mathbb{N}\}$. But then F_n is a closed and discrete set (Prob. 6O). Hence if $g_0(r_0) = 1/n$, there is an interval $(r_0 - \delta, r_0 + \delta)(\delta > 0)$ about r_0 such that no element of F_n other than r_0 lies in that interval, and it follows that

$$\limsup_{\substack{r \to r_0 \\ r \in \mathbb{Q}_+}} g_0(r) = g_0(r_0)$$

at every point r_0 of $\mathbb{Q}_+$. In other words, g_0 is upper semicontinuous on $\mathbb{Q}_+$ (relative to $\mathbb{Q}_+$). Moreover, if t_0 is an arbitrary positive irrational number, then the same argument shows that there is an interval $(t_0 - \delta, t_0 + \delta)(\delta > 0)$ containing no point of F_n, so $\limsup_{\substack{r \to t_0 \\ r \in \mathbb{Q}_+}} g_0(r) \leq 1/n$, and since n is quite arbitrary it follows that

$$\limsup_{\substack{r \to t_0 \\ r \in \mathbb{Q}_+}} g_0(r) = 0.$$

Thus the upper envelope M of the function g_0 agrees with g_0 on $\mathbb{Q}_+$ and vanishes identically elsewhere on X.

We next observe that for an arbitrary positive number ε every nonempty open interval in X contains points r at which $g_0(r)$ is less than ε. Hence the lower envelope m of g_0 is identically zero on X. Thus, in summary, the given function g_0 is upper semicontinuous at every point of $\mathbb{Q}_+$ and lower semicontinuous at no point of $\mathbb{Q}_+$ (relative to $\mathbb{Q}_+$), while the upper envelope M of g_0 is obtained by extending g_0 to be zero at every point of $X \backslash \mathbb{Q}_+$, and the lower envelope m of g_0 is identically zero on X. Incidentally, it should be noted that in this example the set $\widetilde{A}$ of points a at which $\widetilde{g}(a) = \lim_{\substack{r \to a \\ r \in \mathbb{Q}_+}} g_0(r)$ exists is precisely the set $X \backslash \mathbb{Q}_+$ of positive irrational numbers, and is therefore actually disjoint from the domain of definition of g_0. Note too that the function $M = g_0 \cup \widetilde{g}$, which is upper semicontinuous on X (Prop. 7.27), is continuous precisely at the irrational points of X. (Curiously enough, there exists *no* real-valued function on X that is continuous at precisely the rational points of X. These matters are examined further in Chapter 8.)

Problems

A. (i) A mapping ϕ of a metric space X into a metric space Y is continuous at every isolated point of X (cf. Problem 6O). In particular, if X is discrete, then ϕ is continuous. Describe the continuous mappings of Y into X when X is discrete.

(ii) Let ϕ be a one-to-one mapping of $\mathbb{N}$ onto $\mathbb{Q}$. Show that while ϕ is continuous (as has just been seen), the inverse ϕ^{-1} is discontinuous at every point of $\mathbb{Q}$.

B. Let X and Y be metric spaces, and let ϕ be a mapping of X into Y. Show that each of the following is a necessary and sufficient condition in order that ϕ be continuous on X:

(1) The sequence $\{\phi(x_n)\}$ is convergent in Y whenever $\{x_n\}$ is a convergent sequence in X (briefly, ϕ *preserves convergence of sequences*),

(2) $\phi(A^-) \subset \phi(A)^-$ for every subset A of X.

C. If $\phi : A \to Y$ is a mapping of a subset A of a metric space X into a metric space Y that is continuous at some point x_0 of A relative to A, and if B is a subset of A such that $x_0 \in B$, then ϕ is also continuous at x_0 relative to B. In particular, if ϕ is continuous on A, and if $B \subset A$, then ϕ is also continuous on B.

D. Let X and Y be metric spaces. A mapping ϕ of X into Y is *locally open* [*locally closed, locally bounded*] if for every point x of X there is an open subset U of X containing x such that $\phi|U$ is an open [closed, bounded] mapping of U into Y.

(i) Verify that every continuous mapping of X into Y is locally bounded.

(ii) Prove that every locally open mapping is open.

(iii) Clearly a locally bounded mapping need not be bounded. Is it true that a locally closed mapping is necessarily closed?

(iv) A mapping ϕ of X into Y is a *local homeomorphism* if for each point x of X there are open subsets U and V of X and Y, respectively, such that $x \in U$ and $\phi|U$ is a homeomorphism of U onto V. Give necessary and sufficient conditions in order that such a local homeomorphism be a homeomorphism between X and Y.

E. Let X be a metric space and let E and F be nonempty disjoint closed subsets of X. Show first that $d(x,E) + d(x,F) > 0$ for every x in X, so that it makes sense to define

$$f_{E,F}(x) = \frac{d(x,E)}{d(x,E) + d(x,F)}, \qquad x \in X.$$

Then prove that the function $f_{E,F}$ so defined is a continuous mapping of X into the unit interval $[0,1]$, and that

$$E = \{x \in X : f_{E,F}(x) = 0\} \text{ and } F = \{x \in X : f_{E,F}(x) = 1\}.$$

Show, too, that $f_{E,F}$ is Lipschitzian if and only if $d(E,F) > 0$. Show, finally, that $f_{E,F}$ is monotone decreasing as a function of its first variable E and monotone increasing as a function of F. (Hint: Recall Examples B and M.)

F. Let U be an open set in a metric space X. Show that there exists a monotone increasing sequence $\{f_n\}_{n=1}^{\infty}$ of continuous mappings of X into $[0,1]$ with the property that $\chi_U(x) = \lim_n f_n(x), x \in X$. What is the dual of this assertion? (Hint: Recall Problem 6P.)

G. The real-valued function $f(t)$ defined on $\mathbb{R}$ by

$$f(t) = \begin{cases} t, & t \in \mathbb{Q}, \\ 0, & t \in \mathbb{R}\backslash\mathbb{Q}, \end{cases}$$

is continuous at the point $t = 0$ and at no other point (proof?). Likewise the function

$$g(t) = \begin{cases} t - t^3, & t \in \mathbb{Q}, \\ 0, & t \in \mathbb{R}\backslash\mathbb{Q}, \end{cases}$$

is continuous at the points $t = 0, \pm 1$, and at no other point. Construct a real-valued function on $\mathbb{R}$ that is continuous at every point of the Cantor set C and discontinuous at every point of $\mathbb{R}\backslash C$. (Hint: On an open interval $(a,b), a < b$, construct a real-valued function f that is discontinuous at every point of (a,b) but for which $\lim_{t\downarrow a} f(t)$ and $\lim_{t\uparrow b} f(t)$ exist. A precise description of the set of points at which a function of a real variable can be discontinuous will be found in Example 8S; cf. also Problem P and Example X.)

H. Verify that $(s,t) \to s \vee t$ and $(s,t) \to s \wedge t$ are contractive mappings of $\mathbb{R}^2$ into $\mathbb{R}$. Use this observation to show that if f and g are real-valued functions on a metric space X, then $f \vee g$ and $f \wedge g$ are continuous at every point of X at which f and g are both continuous. Conclude that if f is a [uniformly] continuous real-valued function on X, then f^+, f^- and $|f|$ are also [uniformly] continuous.

I. Let X be a set equipped with the discrete metric of Example 6C. Redo Problem A(i) with simple continuity replaced by uniform continuity.

J. Let $\mathcal{E}$ be a normed space, and let ϕ_0 be a continuous mapping of the Cantor set C into $\mathcal{E}$. Suppose ϕ_0 is extended to a mapping ϕ of $[0,1]$ into $\mathcal{E}$ by defining ϕ to be linear on the closure U^- of each (bounded) interval U contiguous to C (see Example N). Show that the extension ϕ is a curve in $\mathcal{E}$ (in other words, that ϕ is continuous on $[0,1]$). (Hint: Show first that if two points x_0 and x_1 belong to some ball D in $\mathcal{E}$, then the entire line segment $\ell(x_0, x_1)$ lies in D (Prob. 3S).)

K. Let (X, ρ) and (Y, σ) be metric spaces, let ϕ be a mapping of X into Y, and let a be a real number such that $0 < a \leq 1$. Then ϕ is said to be *Lipschitz–Hölder continuous*, or to satisfy a *Lipschitz–Hölder condition*, with respect to the *exponent* a if

$$\sigma(\phi(x), \phi(y)) \leq M\rho(x,y)^a$$

for all points x, y of X and some positive constant M, known as a *Lipschitz–Hölder constant* (or *"L.–H." constant* as we shall frequently abbreviate the somewhat unwieldy phrase "Lipschitz–Hölder"). (Note that when the exponent a assumes the value one in this definition, the condition reduces to the simple Lipschitz condition. Exponents greater than one are without analytic interest; see (v) below.) Likewise, ϕ is *locally Lipschitz–Hölder continuous* on X with respect to a if every point of X is contained in an open set U such that $\phi|U$ is L.–H. continuous with respect to a and some constant M. (Note that in this definition the L.–H. constant M depends in general on the set U, but the exponent a does not.)

(i) If ϕ is L.–H. continuous on X with respect to some exponent a and constant M, then ϕ is uniformly continuous on X. If ϕ is locally L.–H. continuous on X with respect to an exponent a, then ϕ is continuous on X.

(ii) If ϕ is L.–H. continuous with respect to some exponent b and constant M on a bounded subset B of X, then ϕ is also L.–H. continuous on B with respect to every exponent a such that $0 < a \leq b$ (and some constant M_a that depends in general on B as well as on a).

(iii) Let $\{\phi_n\}$ be a sequence of mappings of X into Y that converges (pointwise) to ϕ. If for some exponent a and constant M the mappings ϕ_n are all L.–H. continuous with respect to a and M, then the same is true of ϕ.

(iv) The function f defined on the interval $(-1, +1)$ by

$$f(t) = \begin{cases} 0, & t = 0, \\ 1/\log|t|, & 0 < |t| < 1, \end{cases}$$

is locally Lipschitzian on the set $(-1,+1)\setminus\{0\}$ but is not Lipschitz–Hölder continuous with respect to a positive exponent on any open interval containing zero.

(v) If f is a real-valued function defined on a real interval I, and if f satisfies a condition of the form $|f(s) - f(t)| \le M|s-t|^a$ for all s, t in I and for some positive constant M and exponent $a > 1$, then f is constant on I. (Hint: Consider the derivative f'.)

L. Let c be a real number such that $0 < c \le 1$ and set $f_c(t) = t^c, t \ge 0$.

(i) Verify that f_c is L.–H. continuous on $[0,+\infty)$ with respect to the exponent c (and constant one) and *no other* exponent. (Hint: If p and q are any two real numbers such that $p < q$, and A and B are positive numbers, then there is a unique positive real number t_0 such that

$$At^p \gtreqqless Bt^q \quad \text{for} \quad t \lesseqqgtr t_0$$

on $(0,+\infty)$.)

(ii) For any positive number L the function f_c is L.–H. continuous on $[0, L]$ with respect to those exponents a such that $0 < a \le c$ (with constant $M_a = L^{c-a}$) and with respect to no other exponents. Likewise f_c is L.–H. continuous on $[L,+\infty)$ with respect to those exponents a such that $c \le a \le 1$ (with constant $M_a = (c/a)L^{c-a}$) and with respect to no other exponents. (Hint: For any one $\delta > 0$ the increment $f_c(t+\delta) - f_c(t)$ is a monotone decreasing function of t.)

(iii) The functions in the sequence $\{g_n\}$ in Example P are L.–H. continuous with respect to all those exponents a such that $1/2 \le a \le 1$. For which values of a are the functions g_n all L.–H. continuous with respect to a and a common constant M?

M. (i) The Cantor–Lebesgue function (Ex. Q) is Lipschitz–Hölder continuous with respect to the exponent $a = \log 2/\log 3$. Find a suitable L.–H. constant.

(ii) The space-filling curve of Example R is also L.–H. continuous with respect to suitable exponents. Find the largest such exponent and a corresponding L.–H. constant.

N. Let X be a metric space and let f be an extended real-valued function on X. Show that f is upper [lower] semicontinuous at a point x_0 of X if and only if $-f$ is lower [upper] semicontinuous at x_0. (The value of $-f$ is taken to be $\mp\infty$ at a point x where $f(x) = \pm\infty$.) Show, too, that if g is another extended real-valued function on X, if a and b denote nonnegative real numbers, and if X_0 is the subset of X on which the positive linear combination $af+bg$ is defined, then $af+bg$ is upper [lower] semicontinuous relative to X_0 at any point of X_0 where both f and g are upper [lower] semicontinuous. Show, in the same vein, that $f \vee g$ and $f \wedge g$ are upper [lower] semicontinuous at any point where both f and g are upper [lower] semicontinuous.

O. Let I be an interval of real numbers, let ψ be a real-valued function defined on I, and let ϕ be a mapping of a metric space X into I.

(i) Suppose first that ψ is monotone increasing. Show that if ϕ is upper semicontinuous at a point x_0 of X and ψ is also upper semicontinuous at $\phi(x_0)$, then $\psi \circ \phi$ is upper semicontinuous at x_0. Dually, if ϕ is lower semicontinuous at x_0 and ψ is lower semicontinuous at $\phi(x_0)$, then $\psi \circ \phi$ is lower semicontinuous at x_0. Thus, in particular, if ψ is monotone increasing and continuous, then $\psi \circ \phi$ is upper and/or lower semicontinuous along with ϕ.

(ii) Formulate and prove appropriate versions of these results for monotone decreasing ψ.

(iii) Show by example that the assumption that ψ is monotone cannot be dropped.

P. Let X be a metric space and let A be a subset of X. Where is the characteristic function χ_A upper semicontinuous? Where is χ_A lower semicontinuous? At which points of X is χ_A continuous? For which sets A is χ_A upper semicontinuous and for which sets A is χ_A lower semicontinuous? For which sets A is χ_A continuous?

Q. Let X be a metric space, let f be an upper semicontinuous extended real-valued function defined on X, and for each finite real number t, set $U_t = \{x \in X : f(x) < t\}$. Then $\{U_t\}_{t \in \mathbb{R}}$ is a nested family of open sets in X satisfying the following three conditions:

(1) $\bigcup_{t \in \mathbb{R}} U_t = \{x \in X : f(x) < +\infty\}$,

(2) $\bigcap_{t \in \mathbb{R}} U_t = \{x \in X : f(x) = -\infty\}$,

(3) $U_t = \bigcup_{s<t} U_s$ for each t in $\mathbb{R}$.

Suppose, conversely, that M is a dense subset of $\mathbb{R}$ and that $\{V_t\}_{t \in M}$ is a given monotone increasing family of open subsets of X indexed by M. For each point x in X we set

$$f(x) = \inf\{t \in M : x \in V_t\}.$$

(This infimum is taken in $\mathbb{R}^\natural$; recall that the infimum of the empty set of extended real numbers is $+\infty$.) Show that f is an upper semicontinuous extended real-valued function on X satisfying the conditions

(1′) $\bigcup_{t \in M} V_t = \{x \in X : f(x) < +\infty\}$,

(2′) $\bigcap_{t \in M} V_t = \{x \in X : f(x) = -\infty\}$.

Show also that for each element t of M we have

$$\{x \in X : f(x) < t\} \subset V_t \subset \{x \in X : f(x) \leq t\},$$

and that $V_t = \{x \in X : f(x) < t\}$ for each t in M if and only if the given family $\{V_t\}_{t \in M}$ satisfies the additional condition

(3′) $V_t = \bigcup_{\substack{s<t \\ s \in M}} V_s$ for each t in M.

Show, finally, that if (3′) holds, then f is the unique upper semicontinuous extended real-valued function on X such that $V_t = \{x \in X : f(x) < t\}$ for each t in M.

R. Let ϕ be a mapping of a subset A of a metric space X into a metric space Y, let a_0 be a point of A^-, and let B be a subset of A such that $a_0 \in B^-$. Prove that if $\lim_{\substack{x \to a_0 \\ x \in A}} \phi(x) = y_0$, then $\lim_{\substack{x \to a_0 \\ x \in B}} \phi(x) = y_0$. Verify, too, that if B contains a set of the form $D_\varepsilon(a_0) \cap A, \varepsilon > 0$, then the converse implication holds as well. Show, finally, that if $a_0 \in (A \backslash B)^-$ too, then $\lim_{\substack{x \to a_0 \\ x \in A}} \phi(x)$ exists if and only if $\lim_{\substack{x \to a_0 \\ x \in B}} \phi(x) = \lim_{\substack{x \to a_0 \\ x \in A \backslash B}} \phi(x)$, in which case $\lim_{\substack{x \to a_0 \\ x \in A}} \phi(x)$ coincides with this common limit.

S. Let X be a set. A nonempty collection $\mathcal{F}$ of nonempty subsets of X is a *filter* in X if $\mathcal{F}$ satisfies the following two conditions: (a) $\mathcal{F}$ is closed with respect to the formation of finite intersections (that is, if $\{E_1, \ldots, E_p\}$ is an arbitrary nonempty finite subset of $\mathcal{F}$, then $E_1 \cap \ldots \cap E_p \in \mathcal{F}$), (b) if $E \in \mathcal{F}$ and $E \subset F$, then $F \in \mathcal{F}$.

(i) If $X \neq \varnothing$ then the collection Φ of all filters in X is a nonempty partially ordered set in the inclusion ordering. (The elements of Φ are subsets of the power class 2^X; by definition there is no filter in the empty set.) If $\mathcal{F}_1$ and $\mathcal{F}_2$ are filters in X such that $\mathcal{F}_1 \subset \mathcal{F}_2$, one says that $\mathcal{F}_2$ is *finer than* $\mathcal{F}_1$ (or that $\mathcal{F}_2$ is a *refinement* of $\mathcal{F}_1$) and that $\mathcal{F}_1$ is *coarser than* $\mathcal{F}_2$. Prove that if $X \neq \varnothing$, then the singleton $\{X\}$ is the coarsest filter in X (the least element of Φ). Under what circumstances does Φ have a greatest element?

(ii) Show that if Φ_0 is an arbitrary nonempty collection of filters in X, then the intersection $\bigcap \Phi_0$ is again a filter in X. Conclude that if $\mathcal{C}$ is an arbitrary collection of subsets of X, then either there exists no filter $\mathcal{F}$ in X such that $\mathcal{C} \subset \mathcal{F}$, or there exists a unique coarsest filter $\mathcal{F}_0$ in X such that $\mathcal{C} \subset \mathcal{F}_0$. (The filter $\mathcal{F}_0$ is then called the filter *generated* by $\mathcal{C}$, and $\mathcal{C}$ is a *system of generators* for $\mathcal{F}_0$.) Show that if $x \in X$, then the collection $\{\{x\}\}$ consisting of the singleton $\{x\}$ alone generates a filter $\mathcal{F}_x$ in X, and that the filter $\mathcal{F}_x$ is a maximal element of the set Φ of all filters in X.

(iii) For an arbitrary collection $\mathcal{C}$ of subsets of X, let $\mathcal{C}^s$ denote the collection of all those subsets E of X with the property that there exists a set C

in $\mathcal{C}$ such that $C \subset E$ (briefly: "supersets" of the sets in $\mathcal{C}$). Verify that $\mathcal{C}$ is a filter base in X if and only if $\mathcal{C}^s$ is a filter. In particular, every filter in X is also a filter base, and every filter base $\mathcal{B}$ in X generates a filter $\mathcal{F}_\mathcal{B} = \mathcal{B}^s$. (The set $\mathcal{B}$ is said to be a base for the filter $\mathcal{F}_\mathcal{B}$.)

(iv) A nonempty collection $\mathcal{S}$ of subsets of X is a system of generators for a filter in X if and only if it possesses the *finite intersection property*, that is, the property that the intersection of every nonempty finite subcollection of $\mathcal{S}$ is nonempty. Show that if $\mathcal{S}$ is a nonempty collection of subsets of X possessing the finite intersection property, and if $\mathcal{B}$ denotes the collection of all intersections of nonempty finite subcollections of $\mathcal{S}$, then $\mathcal{B}$ is a base for the filter generated by $\mathcal{S}$.

(v) Let $\mathcal{B}$ be a filter base in X and let ϕ be a mapping of X into a second set Y. Show that $\phi(\mathcal{B})(= \{\phi(B) : B \in \mathcal{B}\})$ is a filter base in Y. In particular, if $\mathcal{F}$ is a filter in X, then $\phi(\mathcal{F})$ is a filter base in Y.

T. (i) A filter $\mathcal{F}$ in a metric space X is said to *converge* to a point a_0 of X, or to have *limit* a_0, if $\mathcal{F}$ converges to a_0 as a filter base, that is, if for each positive number ε the ball $D_\varepsilon(a_0)$ contains some set F belonging to $\mathcal{F}$. Verify that if $\mathcal{B}$ is a filter base in X, and if $\mathcal{F}_\mathcal{B}$ denotes the filter generated by $\mathcal{B}$, then $\mathcal{F}_\mathcal{B}$ converges to a limit a_0 in X if and only if $\mathcal{B}$ does so.

(ii) If $\{x_\lambda\}_{\lambda\in\Lambda}$ is a net in an arbitrary set X, and if for each element λ of Λ we set T_λ equal to the tail in that net determined by λ (see Example 6L), then the set $\mathcal{T} = \{T_\lambda : \lambda \in \Lambda\}$ is a filter base in X. (The set $\mathcal{T}$ is the *tail filter base* associated with the given net; the filter $\mathcal{F}_\mathcal{T}$ generated by $\mathcal{T}$ is the *tail filter* associated with it.) Prove that if $\{x_\lambda\}$ is a net in a metric space X, and a_0 is a point of X, then $\lim_\lambda x_\lambda = a_0$ if and only if $\lim \mathcal{T} = a_0$.

(iii) If $\{x_\lambda\}$ is a net in a metric space X, then a point a of X is a *cluster point* of $\{x_\lambda\}$ if for arbitrary positive radius ε and arbitrary index λ_0, there exists an index λ such that $\lambda \geq \lambda_0$ and $x_\lambda \in D_\varepsilon(a)$. Verify that a is a cluster point of $\{x_\lambda\}$ if and only if a is an adherent point of the tail filter (base) associated with the net $\{x_\lambda\}$.

(iv) If $\mathcal{B}$ is a filter base in a set X, then $\mathcal{B}$ is a directed set in the inverse inclusion ordering (Ex. 1Q), and there exist nets $\{x_B\}_{B\in\mathcal{B}}$ in X indexed by $\mathcal{B}$ such that $x_B \in B$ for each B in $\mathcal{B}$. (Such a net will be called a net *along* $\mathcal{B}$.) Show that if $\mathcal{B}$ is a filter base in a metric space X that converges to a point a_0 of X, then every net along $\mathcal{B}$ also converges to a_0, and that if $\mathcal{B}$ does not converge to a_0, then there exists a net along $\mathcal{B}$ that also fails to converge to a_0.

U. Prove that if $\mathcal{B}$ is a filter base in X such that $\lim \mathcal{B} = a_0$, then a_0 is the unique adherent point of $\mathcal{B}$. Is it true, conversely, that if a filter base $\mathcal{B}$ has a unique adherent point a_0, then $\lim \mathcal{B} = a_0$?

V. Let X and Y be metric spaces and let ϕ be a mapping of X into Y. Show that ϕ is continuous at a point x_0 of X if the filter base $\{\phi(D_r(x_0))\}_{r>0}$ is convergent in Y to the point $\phi(x_0)$. Show conversely that if ϕ is continuous at x_0 and $\mathcal{B}$ is an arbitrary filter base in X converging to x_0, then $\phi(\mathcal{B})$ converges to $\phi(x_0)$.

W. Let U be an open subset of Euclidean space $\mathbb{R}^n$. A real-valued function f defined on U is said to be *continuously differentiable* on U if the partial derivatives $\frac{\partial f}{\partial x_i}, i = 1, \ldots, n$, exist and are continuous at every point of U. The collection of all continuously differentiable functions on U is denoted by $\mathcal{C}^{(1)}(U)$; cf. Example 3L.

(i) Prove that a function f belonging to $\mathcal{C}^{(1)}(U)$ is locally Lipschitzian (and therefore continuous) on U. (Hint: Fix a point $x^0 = (x_1^0, \ldots, x_n^0)$ of U and let $x = (x_1, \ldots, x_n)$ be an arbitrary point of the ball $D_\delta(x^0)$, where $\delta > 0$ is chosen small enough so that $D_\delta(x^0) \subset U$. If we write $\widehat{x}_i = (x_1^0, \ldots, x_i^0, x_{i+1}, \ldots, x_n), i = 0, \ldots, n$, then $x = \widehat{x}_0, x^0 = \widehat{x}_n$ and

$$f(x) - f(x^0) = [f(\widehat{x}_0) - f(\widehat{x}_1)] + \ldots + [f(\widehat{x}_{n-1}) - f(\widehat{x}_n)].$$

All of the line segments $\ell(\widehat{x}_{i-1}, \widehat{x}_i)$ belong to U, and by the mean value theorem there exist n points $\widetilde{x}_i$ such that $\widetilde{x}_i \in \ell(\widehat{x}_{i-1}, \widehat{x}_i)$ and such that

$$f(\widehat{x}_{i-1}) - f(\widehat{x}_i) = \left.\frac{\partial f}{\partial x_i}\right|_{\widetilde{x}_i} (x_i - x_i^0),$$

$i = 1, \ldots, n$. Use the fact that the functions $\frac{\partial f}{\partial x_i}$ are locally bounded on U (Prob. D).)

(ii) Prove also that a continuously differentiable function f on U is, in fact, differentiable at every point x^0 of U. (Here and elsewhere in this problem use is made of the terminology and notation introduced in Examples U and V.) (Hint: In the notation introduced in (i) the points $\widetilde{x}_i$ are all at least as close to x^0 as is x. Hence as $r = \|x - x^0\|_2$ tends to zero we have

$$\left.\frac{\partial f}{\partial x_i}\right|_{\widetilde{x}_i} = \left.\frac{\partial f}{\partial x_i}\right|_{x^0} + o(1), \quad i = 1, \ldots, n,$$

and therefore

$$\begin{aligned} f(x) - f(x^0) &= \sum_{i=1}^{n} \left[\left.\frac{\partial f}{\partial x_i}\right|_{x^0} + o(1) \right] (x_i - x_i^0) \\ &= \left(\sum_{i=1}^{n} \left.\frac{\partial f}{\partial x_i}\right|_{x^0} (x_i - x_i^0) \right) + o(r). \;) \end{aligned}$$

(iii) A mapping ϕ of U into a Euclidean space $\mathbb{R}^m$ is defined to be *continuously differentiable* on U if all m of the coordinate functions of ϕ

belong to $\mathcal{C}^{(1)}(U)$. Verify that a continuously differentiable mapping ϕ of U into $\mathbb{R}^m$ is locally Lipschitzian (and therefore continuous) on U, and is also differentiable at every point of U.

X. (i) Let ϕ be a mapping of a set X into a metric space Y. For each subset E of X the extended real number

$$\omega(\phi; E) = \text{ diam } \phi(E)$$

is called the *oscillation* of ϕ on E. Show that the mapping $E \to \omega(\phi; E)$ is a monotone increasing extended real-valued function on the power class 2^X.

(ii) Let $[a, b]$ be a closed interval in $\mathbb{R}(a < b)$, let ϕ be a mapping of $[a, b]$ into a metric space Y, and let $\mathcal{P} = \{a = t_0 < t_1 < \ldots < t_N = b\}$ be a partition of $[a, b]$. The extended real number

$$\omega(\phi; \mathcal{P}) = \sup\{\omega(\phi; [t_{i-1}, t_i]) : i = 1, \ldots, N\}$$

is called the *oscillation of ϕ over $\mathcal{P}$*. Verify that for any one fixed mapping ϕ the mapping $\mathcal{P} \to \omega(\phi; \mathcal{P})$ is a monotone decreasing mapping of the directed set of all partitions of $[a, b]$ into $\mathbb{R}^\natural$ (Prob. 2S).

(iii) Let A be a subset of a metric space X and let ϕ be a mapping of A into a metric space Y. For each point a_0 of A^- and each positive real number ε we set

$$\omega(\phi; a_0, \varepsilon) = \omega(\phi; D_\varepsilon(a_0) \cap A)$$

and define the *oscillation $\omega(\phi; a_0)$ of ϕ at a_0* by setting

$$\omega(\phi; a_0) = \inf_{\varepsilon > 0} \omega(\phi; a_0, \varepsilon).$$

Show that $\omega(\phi; a_0) = 0$ is a necessary condition for the existence of $\lim_{\substack{x \to a_0 \\ x \in A}} \phi(x)$. Show too that the nonnegative extended real-valued mapping $a \to \omega(\phi; a)$ is upper semicontinuous on A^-, and hence that the set of points $\{x \in A^- : \omega(\phi; x) = 0\}$ is a G_δ. Conclude that the set of points of continuity of an arbitrary mapping of X into Y is a G_δ in X, and that the set of points of discontinuity is, accordingly, an F_σ.

(iv) Let f be a real-valued function defined on a subset A of a metric space X, and let a_0 be a point of A^-. Show that the oscillation $\omega(f; a_0)$ is given by the formula

$$\omega(f; a_0) = \limsup_{\substack{x \to a_0 \\ x \in A}} f(x) - \liminf_{\substack{x \to a_0 \\ x \in A}} f(x).$$

8 Completeness and compactness

In classical analysis the *Cauchy criterion* plays a critically important role as a test for convergence. In the theory of abstract metric spaces its role is no less important, but here it serves as a basis for classifying metric spaces.

Definition. An infinite sequence $\{x_n\}$ (indexed by either $\mathbb{N}$ or $\mathbb{N}_0$) in a metric space (X, ρ) is a *Cauchy sequence*, or satisfies the *Cauchy criterion*, if $\lim_{m,n} \rho(x_m, x_n) = 0$, i.e., if for any positive number ε there exists an index N such that $\rho(x_m, x_n) < \varepsilon$ for all $m, n \geq N$. (Equivalently, if T_n denotes the tail $\{x_k : k \geq n\}$ of $\{x_n\}$, then $\{x_n\}$ is Cauchy if and only if $\lim_n \operatorname{diam} T_n = 0$.)

The basic facts about Cauchy sequences are quickly established.

Proposition 8.1. *Let (X, ρ) be a metric space. Every convergent sequence in X is Cauchy, and every Cauchy sequence in X is bounded. Furthermore, every Cauchy sequence in X that possesses a convergent subsequence is itself convergent to the limit of that subsequence. Consequently, a Cauchy sequence can have at most one cluster point, and if it possesses a cluster point, it must converge to that cluster point.*

PROOF. Suppose first that $\{x_n\}$ is a convergent sequence in X and that $\lim_n x_n = a_0$. If $\varepsilon > 0$ is given, then there exists an index n_0 such that $\rho(x_n, a_0) < \varepsilon/2$ for all $n \geq n_0$, whence it follows by the triangle inequality that $\rho(x_m, x_n) < \varepsilon$ for all $m, n \geq n_0$. Thus every convergent sequence is Cauchy. On the other hand, if $\{x_n\}$ is Cauchy, then there exists an index N such that $\rho(x_m, x_n) < 1$ for all $m, n \geq N$. But then, if we set $K = \sup_{n<N} \rho(x_n, x_N)$, the entire sequence $\{x_n\}$ is contained in the ball $D_{K+1}(x_N)$, and is therefore bounded. To complete the proof, it suffices to show that a Cauchy sequence converges to the limit of any of its convergent subsequences (see Proposition 6.5). Suppose then that $\{x_n\}$ is Cauchy and $\{y_k = x_{n_k}\}$ is a subsequence of $\{x_n\}$ that converges to a point a_1. If $\varepsilon > 0$ is given, then there exists an index N such that $\rho(x_m, x_n) < \varepsilon/2$ whenever $m, n \geq N$, and an index k_0 such that $\rho(y_k, a_1) < \varepsilon/2$ for all $k \geq k_0$. But then, if k_1 is any index such that $k_1 \geq k_0$ and $n_{k_1} \geq N (k_1 = N \vee k_0$, for example), we have

$$\rho(x_n, a_1) \leq \rho(x_n, y_{k_1}) + \rho(y_{k_1}, a_1) < \varepsilon/2 + \varepsilon/2 = \varepsilon$$

for all $n \geq N$. □

Definition. A metric space X is *complete* if every Cauchy sequence in X is convergent to some point of X.

It should be acknowledged at once that completeness is not a rare or exceptional property. Indeed, all of the metric spaces of classical analysis are complete.

Example A. The metric space $\mathbb{R}$ is complete. If $\{t_n\}$ is a Cauchy sequence in $\mathbb{R}$, then $\{t_n\}$ is bounded and therefore possesses a convergent subsequence (Ex. 6L). But then $\{t_n\}$ is itself convergent. The metric space $\mathbb{C}$ is also complete. (If $\{\alpha_n\}$ is a Cauchy sequence in $\mathbb{C}$, then both of the sequences $\{\text{Re } \alpha_n\}$ and $\{\text{Im } \alpha_n\}$ are Cauchy in $\mathbb{R}$ and therefore convergent in $\mathbb{R}$. But then $\{\alpha_n\}$ is convergent in $\mathbb{C}$.) More generally, for much the same reasons, the spaces $\mathbb{R}^n$ and $\mathbb{C}^n$ of Examples 6A and 6B are complete. A normed space that is complete as a metric space (Prob. 6A) is a *Banach space*. Thus $\mathbb{R}^n$ and $\mathbb{C}^n$ are, respectively, real and complex Banach spaces.

Example B. If Z is a set and (X, ρ) is a complete metric space, then the space $\mathcal{B}(Z; X)$ of all bounded mappings of Z into X is complete in the metric ρ_∞ of uniform convergence (Ex. 6H). Indeed, if $\{\phi_n\}$ is a sequence in $\mathcal{B}(Z; X)$ that is Cauchy with respect to ρ_∞, then the sequence $\{\phi_n(z)\}$ is Cauchy, and therefore convergent, in X for each z in Z. Let us write $\phi(z) = \lim_n \phi_n(z)$ for all z in Z. We shall show that $\phi \in \mathcal{B}(Z; X)$ and that $\lim_n \rho_\infty(\phi, \phi_n) = 0$, thus proving the assertion. To this end let ε be an arbitrarily prescribed positive number, and let N be an index such that $\rho_\infty(\phi_m, \phi_n) < \varepsilon$ whenever $m, n \geq N$. Then for an arbitrary element z of Z we have $\rho(\phi_m(z), \phi_n(z)) < \varepsilon$ for all $m, n \geq N$ and, letting n tend to infinity, we conclude that

$$\rho(\phi_m(z), \phi(z)) \leq \varepsilon, \quad m \geq N,$$

for each z in Z, since the closed ball $\{x \in X : \rho(\phi_m(z), x) \leq \varepsilon\}$ is a closed set (Ex. 6M), and must therefore contain $\phi(z)$. But from this we see at once that ϕ is indeed bounded (if $M =$ diam $\phi_N(Z)$, then diam $\phi(Z) \leq M+2\varepsilon$), and also that $\rho_\infty(\phi, \phi_m) \leq \varepsilon$ whenever $m \geq N$. Thus $\phi_m \to \phi$ in the metric ρ_∞.

Conversely, if the space (X, ρ) is *not* complete, and if—say—$\{x_n\}$ is a Cauchy sequence in X that fails to converge, then the sequence $\{\phi_n\}$ of constant mappings obtained by setting $\phi_n(z) = x_n, z \in Z$, is also Cauchy in the metric ρ_∞, and it is readily seen that the sequence $\{\phi_n\}$ is not convergent in that metric. Thus if Z is nonempty, *the space $\mathcal{B}(Z; X)$ is complete in the metric of uniform convergence if and only if the space X is complete.*

Example C. It is an immediate consequence of the foregoing example that the space (ℓ_∞) of all bounded complex sequences introduced in Problem 6D is complete. Indeed, (ℓ_∞) in the metric defined by the norm $\| \ \|_\infty$ coincides with the space $\mathcal{B}(\mathbb{N}_0; \mathbb{C})$ in the metric of uniform convergence. It is also true that the metric spaces $(\ell_p), 1 \leq p < +\infty$, introduced in Problem 6D are complete. The argument goes as follows. Fix $p, 1 \leq p < +\infty$, and suppose $\{x_n\}_{n=1}^\infty$ is a Cauchy sequence in (ℓ_p), where $x_n = \{\xi_m^{(n)}\}_{m=0}^\infty, n \in \mathbb{N}$. Then $\{x_n\}$ is certainly Cauchy, and therefore convergent, termwise, to some sequence $x = \{\xi_m\}_{m=0}^\infty$. Let ε be an arbitrary positive number, and let N be a positive integer such that $\|x_m - x_m\|_p < \varepsilon$ for all $m, n \geq N$. Then for each index k it is the case that

$$\sum_{i=0}^{k} \left|\xi_i^{(m)} - \xi_i^{(n)}\right|^p < \varepsilon^p$$

for all $m, n \geq N$. Hence, letting n tend to infinity, we see that

$$\sum_{i=0}^{k} \left|\xi_i^{(m)} - \xi_i\right|^p \leq \varepsilon^p$$

for every index k and all indices m such that $m \geq N$. But then, letting k tend to infinity, we have

$$\sum_{i=0}^{\infty} \left|\xi_i^{(m)} - \xi_i\right|^p \leq \varepsilon^p \tag{1}$$

for every positive integer m such that $m \geq N$. This shows, in the first place, that $x_m - x$ belongs to (ℓ_p) for all $m \geq N$, and hence that x belongs to (ℓ_p) as well. In the second place, (1) shows that $\|x_m - x\|_p \leq \varepsilon$ whenever $m \geq N$, and hence that $\{x_n\}$ converges to x in the metric on (ℓ_p) defined by the norm $\| \ \|_p$. Thus all of the spaces $(\ell_p), 1 \leq p \leq +\infty$, are Banach spaces.

Even if a given metric space fails to be complete it can always be embedded in one that is. The details are set forth in the following theorem.

Theorem 8.2. *For any metric space (X, ρ) there exists a complete metric space $(\widehat{X}, \widehat{\rho})$ and an isometry α of X onto a dense subspace of $\widehat{X}$. Moreover, the pair $\{(\widehat{X}, \widehat{\rho}), \alpha\}$ is unique in the sense that if $\{(\widetilde{X}, \widetilde{\rho}), \widetilde{\alpha}\}$ is another pair with the same properties (that is, if $\widetilde{X}$ is complete and $\widetilde{\alpha}$ is an isometry of X onto a dense subspace of $\widetilde{X}$), then there exists a unique isometry Φ of $\widehat{X}$ onto $\widetilde{X}$ such that $\Phi \circ \alpha = \widetilde{\alpha}$. (The essentially unique pair $(\widehat{X}, \alpha)$ is called the* completion *of X.)*

PROOF. If $\{x_n\}_{n=1}^\infty$ and $\{y_n\}_{n=1}^\infty$ are any two Cauchy sequences in (X, ρ), then the sequence $\{\rho(x_n, y_n)\}_{n=1}^\infty$ is Cauchy—and therefore convergent—in $\mathbb{R}$ (Prob. 6I). Thus if $\mathcal{C}$ denotes the collection of all Cauchy sequences in X, we may, and do, define a nonnegative real-valued function σ on $\mathcal{C} \times \mathcal{C}$ by setting

$$\sigma(\{x_n\}, \{y_n\}) = \lim_n \rho(x_n, y_n),$$

and it is readily verified that σ is a pseudometric on $\mathcal{C}$. Hence if we factor out the equivalence relation induced by σ (that is, the relation $\sim$ defined by setting $\{x_n\} \sim \{y_n\}$ when and only when $\sigma(\{x_n\}, \{y_n\}) = 0$, which is clearly just the relation of equiconvergence on $\mathcal{C}$ (Prob. 6M)), and define, for an arbitrary pair $[\{x_n\}]$ and $[\{y_n\}]$ of equivalence classes of sequences,

$$\widehat{\rho}\left([\{x_n\}], [\{y_n\}]\right) = \sigma\left(\{x_n\}, \{y_n\}\right),$$

then $\widehat{\rho}$ is a metric—the metric associated with σ—on the quotient space $\widehat{X} = \mathcal{C}/\sim$ (Prop. 6.20).

We next define a mapping $\alpha : X \to \widehat{X}$ by setting $\alpha(x) = [\{x, x, \ldots\}]$—the equivalence class of the constant sequence at x—for each point x of X, and observe that α is obviously an isometry of (X, ρ) into $(\widehat{X}, \widehat{\rho})$. (Equivalently, the value of α at a point x of X may be described as the element of $\widehat{X}$ consisting of all those sequences in X that converge to x.) According to this definition, if x is a point of X and $\{x_n\}$ an arbitrary element of $\mathcal{C}$, then $\widehat{\rho}(\alpha(x), [\{x_n\}]) = \sigma(\{x, x, \ldots\}, \{x_n\}) = \lim_n \rho(x, x_n)$. In particular, if for a positive number ε an index N is chosen so that $\rho(x_m, x_n) < \varepsilon$ for all $m, n \geq N$, then

$$\rho(\alpha(x_m), [\{x_n\}]) \leq \varepsilon, \quad m \geq N.$$

Thus $\lim_m \alpha(x_m) = [\{x_n\}]$, which shows that every point of $\widehat{X}$ is an adherent point of $\alpha(X)$, and hence that the latter is dense in $\widehat{X}$.

Moreover, this same observation shows that $\widehat{X}$ is complete. Indeed if $\{y_n\}_{n=1}^\infty$ is an arbitrary Cauchy sequence in $\alpha(X)$, then there exists a unique sequence $\{x_n\}$ in X such that $\alpha(x_n) = y_n, n \in \mathbb{N}$, and it is obvious that this sequence is Cauchy. But then, as we have just seen, $\lim_n y_n = [\{x_n\}]$. Thus every Cauchy sequence in $\alpha(X)$ converges in $\widehat{X}$, whence it follows that $\widehat{X}$ is complete (see Problem B).

To complete the proof, suppose $(\widetilde{X}, \widetilde{\rho})$ is another complete metric space and that $\widetilde{\alpha}$ is an isometry of X onto a dense subspace of $\widetilde{X}$. In the first place, if Φ is an isometry of $\widehat{X}$ into $\widetilde{X}$ such that $\Phi \circ \alpha = \widetilde{\alpha}$, then Φ agrees with $\widetilde{\alpha} \circ \alpha^{-1}$ on the dense subspace $\alpha(X)$, and is therefore unique if it exists (Cor. 7.6). But also, if we write $\Phi_0 = \widetilde{\alpha} \circ \alpha^{-1}$, then Φ_0 is an isometry of the dense subspace $\alpha(X)$ of $\widehat{X}$ onto the dense subspace $\widetilde{\alpha}(X)$ of $\widetilde{X}$. Moreover, if y is an arbitrary point of $\widehat{X}$, and if $\{y_n\}$ and $\{y'_n\}$ are any two sequences in $\alpha(X)$ converging to y, then $\{\Phi_0(y_n)\}$ and $\{\Phi_0(y'_n)\}$ are convergent and equiconvergent in $\widetilde{X}$. Thus to each point y of $\widehat{X}$ there corresponds a unique point $\widetilde{y}$ in $\widetilde{X}$ such that if $\{y_n\}$ is an arbitrary sequence in $\alpha(X)$ that converges to y, then $\{\Phi_0(y_n)\}$ converges in $\widetilde{X}$ to $\widetilde{y}$. The mapping Φ assigning $\widetilde{y}$ to $y, y \in \widehat{X}$, is an extension of Φ_0 to all of $\widehat{X}$, and is easily seen to be an isometry of $\widehat{X}$ into $\widetilde{X}$. Finally, if $\widetilde{z}$ is an arbitrary point of $\widetilde{X}$, and if $\{z_n\}$ is a sequence in $\widetilde{\alpha}(X)$ that converges to $\widetilde{z}$, then $\{\Phi_0^{-1}(z_n)\}$ is a Cauchy sequence in $\alpha(X)$, and if $y_0 = \lim_n \Phi_0^{-1}(z_n)$, then $\Phi(y_0)$ clearly coincides with $\widetilde{z}$. Thus Φ maps $\widehat{X}$ onto $\widetilde{X}$, and the proof is complete. (The mapping Φ is clearly the result of extending Φ_0 by continuity. In this context see also Example M.) □

Note. While it is sometimes advantageous to maintain the above point of view—namely, that the completion of a metric space is a pair $(\widehat{X}, \alpha)$ consisting of an abstract complete metric space $\widehat{X}$ and an isometry α of X onto a dense subspace of $\widehat{X}$, it is more usual to choose one such pair—say the one constructed from Cauchy sequences as in the proof of the foregoing theorem (but in this context see Problem C)—and use the isometry α to identify each point of the subspace $\alpha(X)$ with the corresponding point of X. Once this is done, the metric space X itself is literally a dense subspace of its completion, and the isometry α is replaced by the inclusion mapping of X into $\widehat{X}$. It is this latter point of view that we shall ordinarily take in the sequel.

Example D. The real number system $\mathbb{R}$ may be viewed as the completion of the system $\mathbb{Q}$ of rational numbers. Indeed, this notion may be pursued to obtain an alternate method of constructing $\mathbb{R}$ from $\mathbb{Q}$, with Cauchy sequences of rational numbers taking the place of Dedekind cuts.

The following simple result characterizes those subspaces of a complete metric space that are themselves complete.

Proposition 8.3. *A closed subspace of a complete metric space is complete. Conversely, a complete subspace of an arbitrary metric space X is necessarily closed in X.*

PROOF. If X is complete and if F is closed in X, then any Cauchy sequence in F is convergent in X and therefore in F. If A is a complete subspace of an arbitrary metric space X, and if $\{x_n\}$ is a sequence in A that converges in X, then $\{x_n\}$ is Cauchy—and therefore convergent—in A, so A is closed in X. □

Example E. Let f be a homeomorphism of the real line onto some bounded open interval (a, b)—say the standard mapping $f(t) = t/(1 + |t|)$ of $\mathbb{R}$ onto $(-1, +1)$ (cf. Example 7M). If we use f to define

$$\sigma(f(s), f(t)) = |s - t|, \quad s, t \in \mathbb{R},$$

then it is obvious that σ is a new metric on (a, b) with respect to which f becomes an isometry, so that $((a, b), \sigma)$ inherits all of the properties of $\mathbb{R}$ as a metric space. In particular, (a, b) is complete with respect to the metric σ. But also, f preserves the convergence of sequences, whence it is clear that σ is equivalent to the standard (relative) metric on (a, b), with respect to which (a, b) is not complete. This example shows that completeness is emphatically not a topological property of a metric space. (For a much deeper and more refined discussion of these ideas see Example O below.)

The points made here can be stated more generally. If ϕ is a one-to-one mapping of any set X onto a metric space (Y, ρ), and if ϕ is used to define a metric σ on X as above, then ϕ becomes an isometry between (X, σ) and (Y, ρ), and (X, σ) automatically inherits all of the properties of the metric space (Y, ρ). Moreover, if X comes equipped with a given metric, then the new metric σ is equivalent to that given metric if and only if ϕ was a homeomorphism to begin with.

There are several other versions of the Cauchy criterion that are frequently useful. We begin with one phrased in terms of filter bases.

Definition. A filter base $\mathcal{B}$ in a metric space X is *Cauchy*, or satisfies the *Cauchy criterion*, if $\inf\{\operatorname{diam} B : B \in \mathcal{B}\} = 0$.

Proposition 8.4. *A metric space (X, ρ) is complete if and only if every Cauchy filter base in X is convergent in X.*

PROOF. Suppose first that the stated criterion is satisfied and let $\{x_n\}_{n=1}^{\infty}$ be a Cauchy sequence in X. If we write $T_k = \{x_n : n \geq k\}$, then the system $\mathcal{T} = \{T_k : k \in \mathbb{N}\}$ of all such tails is a filter base in X that is clearly

Cauchy along with the given sequence $\{x_n\}$. Hence there is a point a_0 in X such that $\lim \mathcal{T} = a_0$. But then, according to the definition, the sequence $\{x_n\}$ is eventually in every open ball $D_\varepsilon(a_0)$, $\varepsilon > 0$, so $x_n \to a_0$. Thus X is complete.

Suppose next that X is complete and that $\mathcal{B}$ is a Cauchy filter base in X. For each positive integer n let B_n be a set belonging to $\mathcal{B}$ such that diam $B_n < 1/n$, and let x_n be a point of B_n. Then $\rho(x_m, x_n) < 1/m + 1/n$ (since B_m and B_n must intersect) for all pairs m, n of positive integers. Hence the sequence $\{x_n\}$ is Cauchy, and therefore convergent, in X. Let $\lim_n x_n = a_0$, and let ε be a positive number. There exists a positive integer N such that $\rho(x_n, a_0) < \varepsilon/2$ for all $n \geq N$, and if n is any positive integer greater than $N \vee (2/\varepsilon)$, we have $x_n \in D_{\varepsilon/2}(a_0)$ and also diam $B_n < \varepsilon/2$, whence it follows from the triangle inequality that $B_n \subset D_\varepsilon(a_0)$. Thus $\lim \mathcal{B} = a_0$, and the proof is complete. □

Example F. Let X be a complete metric space and let $\{F_n\}_{n=1}^\infty$ be a decreasing sequence of nonempty closed sets in X such that diam $F_n \to 0$. The collection $\mathcal{F}$ of sets belonging to the sequence $\{F_n\}$ is a filter base in X (Ex. 7T) and that filter base is obviously Cauchy. Hence there exists a point x_0 in X such that $\lim \mathcal{F} = x_0$, and it follows that x_0 is an adherent point of each set F_n (Prob. 7U). Thus

$$x_0 \in \bigcap_{n=1}^\infty F_n$$

since the sets F_n are all closed. But then

$$\bigcap_{n=1}^\infty F_n = \{x_0\}$$

since the intersection $\bigcap_n F_n$ clearly has diameter zero, and therefore cannot contain more than one point. Thus we have the following theorem: *If $\{F_n\}$ is an arbitrary decreasing sequence of nonempty closed subsets of a complete metric space, and if* $\lim_n$ diam $F_n = 0$, *then $\bigcap_n F_n$ is necessarily a singleton.* (In this context it is instructive to consider the sequence $F_n = [n, +\infty), n \in \mathbb{N}$, in the metric space $\mathbb{R}$. The sets F_n are closed and the sequence $\{F_n\}$ is nested, but $\bigcap_n F_n$ is empty. Thus dropping the assumption diam $F_n \to 0$ in this result may allow the intersection $\bigcap_n F_n$ to have more than one element, or to have no elements at all. It is also worth noting that the condition here set forth is, in fact, equivalent to the completeness of the metric space X. For if $\{x_n\}$ is a Cauchy sequence in X and $T_n = \{x_k : k \geq n\}$ for each index n, then $\{T_n^-\}$ is a nested sequence of closed sets with diam $T_n^- \to 0$, and if $\{x_0\} = \bigcap_{n=1}^\infty T_n^-$, then $x_n \to x_0$.)

We next consider two additional sequential criteria for completeness.

Definition. Let (X,ρ) be a metric space and let $\{x_n\}$ be a sequence of points in X. The sequence $\{x_n\}$ is said to be of *bounded variation* in X if $\sum_n \rho(x_n,x_{n+1}) < +\infty$. Likewise, we shall say that $\{x_n\}$ *satisfies condition* (C) with respect to a positive number M and a number r such that $0<r<1$ if $\rho(x_n,x_{n+1}) \le \mathrm{M}\mathrm{r}^n$ for each index n.

Lemma 8.5. *Every sequence in a metric space (X,ρ) that satisfies condition (C) with respect to some positive number M and some number r such that $0<r<1$ is of bounded variation in X. Likewise, every sequence of bounded variation in X is Cauchy.*

PROOF. The first assertion is an immediate consequence of the fact that $\sum_{n=1}^{\infty} Mr^n = Mr/(1-r) < +\infty$. To prove the second, let $\{x_n\}_{n=1}^{\infty}$ be of bounded variation, let $A = \sum_{n=1}^{\infty} \rho(x_n,x_{n+1})$, and let ε be an arbitrary positive number. There exists an index N such that

$$\sum_{n=1}^{N} \rho(x_n,x_{n+1}) > A-\varepsilon,$$

and if m is an index such that $m>N$ then

$$\sum_{k=m}^{m+p} \rho(x_k,x_{k+1}) \le \sum_{k=N+1}^{m+p} \rho(x_k,x_{k+1}) < \varepsilon$$

for every positive integer p. But then if $m,n>N$ and if, say, $m<n$, we have

$$\rho(x_m,x_n) \le \sum_{k=m}^{n-1} \rho(x_k,x_{k+1}) < \varepsilon,$$

so $\{x_n\}$ is a Cauchy sequence. □

Proposition 8.6. *The following conditions are equivalent for any metric space (X,ρ):*

(1) *X is complete,*
(2) *Every sequence of bounded variation in X converges in X,*
(3) *Every sequence in X that satisfies condition (C) with respect to some $M>0$ and some r such that $0<r<1$ converges in X,*
(4) *For some one positive number M_0 and some particular $r_0, 0<r_0<1$, every sequence in X satisfying condition (C) with respect to M_0 and r_0 converges in X.*

PROOF. It is an immediate consequence of Lemma 8.5 that (1) implies (2) and (2) implies (3), and it is obvious that (3) implies (4). To complete the proof suppose (4) is satisfied and let $\{x_n\}$ be a Cauchy sequence in X. For each positive integer k there is an index N_k such that $\rho(x_m, x_n) < M_0 r_0^k$ whenever $m, n \geq N_k$, and if we define the sequence $\{k_n\}_{n=1}^{\infty}$ inductively, setting $k_1 = N_1$ and $k_{n+1} = N_{n+1} \vee (k_n + 1)$, then the subsequence $\{x_{k_n}\}$ satisfies the condition $\rho(x_{k_n}, x_{k_{n+1}}) < M_0 r_0^n, n \in \mathbb{N}$, and is therefore convergent. But then the original sequence $\{x_n\}$ must be convergent too (Prop. 8.1). □

Example G (Method of Successive Approximation). A mapping ϕ of a metric space (X, ρ) into itself is *strongly contractive* if it is Lipschitzian with Lipschitz constant r for some $r < 1$. Let ϕ be a strongly contractive mapping of X into itself and suppose X is complete. Starting with an arbitrary point x_0 of X we define a sequence $\{x_n\}_{n=0}^{\infty}$ inductively, setting $x_1 = \phi(x_0)$, $x_2 = \phi(x_1)$, etc. Then for each positive integer n, $\rho(x_n, x_{n+1}) \leq r\rho(x_{n-1}, x_n)$, and it follows at once by mathematical induction that

$$\rho(x_n, x_{n+1}) \leq r^n \rho(x_0, x_1), \quad n \in \mathbb{N}.$$

Thus $\{x_n\}$ satisfies condition (C) with respect to r and any positive number M exceeding $\rho(x_0, x_1)$. Since X is complete, it follows from the preceding result that $\{x_n\}$ converges in X. Furthermore, if $a_0 = \lim_n x_n$, then

$$\phi(a_0) = \lim_n \phi(x_n) = \lim_n x_{n+1} = a_0,$$

so a_0 is a fixed point for ϕ. Moreover, if a_1 is some other fixed point for ϕ, then $\rho(a_0, a_1) = \rho(\phi(a_0), \phi(a_1)) \leq r\rho(a_0, a_1)$, which implies that $\rho(a_0, a_1) = 0$, and hence that $a_0 = a_1$. Thus *a strongly contractive mapping of a nonempty complete metric space X into itself possesses a unique fixed point a_0, and a_0 is the limit of every sequence of the form $\{x, \phi(x), \phi(\phi(x)), \ldots\}$.*

Example H. A linear transformation T of a normed space $\mathcal{E}$ into itself is strongly contractive if and only if $\|T\| < 1$ (recall Example 7D). Suppose that T is such a strongly contractive linear transformation of $\mathcal{E}$ into itself. Then for an arbitrary fixed vector x_0 in $\mathcal{E}$ the *affine* mapping

$$Ax = Tx + x_0, \quad x \in \mathcal{E},$$

is also strongly contractive ($\|Ax - Ay\| = \|Tx - Ty\| \leq \|T\| \, \|x - y\|$). Hence if $\mathcal{E}$ is complete, there exists a unique vector y_0 in $\mathcal{E}$ such that $y_0 = Ty_0 + x_0$, or, in other words, such that

$$(1_{\mathcal{E}} - T)y_0 = x_0.$$

Thus the linear transformation $1_{\mathcal{E}} - T$ is a linear isomorphism of the Banach space $\mathcal{E}$ onto itself whenever T is strongly contractive. In particular, then, this is true when T is a linear transformation of $\mathbb{R}^n$ into itself that is defined by a matrix A with Hilbert-Schmidt norm $N(A) < 1$ (see Example 7E). (It should be noted that the condition $N(A) < 1$ is merely *sufficient* in order for T to be strongly contractive. Moreover, the same conclusion holds if a linear transformation T on $\mathbb{R}^n$ is strongly contractive with respect to some metric equivalent to the Euclidean metric; see Problem F.) Other applications of the method of successive approximation are touched on in the problems.

Notation and terminology. When X and Y are metric spaces, it is natural to consider the collection $\mathcal{C}(X;Y)$ of all continuous mappings of X into Y, as well as the collection $\mathcal{C}_b(X;Y)$ of the bounded mappings in $\mathcal{C}(X;Y)$. According to Corollary 7.14 this latter *space of bounded continuous mappings* of X into Y coincides with the closed set of continuous mappings in the metric space $\mathcal{B}(X;Y)$ of Proposition 6.2.

It is an important fact in mathematical analysis that the completeness of a metric space Y implies that of various spaces of functions taking their values in Y. In this connection the following result is basic.

Proposition 8.7. *If X and Y are metric spaces ($X \neq \varnothing$), then the space $\mathcal{C}_b(X;Y)$ of bounded continuous mappings of X into Y is complete (in the metric of uniform convergence) if and only if Y is complete.*

PROOF. The space $\mathcal{B}(X;Y)$ of all bounded mappings of X into Y is complete if Y is (Ex. B), and $\mathcal{C}_b(X;Y)$ is a closed subset of that complete space (Cor. 7.14). Thus the condition is sufficient. Its necessity follows from the same construction used in Example B to show that $\mathcal{B}(X;Y)$ is not complete unless Y is. □

Example I. We continue the discussion of the generalized Cantor set of Example 6P. Let $[a,b]$ ($a < b$) be a real interval, let c and d be real numbers, and let θ be a proper ratio ($0 < \theta < 1$). Starting with the real-valued function ℓ that varies linearly from c to d across $[a,b]$, we construct a new function ℓ_θ as follows. Let $\{a < a' < b' < b\}$ be the partition of $[a,b]$ in which $[a',b']$ is the central θth part of $[a,b]$ ($a' = (1/2)[a+b-\theta(b-a)]$, $b' = (1/2)[a+b+\theta(b-a)]$). Then ℓ_θ varies linearly from c to $(c+d)/2$ across $[a,a']$, from $(c+d)/2$ to d across $[b',b]$, and is constantly equal to $(c+d)/2$ across $[a',b']$. (The function ℓ_θ is said to be obtained from ℓ by *flattening the central* θth *portion* of ℓ. Figure 4 shows the graphs of ℓ and $\ell_{1/2}$ in the case $[a,b] = [c,d] = [0,1]$.) It is apparent that ℓ_θ is continuous along with ℓ, and that ℓ and ℓ_θ agree at a, b and the midpoint $(a+b)/2$ (and nowhere else, unless $c = d$, in which case $\ell_\theta = \ell$). It is also readily seen that $\|\ell - \ell_\theta\|_\infty = \theta|d-c|/2$.

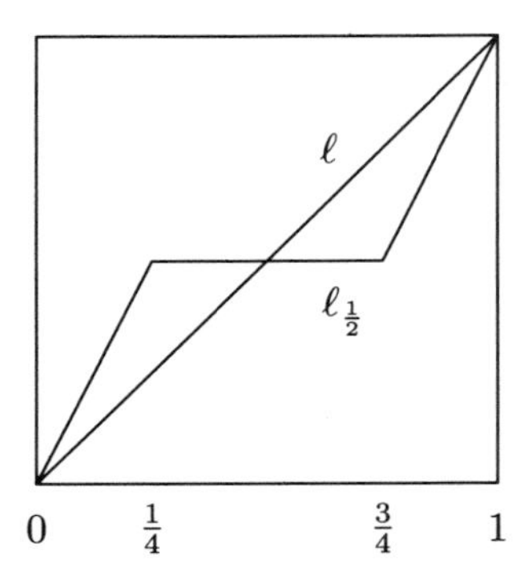

Figure 4

Next, starting from a continuous function p that is piecewise linear with respect to a partition $\mathcal{P} = \{a = t_0 < \ldots < t_N = b\}$ of $[a, b]$ (so that $p_i = p|[t_{i-1}, t_i]$ is linear for each $i = 1, \ldots, N$; see Example 7N), we construct a new function p_θ by replacing each linear segment p_i of p by the corresponding function $p_{i,\theta}$ obtained by flattening the central θth portion of $p_i, i = 1, \ldots, N$. It is again clear that p_θ is continuous along with p, that p_θ is piecewise linear with respect to a refinement $\mathcal{P}'$ of $\mathcal{P}$, and that p and p_θ agree at the partition points of $\mathcal{P}$ as well as on any subinterval of $\mathcal{P}$ on which p is constant. Moreover, if $\omega(p; \mathcal{P})$ denotes the oscillation of p over $\mathcal{P}$ (Prob. 7X), then

$$\omega(p_\theta; \mathcal{P}') \le \omega(p; \mathcal{P})/2 \quad \text{and} \quad \|p - p_\theta\|_\infty \le (\theta/2)\omega(p; \mathcal{P}).$$

Figure 5 illustrates the case $\theta = 1/3$, where $p = \ell_{1/3}$ and ℓ denotes, once again, the identity mapping on $[0, 1]$.

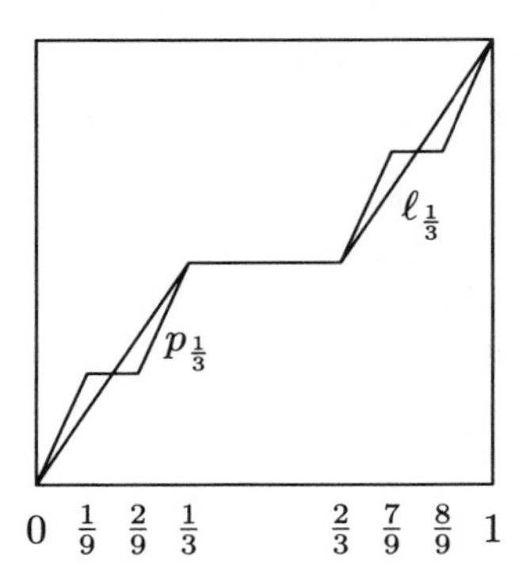

Figure 5

Suppose now that $\{\theta_n\}_{n=0}^{\infty}$ is a sequence of proper ratios. We define inductively a corresponding sequence $\{p_n\}_{n=0}^{\infty}$ of functions on $[0, 1]$, setting $p_0(t) \equiv t, 0 \leq t \leq 1$, and supposing p_n defined,

$$p_{n+1} = (p_n)_{\theta_n}, \quad n \in \mathbb{N}_0.$$

To see how the sequence $\{p_n\}$ develops, let us write $\mathcal{P}_n$ for the partition of $[0, 1]$ consisting of the 2^{n+1} endpoints of the 2^n intervals constituting the system $\mathcal{F}_n$ of Example 6P. Then, in the first place, p_n is continuous, monotone increasing and piecewise linear with respect to $\mathcal{P}_n$. Moreover if for each index n we denote by $\mathcal{U}_n$ the system of $(2^n - 1)$ bounded open intervals contiguous to $F_n = \bigcup \mathcal{F}_n$ (these are just the interiors of the even numbered subintervals of $\mathcal{P}_n$), then $\mathcal{U}_n$ is in one-to-one correspondence with the system of dyadic fractions $j/2^n, j = 1, \ldots, 2^n - 1$, in a unique order preserving fashion, and, for each index j, p_n is constantly equal to $j/2^n$ on the corresponding interval in $\mathcal{U}_n$. (All of these facts are readily established by mathematical induction.) Finally we have $\omega(p_n; \mathcal{P}_n) = 1/2^n$, and therefore

$$\|p_n - p_{n+1}\|_\infty = (\theta_n/2)\, \omega\,(p_n; \mathcal{P}_n) = \frac{\theta_n}{2^{n+1}} < \frac{1}{2^{n+1}},$$

for all n. Thus the sequence $\{p_n\}$ is of bounded variation and is therefore uniformly convergent to a limit $h_{\{\theta_n\}}$ on $[0, 1]$.

Concerning the limit $h_{\{\theta_n\}}$ we already know that it is continuous and it is obviously monotone increasing. As for the rest, the order isomorphisms between the systems $\mathcal{U}_n$ and dyadic fractions extend (being coherent) to an order isomorphism between the collection $\mathcal{U}(= \bigcup_n \mathcal{U}_n)$ of all the bounded open intervals contiguous to the Cantor set $C_{\{\theta_n\}}$ and the system of *all* dyadic fractions in $(0, 1)$, and if U is an interval in $\mathcal{U}$ corresponding to a fraction $k/2^m$ in this way, then p_n is constantly equal to $k/2^m$ on U for all $n \geq m$, so $h_{\{\theta_n\}}$ is too. But $h_{\{\theta_n\}}$ is then nothing but the extension to $[0, 1]$ by continuity of the restriction h_0 of $h_{\{\theta_n\}}$ to the union $\bigcup \mathcal{U}$.

We note, finally, that the classical Cantor-Lebesgue function of Example 7Q is obtained via this construction for the sequence $\{\theta_n\} = \{1/3, 1/3, \ldots\}$. The gist of the present example is that the only thing special about the classical Cantor set C in the construction of the Cantor-Lebesgue function was its extremely simple arithmetic structure. (Still other constructions of this sort can be based on Example 7J.)

Example J. In much the same fashion the Peano space-filling curve of Example 7R may be realized as the limit of a uniformly convergent sequence of piecewise linear functions. To begin with, let $x = (s_1, s_2)$ and $y = (t_1, t_2)$ be points in $\mathbb{R}^2$ and let ℓ denote the linear parametrization on the parameter interval $[a, b]$ of the line segment joining x to y. If $s_1 < t_1$ and $s_2 < t_2$, we shall say that ℓ *slants upward to the right*. In this case (and in this case only) we define two modifications of ℓ as follows. Let a' and b' be the points in $[a,b]$ that partition it into three congruent subintervals ($a' = (1/3)(2a+b)$ and $b' = (1/3)(a+2b)$). Then $\ell^{(x)}$ is defined to be the piecewise linear extension of the mapping of the partition $\{a < a' < b' < b\}$ (Ex. 7N) that agrees with ℓ at a and b (so that $\ell^{(x)}(a) = x$ and $\ell^{(x)}(b) = y$) and that carries a' to $((s_1+t_1)/2, t_2)$ and b' to $((s_1+t_1)/2, s_2)$. Similarly, $\ell^{(y)}$ is defined to be the piecewise linear extension of the mapping of the partition $\{a < a' < b' < b\}$ into $\mathbb{R}^2$ that carries these points into $x, (t_1, (s_2+t_2)/2)$, $(s_1, (s_2+t_2)/2)$, y, respectively. (If ℓ does not slant upward to the right, then, by definition, $\ell^{(x)} = \ell^{(y)} = \ell$.)

Next let π be a continuous mapping of $[a,b]$ into $\mathbb{R}^2$ that is piecewise linear with respect to some partition $\mathcal{P} = \{a = u_0 < \ldots < u_N = b\}$. We first define $\pi^{(x)}$ to be the mapping obtained by replacing each of the linear segments $\pi_i = \pi|[u_{i-1}, u_i], i = 1, \ldots, N$, by the curve $\pi_i^{(x)}$, and we define $\pi^{(y)}$, similarly, to be the curve obtained by replacing each π_i by $\pi_i^{(y)}$. Finally, we denote by $\pi^\ddagger$ the result $\pi^{(x)(y)}$ of applying these two modifications to π one after the other.

Concerning $\pi^\ddagger$ we observe that, for each subinterval $[u_{i-1}, u_i]$ of $\mathcal{P}$, the curve $(\pi^\ddagger)_i = \pi^\ddagger|[u_{i-1}, u_i]$ coincides with $(\pi_i)^\ddagger = (\pi_i)^{(x)(y)}$—the result of applying the two basic operations to the single line segment π_i. It follows that if some segment π_i of π fails to slant upward to the right, then $(\pi^\ddagger)_i = \pi_i$, and, in general, that the behavior of $\pi^\ddagger$ may be determined by investigating the case in which π consists of a single line segment.

Suppose then, once again, that ℓ is the (upward slanting to the right) linear parametrization on the interval $[a,b]$ of the line segment joining x to y in $\mathbb{R}^2$, and let R denote the rectangle having ℓ for a diagonal. If $\mathcal{F} = \{I_1, I_2, I_3, I_4\}$ is the system of subintervals of $[a,b]$ obtained by removing the central third of $[a,b]$ twice (so that $\mathcal{F} = \{[a,b]\}^{**}$ in the notation of Example 6O), and if $\mathcal{P}_0$ is the partition of $[a,b]$ consisting of the eight endpoints of these four subintervals, then $\ell^\ddagger$ is piecewise linear with respect to $\mathcal{P}_0$. Moreover, the subintervals of $\mathcal{P}_0$ on which the restrictions of $\ell^\ddagger$ slant upward to the right are precisely the four intervals I_1, I_2, I_3 and I_4, and the restrictions of $\ell^\ddagger$ to these four intervals are linear parametrizations of the (upward slanting to the right) diagonals of the four rectangles obtained by bisecting the sides of R. Finally, we observe that $\|\ell - \ell^\ddagger\|_\infty \leq \frac{2}{3}\|x-y\|_2$, a fact that is disclosed by direct calculation. (If ℓ does not slant upward to the right, then $\ell^\ddagger = \ell$, so $\|\ell - \ell^\ddagger\|_\infty = 0$.) Figure 6 shows $\ell^\ddagger$ for $[a,b] = [0,1]$ and with vertices labeled with the appropriate values of the parameter.

Suppose now, once again, that π is a continuous mapping of $[a,b]$ into $\mathbb{R}^2$ that is piecewise linear with respect to $\mathcal{P} = \{a = u_0 < \ldots < u_n = b\}$. If we define $N(\pi)$ to be the largest of the numbers $\|\pi(u_i) - \pi(u_{i-1})\|_2$ among all of those subintervals of $\mathcal{P}$ on which π slants upward to the right, it is at once clear that $N(\pi^\ddagger) = \frac{1}{2}N(\pi)$ and also that $\|\pi - \pi^\ddagger\|_\infty \leq (2/3)N(\pi)$. Moreover, π and $\pi^\ddagger$ coincide on any subinterval $[u_{i-1}, u_i]$ of $\mathcal{P}$ on which π

does not slant upward to the right, and if a rectangle contains one of the segments $\pi|[u_{i-1}, u_i]$, it also contains the subcurve $\pi^{\ddagger}|[u_{i-1}, u_i]$.

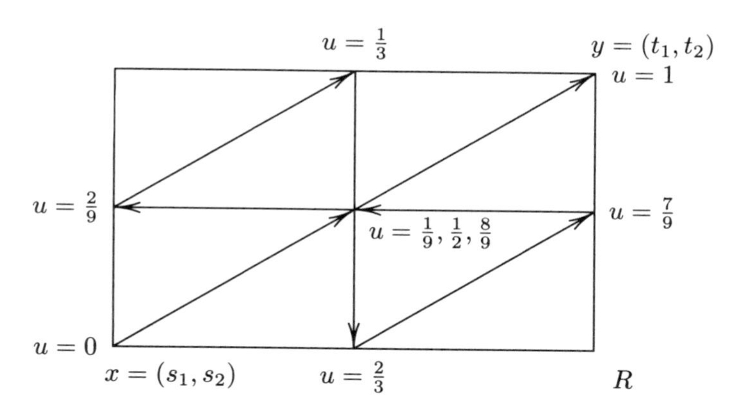

Figure 6

Finally, we define a sequence $\{\pi_n\}_{n=0}^{\infty}$ of piecewise linear mappings of the unit interval into $\mathbb{R}^2$ by mathematical induction, setting $\pi_0(u) = (u, u)$ for $0 \leq u \leq 1$ and, supposing π_n already defined, setting $\pi_{n+1} = (\pi_n)^{\ddagger}$. According to the estimates given above we have $N(\pi_n) = \sqrt{2}/2^n$ and $\|\pi_n - \pi_{n+1}\|_{\infty} \leq \frac{2}{3}N(\pi_n)$ for each index n, and it follows that $\{\pi_n\}$ satisfies condition (C) with respect to $M = 2\sqrt{2}/3$ and $r = \frac{1}{2}$, and therefore converges uniformly to a continuous mapping ψ of $[0, 1]$ into $\mathbb{R}^2$ (Props. 8.6 and 8.7). But the mapping π_n may also be described as the piecewise linear extension of a certain mapping of the partition $\mathcal{P}_{2n}$ of $[0, 1]$ consisting of the endpoints of the various intervals $I_{\varepsilon_1,\ldots,\varepsilon_{2n}}$ employed in the construction of the Cantor set. Moreover a straightforward mathematical induction shows that π_k—and therewith every $\pi_n, n \geq k$, and hence ψ as well—agrees on $\mathcal{P}_{2k}$ with the space-filling curve defined in Example 7R, whence it follows that ψ coincides with that curve.

We next turn our attention to a version of the Cauchy criterion especially tailored for infinite series.

Definition. An infinite series $\sum_{n=0}^{\infty} a_n$ in a normed space $\mathcal{E}$ is said to satisfy *condition* (S) if for each positive number ε there exists a nonnegative integer N such that

$$\left\| \sum_{k=m+1}^{m+p} a_k \right\| < \varepsilon$$

for all p and all m such that $m \geq N$.

Proposition 8.8. *Every convergent infinite series in a normed space $\mathcal{E}$ satisfies condition (S). The space $\mathcal{E}$ is complete if and only if every infinite series in $\mathcal{E}$ that satisfies condition (S) is convergent in $\mathcal{E}$.*

PROOF. Let $s_n = \sum_{k=0}^{n} a_k$ denote the nth partial sum of the series $\sum_{k=0}^{\infty} a_k$. If m and n are nonnegative integers such that $m < n$, then $\|s_n - s_m\| = \|\sum_{k=m+1}^{n} a_k\|$. Thus the stated condition is equivalent to the standard version of the Cauchy criterion for the sequence $\{s_n\}$. □

The following generalization of the well-known fact that an absolutely convergent series of scalars converges is an immediate consequence of the foregoing result. (Here again it is easily seen that the stated condition is also sufficient for the completeness of a normed space $\mathcal{E}$.)

Corollary 8.9. *Let $\mathcal{E}$ be a Banach space and let $\{a_n\}_{n=0}^{\infty}$ be a sequence in $\mathcal{E}$ such that $\sum_{n=0}^{\infty} \|a_n\| < +\infty$. Then the series $\sum_{n=0}^{\infty} a_n$ is convergent in $\mathcal{E}$.*

Example K. Let $\mathcal{E}$ be a complex Banach space, let $\{a_n\}_{n=0}^{\infty}$ be a sequence in $\mathcal{E}$ and suppose, for some positive number r_0, that the sequence $\{\|r_0^n a_n\|\}$ is bounded in $\mathbb{R}$ by a positive number M. If $0 < r_1 < r_0$ and λ and λ_0 are complex numbers such that $|\lambda - \lambda_0| \leq r_1$, then

$$\|(\lambda - \lambda_0)^n a_n\| = \|r_0^n a_n\|(|\lambda - \lambda_0|^n / r_0^n) \leq M(r_1/r_0)^n.$$

Hence the sup norm of the function $(\lambda - \lambda_0)^n a_n$ is dominated by $M(r_1/r_0)^n$, $n \in \mathbb{N}_0$, and it follows that the power series

$$\sum_{n=0}^{\infty} (\lambda - \lambda_0)^n a_n \tag{2}$$

is uniformly convergent on the disc $D_{r_1}(\lambda_0)^- = \{\lambda \in \mathbb{C} : |\lambda - \lambda_0| \leq r_1\}$ (i.e., convergent in the metric of uniform convergence on $\mathcal{B}(D_{r_1}(\lambda_0)^-; \mathcal{E})$). Since r_1 was an arbitrary positive number less than r_0, this also shows that the series (2) is convergent in $\mathcal{E}$ for each complex number λ in the open disc $D_{r_0}(\lambda_0)$. From these observations it follows, via a standard argument (which we omit), that for an arbitrary sequence $\{a_n\}_{n=0}^{\infty}$ of *coefficients* in $\mathcal{E}$, and for an arbitrary *center* λ_0, the $\mathcal{E}$-valued power series (2) possesses a *radius of convergence* $R, 0 \leq R \leq +\infty$, having the property that (2) converges for all λ such that $|\lambda - \lambda_0| < R$ and converges for no λ such that $|\lambda - \lambda_0| > R$. (This statement is to be interpreted appropriately if $R = 0$ or $R = +\infty$; moreover, the behavior of (2) as regards convergence on the circle $|\lambda - \lambda_0| = R$ is not touched on here.)

The theory of limits at a point for a mapping taking values in a metric space Y is also considerably richer when Y is complete.

Proposition 8.10. *Let ϕ be a mapping of a subset A of a metric space X into a complete metric space Y. Then the limit $\lim_{\substack{x \to a \\ x \in A}} \phi(x)$ exists at a point a of A^- if and only if the oscillation $\omega(\phi; a)$ of ϕ at a is zero (Prob. 7X).*

PROOF. The oscillation $\omega(\phi; a)$ vanishes at a point a of A^- if and only if the filter base $\phi(D_r(a) \cap A), r > 0$, is Cauchy; the limit $\lim_{\substack{x \to a \\ x \in A}} \phi(x)$ exists if and only if the same filter base is convergent. Thus the stated result follows at once from Proposition 8.4. □

Corollary 8.11. *Let ϕ, A, X, and Y be as in the preceding proposition, and let $\widetilde{A}$ denote the set of all those points of A^- at which the limit of ϕ through A exists (cf. Proposition 7.26). Then $\widetilde{A}$ is a G_δ in X. In particular, if ϕ is continuous on A, then the result of extending ϕ by continuity is a continuous extension of ϕ to a G_δ in X.*

PROOF. As was just seen, $\widetilde{A} = \{a \in A^- : \omega(\phi; a) = 0\}$, and this set is a G_δ in X (Prob. 7X). □

Example L. For the function g_0 of Example 7X the set $\widetilde{A}$ of those positive real numbers at which the limit exists is the set of all positive irrational numbers, clearly a G_δ in the set of all positive real numbers, since it is the complement of the countable set $\mathbb{Q}_+$ of positive rationals.

When the mapping ϕ is uniformly continuous, one can say considerably more.

Proposition 8.12. *Suppose ϕ is a uniformly continuous mapping of a subset A of a metric space (X, ρ) into a complete metric space (Y, σ). Then the subset of X to which ϕ extends by continuity coincides with A^-, a closed subset of X.*

PROOF. Let ε be a positive number, let $\delta > 0$ be chosen so that $\rho(x, x') < \delta$ implies $\sigma(\phi(x), \phi(x')) < \varepsilon$ for any two points x and x' in A, and let a be a point of A^-. If x and x' belong to $D_{\delta/2}(a)$, then $\rho(x, x') < \delta$, so $\sigma(\phi(x), \phi(x')) < \varepsilon$. Thus diam $\phi(D_{\delta/2}(a) \cap A) \leq \varepsilon$, whence it follows that $\omega(\phi; a) = 0$, and hence that $\lim_{\substack{x \to a \\ x \in A}} \phi(x)$ exists. □

Example M. Suppose, even more particularly, that ϕ is a Lipschitzian mapping with Lipschitz constant M of a subset A of a metric space (X, ρ)

into a complete metric space (Y, σ), and let $\widetilde{\phi}$ be the extension by continuity of ϕ to A^-. If a and a' are any two points of A^-, and if $\{x_n\}$ and $\{x'_n\}$ are sequences in A such that $x_n \to a$ and $x'_n \to a'$, then $\lim_n \sigma(\phi(x_n), \phi(x'_n)) = \sigma(\widetilde{\phi}(a), \widetilde{\phi}(a'))$ (Prob. 6I). But also $\sigma(\phi(x_n), \phi(x'_n)) \leq M\rho(x_n, x'_n)$ for each index n, which shows that $\lim_n \sigma(\phi(x_n), \phi(x'_n)) \leq M \lim_n \rho(x_n, x'_n) = M\rho(a, a')$. Thus $\sigma(\widetilde{\phi}(a), \widetilde{\phi}(a')) \leq M\rho(a, a')$, so the extension $\widetilde{\phi}$ is Lipschitzian with respect to the same Lipschitz constant M. Like considerations show that the extension by continuity of an isometry of a subspace A of a metric space X into a complete metric space Y is an isometry of A^- into Y, and similarly as regards a mapping that is Lipschitz-Hölder continuous on A with respect to some exponent (Prob. 7K).

The completeness of the codomain Y is absolutely essential for the validity of any of the immediately foregoing results.

Example N. Let $X = \mathbb{R}$, let $Y = \mathbb{Q}$, and consider the identity mapping ι of $\mathbb{Q}(\subset X)$ onto Y. The isometric function ι cannot be extended by continuity to a single point of $\mathbb{R} \backslash \mathbb{Q}$, and $\mathbb{Q}$ is not a G_δ in $\mathbb{R}$. (The truth of this latter remark is far from obvious; see Problem J, for example.)

Suppose now that A and B are subsets of complete metric spaces X and Y, respectively, and that ϕ is a homeomorphism of A onto B. Then ϕ can be extended by continuity onto a G_δ in X, and ϕ^{-1}: $B \to A$ can also be extended by continuity onto a G_δ in Y, but there is no reason to suppose that either of these extensions is one-to-one (and simple examples show that both may fail to be one-to-one). Nevertheless, we have the following remarkable result.

Proposition 8.13 (Theorem of Lavrentiev [17]). *Let X and Y be complete metric spaces, and let ϕ be a homeomorphism of a subset A of X onto a subset B of Y. Then there exists an extension $\widehat{\phi}$ of ϕ that is a homeomorphism of a set $\widehat{A} \subset A^-$ in X onto a set $\widehat{B} \subset B^-$ in Y where $\widehat{A}$ and $\widehat{B}$ are G_δs in X and Y, respectively.*

Proof. Let $\widetilde{\phi}$ denote, as usual, the result of extending ϕ by continuity onto $\widetilde{A}$, a G_δ in X (Cor. 8.11), and let us write $\widetilde{\psi}$ for the result of extending ϕ^{-1} by continuity onto $\widetilde{B}$, a G_δ in Y. If $\widetilde{a}$ is a point of $\widetilde{A}$ and $\mathcal{B}$ is an arbitrary filter base in A that converges to $\widetilde{a}$, then $\phi(\mathcal{B})$ is a convergent filter base in B^-, and $\lim \phi(\mathcal{B}) = \widetilde{\phi}(\widetilde{a})$. (Indeed, $\mathcal{B}$ is also a filter base in $\widetilde{A}$, and $\widetilde{\phi}$ is continuous, so $\lim \widetilde{\phi}(\mathcal{B}) = \widetilde{\phi}(\widetilde{a})$ (Prob. 7V), but $\widetilde{\phi}(\mathcal{B}) = \phi(\mathcal{B})$ since $\widetilde{\phi}|A = \phi$.) The point $\widetilde{\phi}(\widetilde{a})$ may belong to $\widetilde{B}$ or not, but if it does so, then $\phi(\mathcal{B})$ is a filter base in B converging to it, so, by symmetry, $\phi^{-1}(\phi(\mathcal{B}))$ must converge to $\widetilde{\psi}(\widetilde{\phi}(\widetilde{a}))$. But $\phi^{-1}(\phi(\mathcal{B})) = \mathcal{B}$ since ϕ is one-to-one, and it follows that $\widetilde{\psi}(\widetilde{\phi}(\widetilde{a})) = \widetilde{a}$. Thus every point $\widetilde{a}$ of $\widehat{A} = \widetilde{\phi}^{-1}(\widetilde{B})$ possesses the property that $(\widetilde{\psi} \circ \widetilde{\phi})(\widetilde{a}) = \widetilde{a}$. Likewise, again by symmetry, $(\widetilde{\phi} \circ \widetilde{\psi})(\widetilde{b}) = \widetilde{b}$ for every point $\widetilde{b}$ of $\widehat{B} = \widetilde{\psi}^{-1}(\widetilde{A})$. Moreover, this shows that $\widetilde{\phi}(\widehat{A}) \subset \widehat{B}$ and $\widetilde{\psi}(\widehat{B}) \subset \widehat{A}$, whence it follows that $\widehat{\phi} = \widetilde{\phi}|\widehat{A}$ and $\widehat{\psi} = \widetilde{\psi}|\widehat{B}$ are mutually inverse homeomorphisms between $\widehat{A}$ and $\widehat{B}$, and the proof will be complete once we verify that $\widehat{A}$ and $\widehat{B}$ are G_δs. But $\widehat{A}$ is the inverse image of a G_δ, namely $\widetilde{B}$, under the continuous mapping $\widetilde{\phi}$, whence it follows at once that $\widehat{A}$ is a G_δ in $\widetilde{A}$ (since the formation of inverse images preserves countable intersections). Thus $\widehat{A}$ is a G_δ in a G_δ in X, and is therefore a G_δ in X itself (Prob. 6P). Similarly $\widehat{B}$ is a G_δ in Y. □

Example O. Suppose a subset A of a metric space X is homeomorphic to some complete metric space Y, and let ϕ be a homeomorphism of A onto Y. Even though X need not be complete as given, we can complete it, and regard A as a subset of the completion $\widehat{X}$ (Prop. 8.2). Then according to the foregoing result (with $B = Y$), the given homeomorphism ϕ can be extended to a homeomorphism of a G_δ subset $\widehat{A}$ of $\widehat{X}$ onto Y. But ϕ is already a one-to-one mapping of A onto *all* of Y and thus cannot be extended to any properly larger one-to-one mapping. Hence $\widehat{A} = A$, so A is a G_δ in $\widehat{X}$. But if $\{U_n\}_{n=1}^\infty$ is a sequence of open sets in $\widehat{X}$ such that $A = \bigcap_{n=1}^\infty U_n$, and if $V_n = U_n \cap X$, $n \in \mathbb{N}$, then the sets V_n are open in X, and $A = \bigcap_{n=1}^\infty V_n$. Thus A is also a G_δ in X.

Conversely, it is not hard to show that if A is a G_δ in a complete metric space (X, ρ), then A can be given a new metric σ such that σ is equivalent to the relative metric $\rho|(A \times A)$ on A, and such that the metric space (A, σ) is complete (see Problem D). But then A is obviously homeomorphic to a complete metric space via the identity mapping. Thus the following three conditions are equivalent for a subset A of a metric space X:

(1) A is a G_δ in the completion of X,
(2) A is homeomorphic to a complete metric space,
(3) A can be given an equivalent metric with respect to which it is itself complete.

Such a metric space is known as an *absolute* G_δ. (If A is such an absolute G_δ, and if A is mapped homeomorphically onto some subset B of another metric space Y, then it follows from what we have just shown that B is a G_δ in Y, no matter whether Y is complete or not.)

We conclude our treatment of the theory of complete metric spaces with a brief introduction to a very important classification of the subsets of metric spaces discovered by R. Baire [1]. The root notion is of a kind of set that is inconsequential in a very precise sense.

Definition. A subset N of a metric space X is said to be *nowhere dense* in X if the closure N^- has no interior points, i.e., if $(N^-)^\circ = \varnothing$; recall Problem 6Q.

Here is an alternate characterization of nowhere dense sets that is frequently useful. The precise formulation of the following proposition was chosen with a view to future application.

Proposition 8.14. *A set N is nowhere dense in a metric space X if and only if, for every nonempty open set U in X, there exists a nonempty open set V such that $V \subset U$ and $V \cap N = \varnothing$. Moreover, if N is nowhere dense, it is always possible to arrange for V to be an open ball with arbitrarily small positive radius satisfying the stronger conditions $V^- \subset U$ and $V^- \cap N = \varnothing$.*

PROOF. If N is not nowhere dense, and U is a nonempty open set contained in N^-, then every nonempty open subset V of U clearly meets N. This proves the sufficiency of the given condition. To prove the necessity, set $V = U \backslash N^-$. If $V = \varnothing$, then N is not nowhere dense. Otherwise V is a nonempty open subset of U that is disjoint from N. Thus the necessity of the given condition is proved. To prove the last assertion observe that if x_0 is any point of V, then there exists a radius $r > 0$ such that $D_r(x_0) \subset V$. But then, if ε is a positive number less than r, the ball $D_\varepsilon(x_0)$ satisfies all the desired conditions. □

On the basis of this result it is easily seen that any finite union of nowhere dense sets is itself nowhere dense. This is far from true, of course, for infinite unions, and we are thus led to the distinction introduced by Baire.

Definition. A subset A of a metric space X is of *first category* in X if it can be expressed as the union of a countable collection of sets each of which is nowhere dense in X. A subset B of X is of *second category* in X if it is not of first category in X.

The following result is an almost trivial consequence of the various definitions.

Proposition 8.15. *If a subset A of a metric space X is of first category in X [nowhere dense in X], and if M is a subset of A, then M is also of first category in X [nowhere dense in X]. Dually, if a subset A of X is of second category in X, and if $A \subset B \subset X$, then B is also of second category in X.*

PROOF. If N is nowhere dense in X and $M \subset N$, then $(M^-)^\circ \subset (N^-)^\circ =$

$\varnothing$. Thus M is also nowhere dense in X, and the stated facts follow at once. □

The next result is not quite so obvious as the last, but is still far from deep.

Proposition 8.16. *If a set M is of first category in a subspace A of a metric space X, then M is also of first category in X. Dually, if A is of second category in X, and if B is a subspace of X such that $A \subset B$, then A is also of second category in B.*

PROOF. Suppose N is a subset of A that is not nowhere dense in X, and let $U = (N^-)^\circ$, so that $U \neq \varnothing$. Then $U \subset N^- \subset A^-$, whence it follows that $U \cap A \neq \varnothing$. But $U \cap A$ is open relative to A and is obviously contained in $N^- \cap A$, the closure of N relative to A (Prop. 6.15). Thus N is also not nowhere dense in A, and the result follows easily. □

Corollary 8.17. *If a metric space X is of first category in itself, then X is automatically of first category in any metric space containing it as a subspace. If a subset B of X is of second category in X, then B is also of second category in itself.*

It is important to recognize that each of these three concepts has to do with a relation between a metric space and its various subsets, and not with any intrinsic property of the subsets themselves.

Example P. No nonempty set in a metric space is nowhere dense in itself. A singleton is nowhere dense in any metric space containing it provided it is not an isolated point of that space. Hence an arbitrary countable subset of a metric space X is of first category in X unless it contains one or more isolated points of X. (An isolated singleton is actually of second category in an arbitrary metric space containing it—a good indication that in studying the theory of *Baire categories* we will do well to stick to metric spaces that are dense in themselves (Prob. 6O).) Thus the set $\mathbb{Q}$ of rational numbers is of first category in $\mathbb{R}$; indeed $\mathbb{Q}$ is of first category in itself. By contrast (as we shall see) the set $\mathbb{R}\backslash\mathbb{Q}$ of irrational numbers is of second category in $\mathbb{R}$ and accordingly therefore of second category in itself.

If a subset N of a metric space X is nowhere dense in X, then N has empty interior, and is therefore all boundary ($N \subset \partial N$). The converse is false, of course, but there are a lot of sets with *nowhere dense boundaries*. Thus every closed set F in X has nowhere dense boundary ($((\partial F)^-)^\circ = (\partial F)^\circ \subset F^\circ$), and so therefore does every open set (since a subset and its complement share a common boundary).

A line in the plane $\mathbb{R}^2$ is nowhere dense there. Hence the union of an arbitrary countable collection of points, line segments, rays and straight lines is a first category subset of $\mathbb{R}^2$—and likewise of any $\mathbb{R}^n, n \geq 2$. On the other hand, as we shall see, a nondegenerate interval is of second category in $\mathbb{R}$.

Any subset of a finite union of nowhere dense sets in a metric space X is itself nowhere dense in X, a fact expressed by saying that the nowhere dense sets constitute an *ideal* in 2^X. On the other hand, any subset of a *countable* union of sets of the first category in X is itself of first category in X, which is expressed by saying that the first category subsets of X form a σ*-ideal* in 2^X.

The essential importance of the rather technical distinction between sets of first and second category is the means it provides for establishing a wide assortment of existence and nonexistence theorems. The simplest version of such a *category argument* is presented in the following proposition.

Proposition 8.18. *Let B be a set of second category in a metric space X, and let $\{M_n\}_{n=1}^{\infty}$ be a countable covering of B consisting of subsets of X. Then $B \cap (M_n^-)^\circ \neq \varnothing$ for at least one index n.*

PROOF. If $B \cap (M_n^-)^\circ = \varnothing$ for every n, then B is covered by the sequence $\{M_n^- \backslash (M_n^-)^\circ = \partial(M_n^-)\}$, and each of these sets is closed and nowhere dense. □

Even the following much weaker version of this principle is frequently useful.

Corollary 8.19. *Let the metric space X be of second category in itself, and suppose given a countable closed covering $\{F_n\}_{n=1}^{\infty}$ of X. Then one (at least) of the sets F_n has nonempty interior. Dually, if $\{G_n\}$ is an arbitrary sequence of dense open subsets of X, then $\bigcap_{n=1}^{\infty} G_n$ is of second category in X. (In particular then, $\bigcap_{n=1}^{\infty} G_n$ is not empty.)*

Example Q. A mapping ψ of a metric space (X, ρ) into a metric space (Y, σ) is said to be a mapping of *Baire class one* of X into Y if there exists a sequence $\{\phi_n\}_{n=1}^{\infty}$ of continuous mappings of X into Y that converges to ψ pointwise on X (i.e., such that $\lim_n \phi_n(x) = \psi(x)$ for each x in X). In general there is not much to be said about such a mapping (see Problem K), but if, as we assume henceforth in this example, the domain X is of second category in itself, then a good deal can be said.

Indeed, suppose $\{\phi_n\}$ is a sequence of continuous mappings of X into Y that converges pointwise to ψ. To begin with, according to a standard, and by now familiar, argument, the real-valued function $f_{n,k}$ defined by setting

$$f_{n,k}(x) = \sigma(\phi_n(x), \phi_k(x)), \quad x \in X,$$

is continuous on X for each pair n, k of positive integers. Moreover, the sequence $\{\phi_n(x)\}$ is convergent—and therefore bounded—in Y for each fixed x in X. Hence if we define, for each index n,

$$g_n(x) = \sup_{k \geq n} f_{n,k}(x), \quad x \in X,$$

the function g_n is finite-valued and lower semicontinuous on X, and for each point x of X the sequence $\{g_n(x)\}$ is monotone decreasing to zero in $\mathbb{R}$.

Suppose next that η is some positive number, and let

$$G_n = \{x \in X : g_n(x) > \eta\}, \quad n \in \mathbb{N}.$$

Then, by what has just been observed, $\{G_n\}_{n=1}^{\infty}$ is a decreasing sequence of open sets in X such that $\bigcap_{n=1}^{\infty} G_n = \varnothing$. Hence there is an index N such that G_N is not dense in X, so $V = X \backslash G_N^-$ is a nonempty open subset of X. This set V has the property that if $x \in V$, then $\sigma(\phi_k(x), \phi_N(x)) \leq \eta$ for all $k \geq N$, whence it follows that $\sigma(\psi(x), \phi_N(x)) \leq \eta$ as well.

Now let ε be a second positive number and let a_0 denote a point of V. Since the mapping ϕ_N is continuous at a_0, there exists a positive number δ such that $\rho(x, a_0) < \delta$ implies $\sigma(\phi_N(x), \phi_N(a_0)) < \varepsilon/2$. Moreover, by reducing δ if necessary, we may arrange matters so that $D_\delta(a_0) \subset V$. But then, for any two points x and x' of $D_\delta(a_0)$, we have $\sigma(\phi_N(x), \phi_N(x')) < \varepsilon$ and therefore

$$\begin{aligned}\sigma(\psi(x), \psi(x')) &\leq \sigma(\psi(x), \phi_N(x)) + \sigma(\phi_N(x), \phi_N(x')) + \sigma(\phi_N(x'), \psi(x')) \\ &< \eta + \varepsilon + \eta.\end{aligned}$$

Thus diam $\psi(D_\delta(a_0)) \leq 2\eta + \varepsilon$, and it follows that the oscillation of the mapping ψ satisfies the inequality $\omega(\psi; a_0) \leq 2\eta$. Finally, since a_0 denotes an arbitrary point of V, we have $\omega(\psi; x) \leq 2\eta$ for every point x of V. Thus we have shown that *if ψ is an arbitrary mapping of Baire class one on a metric space that is of second category in itself, then the open set $\{x \in X : \omega(\psi; x) < \varepsilon\}$ is nonempty for each positive number ε.*

The foregoing observations suggest that it would be of major interest to find metric spaces having a substantial number of second category subsets. In this connection the following theorem is of paramount importance.

Theorem 8.20 (Baire Category Theorem). *Let X be a complete metric space, and let U be a nonempty open subset of X. Then U is of second category in X.*

Proof. Let $\{N_n\}_{n=1}^{\infty}$ be an arbitrary sequence of nowhere dense sets in X. It suffices to show that if $A = \bigcup_{n=1}^{\infty} N_n$, then $U \backslash A \neq \varnothing$. Note first that by Proposition 8.14 there exists an open ball $D_1 = D_{r_1}(x_1)$ with radius $r_1 < 1$ such that $D_1^- \subset U$ and $D_1^- \cap N_1 = \varnothing$. Continuing via mathematical induction, one easily constructs a nested sequence

$$D_1 \supset D_2 \supset \ldots \supset D_n \supset \ldots$$

of open balls $D_n = D_{r_n}(x_n)$ such that $r_n < 1/n$ and such that $D_n^- \cap N_n = \varnothing$ for every n. The sequence of centers $\{x_n\}$ is then obviously a Cauchy sequence, and since X is complete, $\{x_n\}$ must converge to some limit x_0. Since x_0 is also the limit of every tail $\{x_n\}_{n=m}^{\infty}$, and since this tail lies in D_m, it follows that $x_0 \in D_m^-$ for $m = 1, 2, \ldots$. Thus $x_0 \notin A$ and since $x_0 \in D_1^-$ and $D_1^- \subset U$, we have $x_0 \in U \backslash A$. □

Note. It is a remarkable fact that, while the hypotheses of the Baire category theorem are metric in nature, its conclusion is purely topological. Thus the theorem remains in force if the metric space X is merely assumed to be an absolute G_δ, i.e., to admit an equivalent metric with respect to which it is complete (recall Example O), and we are led to the following reformulation of Theorem 8.20, for which no proof is needed.

Theorem 8.21 (Baire Category Theorem; Version II). *Let X be a metric space that is an absolute G_δ. Then any nonempty open subset of X is of second category in X.*

Corollary 8.22. *Let X be a nonempty metric space that is an absolute G_δ and let $\{G_n\}_{n=1}^{\infty}$ be a sequence of dense open subsets of X. Then the set*

$$B = \bigcap_{n=1}^{\infty} G_n$$

is dense and of second category in X, and $X \backslash B$ is of first category in X.

Proof. Each complement $F_n = X \backslash G_n$ is nowhere dense in X by hypothesis. Hence $X \backslash B = \bigcup_{n=1}^{\infty} F_n$ is of first category in X, whence it follows at once that B is of second category in X (since X is not of first category in itself by Theorem 8.21). Moreover, if U is an arbitrary nonempty open set in X, then U is not contained in $X \backslash B$, so $U \cap B \neq \varnothing$. Thus B is dense in X. □

Corollary 8.23. *If X is a nonempty metric space that is an absolute G_δ and B is an arbitrary dense G_δ in X, then B is of second category in X and $X \backslash B$ is of first category in X.*

PROOF. If $B = \bigcap_{n=1}^{\infty} U_n$ where each U_n is open, then each U_n is also dense in X. □

Corollary 8.24. *Let X be a nonempty metric space that is an absolute G_δ, and let ϕ be a mapping of X into a metric space Y. If the set B of points of continuity of ϕ is dense in X, then $X \backslash B$ is of first category in X.*

PROOF. The set B is known to be a G_δ; see Problem 7X. □

Example R. Let X be a complete metric space, and let ψ be a mapping of Baire class one of X into a metric space Y, as defined in Example Q. If U is an arbitrary nonempty open subset of X, then $\psi|U$ is obviously of Baire class one as a mapping of U into Y. Hence if ε is an arbitrary positive number, there is a nonempty open subset of U on which the oscillation of $\psi|U$ is less than ε. But U is open, so the oscillation of $\psi|U$ is just the restriction to U of the oscillation of ψ on X. Consequently if we write $G_\varepsilon = \{x \in X : \omega(\psi; x) < \varepsilon\}$, then $G_\varepsilon \cap U \neq \varnothing$, which shows that G_ε is dense in X. In particular, if we set $V_n = G_{1/n}, n \in \mathbb{N}$, then $\{V_n\}_{n=1}^{\infty}$ is a decreasing sequence of dense open subsets of X, and the intersection $\bigcap_{n=1}^{\infty} V_n$ is precisely the set of points of continuity of ψ. Thus we have shown that *the set of points of continuity of a mapping ψ of Baire class one of a nonempty complete metric space X into a metric space Y is a dense G_δ of second category in X, the complementary set of points of discontinuity of ψ being of first category in X.*

Example S. Let X be a complete, separable metric space, let M be a countable dense set in X, and suppose that X is perfect, i.e., dense in itself. Then M is of first category in X, and therefore has empty interior, so the complement $X \backslash M$ is also dense in X.

Suppose next that S is an F_σ subset of X, and let $\{F_n\}_{n=1}^{\infty}$ be an increasing sequence of closed sets in X such that $S = \bigcup_{n=1}^{\infty} F_n$. We write A_n for the nth difference, $A_n = F_n \backslash F_{n-1}$ (setting $F_0 = \varnothing$), and define

$$B_n = (A_n \backslash A_n^\circ) \cup (A_n^\circ \cap M), \quad n \in \mathbb{N}.$$

Since $B_n \subset A_n$ for each n, the sequence $\{B_n\}_{n=1}^{\infty}$ is disjoint. Moreover each set B_n is of first category in X. Indeed, $A_n^\circ \cap M$ is countable, and therefore certainly of first category, while $A_n \backslash A_n^\circ \subset \partial A_n$, and, as a difference of closed sets, A_n has nowhere dense boundary (cf. Example P and Problem 6Q).

Finally, we define a real-valued function on X, setting

$$f(x) = \begin{cases} 1/n, & x \in B_n, \quad n \in \mathbb{N}, \\ 0, & x \notin \bigcup_n B_n. \end{cases}$$

(Equivalently, f is given by the (formally infinite) sum

$$f = \sum_{n=1}^{\infty} (1\backslash n)\chi_{B_n}.)$$

The function f is nonnegative and vanishes on the complement of the first category set $\bigcup_n B_n$, so $f(x) = 0$ on an everywhere dense set, and the lower envelope of f (Prop. 7.27) is accordingly identically zero on X.

On the other hand, as regards the upper envelope of f, we note first that if x_0 is a point of $X \backslash S$, then for each positive integer n there is an open set U containing x_0 such that $U \cap F_n = \varnothing$, and therefore such that $0 \leq f \leq 1/(n+1)$ on U. Thus $\lim_{x \to x_0} f(x) = 0$. But also, if x_0 belongs to S, and hence to some one of the sets A_n, then there is an open set U containing x_0 such that $U \cap F_{n-1} = \varnothing$, and on this set U we have $0 \leq f \leq 1/n$. But also, f actually assumes the extreme value $1/n$ on U, since either $x_0 \in A_n \backslash A_n^\circ$, in which case $f(x_0) = 1/n$, or $x_0 \in A_n^\circ$, in which case f is constantly equal to $1/n$ on the nonempty set $U \cap A_n^\circ \cap M$. Thus

$$\limsup_{x \to x_0} f(x) = \frac{1}{n},$$

which shows that the upper envelope of f is given by the formally infinite sum

$$\sum_{n=1}^{\infty} (1/n)\chi_{A_n}.$$

In particular, we see that f is continuous on $X \backslash S$ and discontinuous at every point of S. In view of Problem 7X, this shows that in a *perfect, separable, complete metric space X a set S is the set of points of discontinuity of a (real-valued) function on X if and only if S is an F_σ*. (It is perhaps worth remarking that if we take $X = \mathbb{R}_+$ and set $F_n = (1/n)\mathbb{N}$ for each index n, then the present example essentially reproduces the construction of Example 7X.)

We conclude this chapter with a discussion of some properties of a metric space that have to do with how "scattered" the space is.

Definition. A subset N of a metric space (X, ρ) is an *r-net* for a subset A of X if for every point x of A there is a point y in N such that $\rho(x, y) < r$.

Proposition 8.25. *The following four conditions are equivalent for an arbitrary subset A of a metric space X:*

(1) *For every positive number ε there is a finite ε-net for A,*
(2) *For every positive number ε there exists a finite covering of A consisting of sets of diameter less than ε,*

(3) *For every positive number ε there exists a finite partition of A into sets of diameter less than ε,*

(4) *For every positive number ε there exists a finite ε-net for A consisting of points of A, in short, a finite ε-net* in *A.*

PROOF. If N is a finite $(\varepsilon/3)$-net for A, then the set of balls $D_{\varepsilon/3}(x), x \in N$, constitutes a finite covering of A consisting of sets of diameter less than ε. Thus (1) implies (2), and it is clear that (2) implies (3) (Prob. 1U), and also that (3) implies (4) and (4) implies (1). □

Definition. A subset A of a metric space X is *totally bounded* if it possesses any one (and therefore all) of the properties set forth in Proposition 8.25.

Example T. A totally bounded subset of an arbitrary metric space X is automatically a bounded subset of X. In Euclidean space $\mathbb{R}^n$ it is the case, conversely, that every bounded set is actually totally bounded. Indeed if B is a bounded subset of $\mathbb{R}^n$, if a is the point $a = (s_1, \ldots, s_n)$, and if $D_R(a)$ is a ball in $\mathbb{R}^n$ large enough so that $B \subset D_R(a)$, then B is also contained in the cube $W = [s_1 - R, s_1 + R\,] \times \ldots \times [s_n - R, s_n + R\,]$ with edge $2R$. If each of the intervals $[s_i - R, s_i + R\,], i = 1, \ldots, n$, is partitioned into M subintervals of equal length, then the various products of all of these subintervals (one from each partition) provide a *cellular partition* of W into M^n subcubes, each of edge $2R/M$, and it is easily seen that the diameter of each of these subcubes is simply $2R\sqrt{n}/M$ (the diameter of a cube of edge e being $e\sqrt{n}$). Thus if, for a given positive ε, M is chosen large enough so that

$$\frac{2R\sqrt{n}}{M} < \varepsilon,$$

then the resulting cellular partition of W provides a finite covering of B satisfying condition (2) of Proposition 8.25.

Proposition 8.26. *A totally bounded metric space X is separable, and therefore satisfies the second axiom of countability.*

PROOF. For each positive integer n let N_n be a finite $(1/n)$-net for X. Then the set $M = \bigcup_{n=1}^{\infty} N_n$ is countable and is clearly dense in X, and the proposition follows (Th. 6.18). □

If A is a subset of a metric space (X, ρ) and if A is not totally bounded, then for some sufficiently small positive number ε_0 there exists no finite ε_0-net in A. Thus, in particular, A is nonempty, and if $x_1 \in A$, there also exists a point x_2 in A such that $\rho(x_1, x_2) \geq \varepsilon_0$. Suppose points $x_1, \ldots, x_k$ have already been chosen in A in such a way that $\rho(x_i, x_j) \geq \varepsilon_0, i \neq j$;

$i, j = 1, \ldots, k$. The set $\{x_1, \ldots, x_k\}$ is not an ε_0-net in A (since no finite ε_0-net exists for A), so there is a point in A—call it x_{k+1}—such that $\rho(x_i, x_{k+1}) \geq \varepsilon_0, i = 1, \ldots, k$. Thus by mathematical induction we construct a sequence $\{x_n\}_{n=1}^{\infty}$ in A with the property that $\rho(x_m, x_n) \geq \varepsilon_0$ for every pair of distinct indices m and n. (Such a sequence will be said to be *uniformly scattered*.) This proves one half of the following basic result.

Theorem 8.27. *A set A in a metric space X is totally bounded if and only if every infinite sequence $\{x_n\}_{n=1}^{\infty}$ in A possesses a Cauchy subsequence.*

Proof. As just noted, if A is not totally bounded, then it contains a uniformly scattered sequence, and it is clear that no subsequence of such a sequence can be Cauchy. Thus the condition is sufficient. To prove its necessity, suppose that A is a totally bounded subset of X, and that $\{x_n\}_{n=1}^{\infty}$ is an arbitrary sequence in A. According to Proposition 8.25, there exists a partition of A into a finite number of subsets, each having diameter less than one. Since these sets are finite in number, it is clear that there is (at least) one of them with the property that $\{x_n\}$ belongs to that subset infinitely often. Let A_1 denote any one such set in the partition, and let $\{x_k^{(1)} = x_{n_k}\}_{k=1}^{\infty}$ be a subsequence of $\{x_n\}$ lying in A_1. The set A_1 is also totally bounded (for it is evident that any subset of a totally bounded set is itself totally bounded). Hence there exists a partition of A_1 into a finite number of subsets each having diameter less than $1/2$. Once again, there must be at least one of these subsets with the property that $\{x_n^{(1)}\}$ belongs to it infinitely often. Let A_2 be such a set, and let $\{x_k^{(2)} = x_{n_k}^{(1)}\}$ be a subsequence of $\{x_n^{(1)}\}$ lying in A_2. Continuing in this fashion we obtain (by mathematical induction) a nested sequence $\{A_n\}_{n=1}^{\infty}$ of subsets of A with the property that diam $A_n \leq 1/n, n \in \mathbb{N}$, and a sequence $\{\{x_k^{(n)}\}_{k=1}^{\infty}\}_{n=1}^{\infty}$ of sequences in A with the properties that (a) all of the sequences $\{x_k^{(n)}\}_{k=1}^{\infty}$ are subsequences of the given sequence $\{x_n\}$, (b) each sequence $\{x_k^{(n)}\}_{k=1}^{\infty}, n \geq 2$, is a subsequence of its predecessor $\{x_k^{(n-1)}\}_{k=1}^{\infty}$, and (c) the subsequence $\{x_k^{(n)}\}_{k=1}^{\infty}$ lies in A_n.

Consider, finally, the *diagonal sequence* $\{x_n^{(n)}\}_{n=1}^{\infty}$ in the infinite array

$$\begin{array}{l}
x_1^{(1)}, x_2^{(1)}, \ldots, x_k^{(1)}, \ldots \\
x_1^{(2)}, x_2^{(2)}, \ldots, x_k^{(2)}, \ldots \\
\vdots \quad\; \vdots \qquad\quad \vdots \\
x_1^{(n)}, x_2^{(n)}, \ldots, x_k^{(n)}, \ldots \\
\vdots \quad\; \vdots \qquad\quad \vdots
\end{array}$$

This sequence is a subsequence of $\{x_n\}$ with the property that, for each positive integer m, some tail of $\{x_n^{(n)}\}$ is a subsequence of $\{x_k^{(m)}\}_{k=1}^{\infty}$, and therefore lies wholly in A_m. Thus $\{x_n^{(n)}\}$ is Cauchy. □

Note. The central feature of this argument—the extraction of ever finer subsequences from a given sequence, followed by the extraction of the diagonal sequence from the resulting infinite array—is known as the *diagonal process*. This is a powerful tool with numerous uses.

There are many contexts in mathematical analysis in which it is of great importance to know that a given sequence possesses a convergent subsequence. Accordingly, the following concept is of major interest.

Definition. A subset K of a metric space X is *compact* if every infinite sequence in K possesses a subsequence that converges to some point of K, or, equivalently, if every infinite sequence in K possesses a cluster point belonging to K (Prop. 6.5).

The following proposition contains interesting and useful information, but it is even more important for the illustration it provides of a couple of typical compactness arguments.

Proposition 8.28. *Let (X, ρ) be a metric space, and let K be a compact subset of X. Then K is closed and a subset of K is compact if and only if it is closed (in either K or X; see Corollary 6.16). Moreover, K is bounded and, if K is nonempty, then there exist points x_0 and y_0 in K such that $\rho(x_0, y_0) = \operatorname{diam} K$.*

PROOF. Suppose first that $\{x_n\}$ is a sequence in K that converges in X to some limit a_0. Since K is compact, there is a subsequence of $\{x_n\}$ that converges to a point of K, and this point must coincide with a_0 (Prop. 6.4). Thus $a_0 \in K$ and K is closed. It follows at once that if L is a compact subset of K, then L is also closed in X (and in K). Suppose, conversely, that F is a closed subset of K and $\{x_n\}$ is a sequence of points in F. Since K is compact, there exists a subsequence of $\{x_n\}$ that converges to a point a_0 of K, and since F is closed we have $a_0 \in F$. Thus F is compact.

To establish the remaining assertions of the proposition, let $\{(x_n, y_n)\}_{n=1}^{\infty}$ be a sequence of pairs of points in K such that $\{\rho(x_n, y_n)\}$ tends upward to diam K. Since K is compact, the sequence $\{x_n\}_{n=1}^{\infty}$ has a subsequence $\{x_{n_m}\}_{m=1}^{\infty}$ that converges to a point x_0 of K. Similarly, the sequence $\{y_{n_m}\}_{m=1}^{\infty}$ has a subsequence $\{y_{n_{m_k}}\}_{k=1}^{\infty}$ that converges to a point y_0 of K, and it is easily seen that $\lim_k x_{n_{m_k}} = x_0$ and $\lim_k \rho(x_{n_{m_k}}, y_{n_{m_k}}) = \rho(x_0, y_0)$. Hence diam $K = \rho(x_0, y_0) < +\infty$. □

It is an elementary, but still noteworthy, fact that compactness is equivalent to the conjunction of two properties introduced earlier.

Theorem 8.29. *A metric space X is compact if and only if it is both totally bounded and complete.*

PROOF. If X is both totally bounded and complete, and if $\{x_n\}$ is an infinite sequence in X, then, in the first place, $\{x_n\}$ possesses a Cauchy subsequence (Th. 8.27), and that subsequence is then convergent because X is complete. Thus X is compact. On the other hand, if X is compact, then every sequence in X possesses a subsequence that is convergent and therefore Cauchy, so X is totally bounded (Th. 8.27). Similarly, if X is compact and if $\{x_n\}$ is a Cauchy sequence in X, then $\{x_n\}$ possesses a subsequence that converges to a point a_0 of X. But then $\{x_n\}$ converges to a_0 also (Prop. 8.1), which shows that X is complete. □

Example U. Let us fix a number $p, 1 \leq p < +\infty$, and consider the Banach space (ℓ_p) (Ex. C). For each index m we define a function h_m on (ℓ_p) by setting, for each $x = \{\xi_n\}_{n=0}^{\infty}$ in (ℓ_p),

$$h_m(x) = \left[\sum_{n=m}^{\infty} |\xi_n|^p\right]^{1/p}$$

Thus $h_m(x)$ is just the norm of the mth tail of x, whence it is apparent that $h_m(x) \leq \|x\|_p$ and $h_m(x+y) \leq h_m(x) + h_m(y)$ for all x, y in (ℓ_p), and also that the sequence $\{h_m\}_{m=0}^{\infty}$ is pointwise monotone decreasing to zero on (ℓ_p). Moreover, if $\{x_n\}_{n=1}^{\infty}$ is a Cauchy sequence in (ℓ_p) and if ε is a positive number, then there is an index N such that $\|x_n - x_N\|_p < \varepsilon/2$ for all $n \geq N$. Since $x_n = (x_n - x_N) + x_N$, this implies that

$$h_m(x_n) \leq h_m(x_N) + \varepsilon/2$$

for $n \geq N$ and every index m. But then, since $\{h_m\}$ tends to zero at each of the vectors $x_1, \ldots, x_N$, it follows that there exists an index M such that $h_m(x_n) < \varepsilon$ for all n and all $m \geq M$. In other words, $\{h_m\}$ tends to zero *uniformly* on any Cauchy sequence in (ℓ_p).

This much said, it becomes a simple matter to determine which subsets of (ℓ_p) are totally bounded. To begin with, if $S \subset (\ell_p)$ and if S is totally bounded, then S must, of course, be bounded (in norm), and it is easy to see that $\{h_m\}$ must tend to zero uniformly on S. Indeed, if this is not the case, then there is a positive number ε_0 so small that for each index m there is a vector x_m in S such that $h_m(x_m) \geq \varepsilon_0$, and it is clear from the foregoing discussion that no subsequence of the sequence $\{x_m\}$ can be Cauchy. On the other hand, if these two conditions are satisfied by S, then it is easy to construct a finite ε-net for S for any prescribed $\varepsilon > 0$. We first choose M large enough so that $h_M(x) < \varepsilon/2^{1/p}$ for all x in S. We then set $\eta = \varepsilon/(2M)^{1/p}$ and select a finite η-net A in $\mathbb{C}$ for the disc $D_K(0)^-$ where K is a positive number large enough so that $\|x\|_p \le K$ for every x in S. Then the set of all sequences $\{\xi_n\}_{n=0}^\infty$ with the property that $\xi_n \in A$ for $n = 0, 1, \ldots, M-1$, and $\xi_n = 0$ for all $n \ge M$ is obviously finite, and is readily perceived to be an ε-net for S. (It may be noted that these two conditions can be paraphrased in the following way: If we define

$$H_m = \sup_{x \in S} h_m(x), \qquad m \in \mathbb{N}_0,$$

then S is totally bounded if and only if $H_0 < +\infty$ and $\lim_m H_m = 0$.)

Suppose now that $\{\xi_n^{(0)}\}_{n=0}^\infty$ is some one fixed vector in (ℓ_p), and we take for S the set of all those sequences $\{\xi_n\}_{n=0}^\infty$ such that $|\xi_n| \le |\xi_n^{(0)}|, n \in \mathbb{N}_0$. Then S is a subset of (ℓ_p), and it is clear from the preceding discussion that S is totally bounded. But S is also closed in (ℓ_p), and it follows that S is, in fact, compact. Thus, for example, the set Q of those sequences $x = \{\xi_n\}_{n=0}^\infty$ such that

$$|\xi_n| \le \frac{1}{n+1}, \qquad n \in \mathbb{N}_0,$$

is a compact subset of (ℓ_2). (The set Q is known as the *Hilbert parallelotope*.)

The property of compactness is of sufficient importance that it is extremely useful to have other criteria for it. The following result provides one such criterion.

Proposition 8.30 (Cantor's Theorem). *A metric space X is compact if and only if every decreasing sequence $\{F_n\}_{n=1}^\infty$ of nonempty closed subsets of X has the property that $\bigcap_{n=1}^\infty F_n$ is nonempty.*

PROOF. Let X be compact and suppose given a decreasing sequence $\{F_n\}_{n=1}^\infty$ of nonempty closed sets in X. For each index n let x_n be a point of F_n, and let a_0 be a cluster point of the resulting sequence $\{x_n\}_{n=1}^\infty$. For each index n this sequence belongs eventually to F_n, and the same is therefore true of any subsequence of $\{x_n\}$. This implies that $a_0 \in F_n$, and since n is arbitrary, it follows that $a_0 \in \bigcap_{n=1}^\infty F_n$. Thus the stated condition is necessary. Suppose next that the condition holds, and let $\{x_n\}_{n=1}^\infty$ be an arbitrary infinite sequence in X. The sequence $\{T_n\}_{n=1}^\infty$ of tails of $\{x_n\}$ ($T_n = \{x_m : m \ge n\}$) is nested, and so therefore is the sequence $\{T_n^-\}_{n=1}^\infty$. Hence by hypothesis $\bigcap_{n=1}^\infty T_n^- \ne \varnothing$. Let $a_0 \in \bigcap_{n=1}^\infty T_n^-$, and let ε be a positive number. The ball $D_\varepsilon(a_0)$ meets every tail T_n, so for every index n there are indices $m \geq n$ such that $x_m \in D_\varepsilon(a_0)$. In other words, $\{x_n\}$ is in $D_\varepsilon(a_0)$ infinitely often, and a_0 is thus a cluster point of $\{x_n\}$. □

There is also a very important criterion for the compactness of a metric space that is stated in terms of open sets.

Theorem 8.31. *A metric space X is compact if and only if every open covering of X contains a finite subcovering.*

PROOF. Suppose first that the condition is satisfied, and let $\{F_n\}_{n=1}^{\infty}$ be a decreasing sequence of nonempty closed sets in X. If $V_n = X \backslash F_n, n \in \mathbb{N}$, then $\{V_n\}_{n=1}^{\infty}$ is an increasing sequence of open sets such that $V_n \neq X$ for all n. But then $\bigcup_{n=1}^{\infty} V_n \neq X$, for otherwise, by hypothesis, there would have to be an index N such that $V_1 \cup \ldots \cup V_N = V_N = X$. Thus $\bigcap_{n=1}^{\infty} F_n \neq \varnothing$, so X is compact by Proposition 8.30.

Suppose next that X is compact, and let $\mathcal{U}$ be an arbitrary open covering of X. Since X satisfies the second axiom of countability (Prop. 8.26), there exists a countable subset $\mathcal{U}_0$ of $\mathcal{U}$ that also covers X (Prob. 6S). If $\mathcal{U}_0$ happens to be finite, then $\mathcal{U}_0$ itself is a finite subcovering of X; otherwise $\mathcal{U}_0$ is countably infinite and may be arranged into an infinite sequence $\{U_n\}_{n=1}^{\infty}$. Set $F_n = X \backslash (U_1 \cup \ldots \cup U_n), n \in \mathbb{N}$. Then $\{F_n\}_{n=1}^{\infty}$ is a decreasing sequence of closed sets, and $\bigcap_{n=1}^{\infty} F_n = \varnothing$. Hence, by Cantor's theorem (Prop. 8.30), some set F_N must be empty, and therefore $U_1 \cup \ldots \cup U_N = X$. Thus in either case $\mathcal{U}$ contains a finite subcovering. □

Corollary 8.32. *A subset K of a metric space X is compact as a subspace of X if and only if every covering of K by means of open subsets of X contains a finite subcovering.*

PROOF. If $\mathcal{U}$ is a covering of K consisting of open subsets of X, then $\widetilde{\mathcal{U}} = \{U \cap K : U \in \mathcal{U}\}$ is an open covering of the subspace K. Conversely, every covering of K consisting of relatively open subsets of K can be obtained in this fashion (Prop. 6.15). □

Corollary 8.33 (Heine-Borel Theorem). *A subset of Euclidean space $\mathbb{R}^n$ is compact if and only if it is closed and bounded. Thus every open covering of a closed and bounded set in $\mathbb{R}^n$ contains a finite subcovering.*

PROOF. A compact subset of any metric space is necessarily closed and bounded (Prop. 8.28). A closed and bounded subset of Euclidean space is complete (Prop. 8.3) and totally bounded (Ex. T), and is therefore compact by Theorem 8.29. The covering property follows by Corollary 8.32. □

Some of the most important theorems of mathematical analysis have to do with continuous mappings on compact metric spaces. The following result is one of those theorems.

Theorem 8.34. *A continuous mapping of a compact metric space (X, ρ) into an arbitrary metric space (Y, ρ') is uniformly continuous.*

PROOF. Let $\phi : X \to Y$ be continuous and suppose ϕ is not uniformly continuous. Then there exists a positive number ε_0 so small that for each positive integer n there exist points x_n and x'_n of X such that $\rho(x_n, x'_n) < 1/n$, while $\rho'(\phi(x_n), \phi(x'_n)) \geq \varepsilon_0$. The sequences $\{x_n\}$ and $\{x'_n\}$ thus obtained are obviously equiconvergent (Prob. 6M), and since X is compact, the sequence $\{x_n\}$ has a subsequence $\{x_{n_k}\}$ that converges to a limit a. But then $\{x'_{n_k}\}$ also converges to a, and it follows that $\lim_k \phi(x_{n_k}) = \lim_k \phi(x'_{n_k}) = \phi(a)$, which is impossible since $\rho'(\phi(x_{n_k}), \phi(x'_{n_k})) \geq \varepsilon_0$ for all k. □

Example V. Let α be a continuous mapping of a closed interval $[a, b]$ into a metric space (Y, ρ'). Then for any given positive number ε there exists a positive number δ such that if $\mathcal{P} = \{a = t_0 < \ldots < t_n = b\}$ is an arbitrary partition of $[a, b]$ with mesh $\mathcal{P} < \delta$ (Prob. 2S), then the oscillation $\omega(\alpha; \mathcal{P})$ of α over $\mathcal{P}$ is less than or equal to ε (see Problem 7X). Indeed, $[a, b]$ is compact by the Heine-Borel theorem, so α is uniformly continuous by the foregoing result. Hence for any given $\varepsilon > 0$ there exists a number $\delta > 0$ such that $|t - t'| < \delta$ (with $a \leq t, t' \leq b$) implies $\rho'(\alpha(t), \alpha(t')) < \varepsilon$. Thus if mesh $\mathcal{P} < \delta$, then $\omega(\alpha; \mathcal{P}) \leq \varepsilon$. More generally, if ϕ is a continuous mapping of any compact metric space X into Y, then for any positive number ε there is a positive number δ such that if $\mathcal{C}$ is an arbitrary collection of subsets C of X with the property that diam $C < \delta$ for each set C in $\mathcal{C}$, then the oscillation of ϕ over each set C in $\mathcal{C}$ is no greater than ε.

The following result may seem too elementary to be important, but that appearance is misleading.

Proposition 8.35. *Let ϕ be a continuous mapping of a metric space X into a metric space Y. If K is a compact subset of X, then $\phi(K)$ is compact and therefore closed in Y. If the entire space X is compact, then ϕ is a closed bounded mapping.*

PROOF. Let $\{y_n\}$ be a sequence in $\phi(K)$, and for each n let x_n be a point of K such that $y_n = \phi(x_n)$. The sequence $\{x_n\}$ possesses a subsequence $\{x_{n_k}\}$ that is convergent to a point x_0 of K. But then $\{\phi(x_{n_k}) = y_{n_k}\}$ converges to $\phi(x_0)$, which belongs to $\phi(K)$. Thus $\phi(K)$ is compact, and is therefore bounded and closed (Prop. 8.28).

Suppose now that X is a compact metric space. If F is a closed set in X, then F is compact, so $\phi(F)$ is also compact. But then $\phi(F)$ is closed, and therefore ϕ is closed. □

Corollary 8.36 (Theorem of Weierstrass). *If f is a continuous real-valued function on a metric space, then f assumes both a greatest and a least value on each nonempty compact subset of X.*

PROOF. If K is a nonempty compact subset of X, then $L = f(K)$ is a compact subset of $\mathbb{R}$, and is therefore both closed and bounded by the Heine-Borel theorem (Cor. 8.33). But then L contains both $\sup L$ and $\inf L$. □

Example W (Rolle's Theorem). Let f be a continuous real-valued function on a closed interval $[a,b]$ in $\mathbb{R}(a < b)$ such that $f(a) = f(b) = 0$, and suppose that the derivative $f'(t)$ exists at each point of the open interval (a,b). If there is a number t between a and b at which $f(t) > 0$, then the maximum of f on $[a,b]$ (which exists by the theorem of Weierstrass above) occurs at a point t_0 in (a,b), and it is easily seen that $f'(t_0)$ must vanish. (The limit

$$\lim_{h\uparrow 0} \frac{f(t_0+h) - f(t_0)}{h}$$

of the difference quotient from the left must be nonnegative; the limit

$$\lim_{h\downarrow 0} \frac{f(t_0+h) - f(t_0)}{h}$$

from the right must be nonpositive.) On the other hand, if there is a number t in (a,b) at which $f(t) < 0$, then the minimum of f on $[a,b]$ (which also exists by the theorem of Weierstrass) occurs at a point t_1 of (a,b), and it is easily seen, once again, that $f'(t_1) = 0$. Finally, if neither of these two possibilities occurs, then $f(t) = 0$ identically on $[a,b]$, in which event $f'(t) = 0$ identically on (a,b). Thus, in any case *there exists at least one point t of (a,b) such that $f'(t) = 0$*. (This result, known as *Rolle's theorem*, yields immediately a proof of the mean value theorem, and therefore stands at the heart of elementary calculus. Its rigorous demonstration, sometimes omitted from calculus courses, is, as we have just seen, an easy consequence of the theorem of Weierstrass.)

Note. It is an immediate consequence of Proposition 8.35 that if X is a compact metric space, then the space $\mathcal{C}(X;Y)$ of continuous mappings of X into a second metric space Y coincides with the space $\mathcal{C}_b(X;Y)$ of all bounded continuous mappings of X into Y. In the sequel we shall make free use of this as well as the consequent fact that $\mathcal{C}(X;Y)$ is complete in the metric of uniform convergence whenever X is compact and Y is complete (Prop. 8.7). (If X is compact and $\mathcal{E}$ is a Banach space, then $\mathcal{C}(X;\mathcal{E})$ is also a Banach space in the metric of uniform convergence with pointwise linear operations; see Problem 6E. In particular, $\mathcal{C}(X;\mathbb{C})$ is a complex Banach space and $\mathcal{C}(X;\mathbb{R})$ is a real Banach space.)

The following result is little more than a corollary of Proposition 8.35. Nevertheless, it is one of the most important theorems of modern analysis, the most basic of the *open mapping theorems*.

Theorem 8.37. *Let X be a compact metric space and let ϕ be a continuous one-to-one mapping of X into a metric space Y. Then ϕ is a homeomorphism of X onto $\phi(X)$.*

PROOF. The mapping ϕ is closed, as we already know (Prop. 8.35), and since ϕ is one-to-one, this implies that it is also open as a mapping of X onto $\phi(X)$ (since open and closed sets in $\phi(X)$ in the relative metric are complements of one another). But then ϕ is a homeomorphism of X onto $\phi(X)$. □

Example X. A compact convex subset K of Euclidean space $\mathbb{R}^n$ (Prob. 3S) is called a *convex body* if its interior K° is nonempty. If K is a convex body and L is a straight line in $\mathbb{R}^n$ that meets K, then $L \cap K$ is a line segment ℓ in L, and if L meets K° at all, then ℓ is nondegenerate and every point of ℓ lies in K° except the endpoints, which must belong to ∂K, of course. (Indeed, if $D_r(x_0) \subset K$ for some positive radius r, if x_1 is a point of K, and if for some $0 < t < 1$ we write $y_0 = tx_0 + (1-t)x_1$, then direct calculation discloses that $D_{tr}(y_0) \subset K$.)

Suppose now that K is a convex body in $\mathbb{R}^n$, and suppose further, for the sake of convenience, that the origin 0 belongs to K°. Then for each point u of the *unit sphere* $S = \{u \in \mathbb{R}^n : \|u\|_2 = 1\}$ there is a unique positive real number t_u such that tu belongs to K° for $0 \le t < t_u$, while tu fails to belong to K for all $t > t_u$, and $t_u u \in \partial K$. We shall show that the real-valued function h defined on S by setting $h(u) = t_u, u \in S$, is continuous. To this end, let $\{u_n\}$ be a sequence of points in S tending to a limit u_0, and suppose $\{h(u_n)\}$ fails to converge to $h(u_0)$ in $\mathbb{R}$. Then there is a positive number ε_0 small enough so that the sequence $\{h(u_n)\}$ lies outside the interval $U = (h(u_0) - \varepsilon_0, h(u_0) + \varepsilon_0)$ infinitely often. But then $\{u_n\}$ possesses a subsequence $\{v_m = u_{n_m}\}$ such that $h(v_m) \notin U$ for every index m. Moreover, the sequence $\{h(v_m)\}$ is bounded in $\mathbb{R}$ (for K is compact, and therefore bounded, in $\mathbb{R}^n$) so $\{v_m\}$ in turn has a subsequence $\{w_k = v_{m_k}\}$ such that $\{h(w_k)\}$ converges to a limit t_0 in $\mathbb{R}$, and t_0 cannot belong to U since $\mathbb{R}\backslash U$ is a closed set. But now, the sequence $\{h(w_k)w_k\}$ also converges in $\mathbb{R}^n$ to t_0u_0 (Ex. 7L), which implies that $t_0u_0 \in \partial K$, since $h(w_k)w_k \in \partial K$ for all k and ∂K is a closed set (Prob. 6Q). But $t_0 \neq h(u_0)$, and $h(u_0)$ is, by its construction, the only positive number t such that

$tu_0 \in \partial K$. Thus we have arrived at a contradiction, and the continuity of h is verified. Moreover (as was just noted), the mapping ϕ_0 of S onto ∂K defined by setting

$$\phi_0(u) = h(u)u, \quad u \in S,$$

is also continuous. Furthermore, ϕ_0 is obviously one-to-one, and is therefore a homeomorphism of S onto ∂K according to Theorem 8.37.

Finally, we use ϕ_0 to define a homeomorphism ϕ of the closed unit ball $D_1(0)^-$ onto the convex body K, setting

$$\phi(x) = \begin{cases} \|x\|_2 \phi_0(x/\|x\|_2), & x \in K\backslash\{0\}, \\ 0, & x = 0. \end{cases}$$

To show that ϕ is a homeomorphism it is enough to prove that it is continuous, since it is obvious that it is one-to-one, and both $D_1(0)^-$ and K are compact. Moreover, it is easy to see that ϕ is continuous at every point of $K\backslash\{0\}$ (see Example 7M), while if $\{x_n\}$ is an arbitrary sequence of nonzero vectors in K tending to 0, then $\|\phi(x_n)\|_2 = \|x_n\|_2 h(x_n/\|x_n\|_2) \to 0$, and it follows that ϕ is also continuous at 0.

According to the foregoing considerations, if K_1 and K_2 are any two convex bodies in $\mathbb{R}^n$, then there exist homeomorphisms of K_1 onto K_2 that carry the boundary of K_1 homeomorphically onto the boundary of K_2. Indeed, it is obvious that suitable translates of K_1 and K_2 have the origin in their interiors, and it follows that both K_1 and K_2 are homeomorphic to $D_1(0)^-$ under homeomorphisms that carry S onto ∂K_1 and ∂K_2, respectively.

As a matter of fact, *any* homeomorphism of a convex body K in $\mathbb{R}^n$ onto a subset A of $\mathbb{R}^n$ must carry ∂K homeomorphically onto ∂A, but this theorem is much deeper than the results developed here and cannot be proved without recourse to the arcana of algebraic topology.

Example Y. A curve α in a metric space X (parametrized on an interval $[a, b]$) is said to be a *simple arc* if it does not cross itself, that is, if it is one-to-one. (Recall (Ex. 7R) that a curve in X is just a continuous mapping of a parameter interval into X.) According to Theorem 8.37, such a curve is actually a homeomorphism of $[a, b]$ into X. A very interesting example of such a homeomorphism of the unit interval into $\mathbb{R}^2$ may be obtained by modifying in a suitable manner the construction of the space-filling curve in Example J. Here are the details. (Throughout all that follows in this example use is made of the notation and terminology introduced in Examples 6O, 6P and 6Q.)

To begin with, once again, let $x = (s_1, s_2)$ and $y = (t_1, t_2)$ be points in $\mathbb{R}^2$, let ℓ be the linear parametrization on the parameter interval $[a, b]$ of the line segment $\ell(x, y)$, and suppose that ℓ slants upward to the right. Let a' and b' be the points in $[a, b]$ that partition it into three congruent subintervals ($a' = (2a+b)/3, b' = (a+2b)/3)$), let θ be a proper ratio ($0 < \theta < 1$), and let s_1' and t_1' denote the endpoints of the central θth part of $[s_1, t_1]$ ($s_1' = (1/2)[s_1 + t_1 - \theta(t_1 - s_1)]$, $t_1' = (1/2)[s_1 + t_1 + \theta(t_1 - s_1)]$). Then $\ell^{(x,\theta)}$ is defined to be the piecewise linear extension (Ex. 7N) of the mapping of the partition $\{a < a' < b' < b\}$ that agrees with ℓ at a and b and that carries a' to (s_1', t_2) and b' to (t_1', s_2). Similarly, if s_2' and t_2' denote the endpoints of the central θth part of $[s_2, t_2]$, then $\ell^{(y,\theta)}$ is defined to be the piecewise linear extension of the mapping of the partition $\{a < a' < b' < b\}$ that agrees with ℓ at a and b and carries a' to the point (t_1, s_2') and b' to the point (s_1, t_2'). (If ℓ does not slant upward to the right, then, by definition, $\ell^{(x,\theta)} = \ell^{(y,\theta)} = \ell$.)

Next let π be a continuous mapping of $[a, b]$ into $\mathbb{R}^2$ that is piecewise linear with respect to some partition $\mathcal{P} = \{a = u_0 < \ldots < u_N = b\}$. We first define $\pi^{(x,\theta)}$ to be the mapping obtained by replacing each of the linear segments $\pi_i = \pi|[u_{i-1}, u_i], i = 1, \ldots, N$, by the curve $\pi_i^{(x,\theta)}$, and we define $\pi^{(y,\theta)}$, similarly, to be the curve obtained by replacing each π_i by $\pi_i^{(y,\theta)}$. Finally, we denote by $\pi^{\ddagger(\theta)}$ the result $\pi^{(x,\theta)(y,\theta)}$ of applying these two modifications to π one after the other.

Concerning $\pi^{\ddagger(\theta)}$ we observe that, for each subinterval $[u_{i-1}, u_i]$ of $\mathcal{P}$, the curve $(\pi^{\ddagger(\theta)})_i = \pi^{\ddagger(\theta)}|[u_{i-1}, u_i]$ coincides with $(\pi_i)^{\ddagger(\theta)} = (\pi_i)^{(x,\theta)(y,\theta)}$ —the result of applying the two basic operations to the single line segment π_i. It follows that if some segment π_i of π fails to slant upward to the right, then $(\pi^{\ddagger(\theta)})_i = \pi_i$, and, in general, that the behavior of $\pi^{\ddagger(\theta)}$ may be determined by investigating the case in which π consists of a single line segment.

Suppose then, once again, that ℓ is the (upward slanting to the right) linear parametrization on the interval $[a, b]$ of the line segment joining x to y in $\mathbb{R}^2$, and let R denote the rectangle having ℓ for a diagonal. If $\mathcal{F} = \{I_1, I_2, I_3, I_4\}$ is the system of subintervals of $[a, b]$ obtained by removing the central third twice (so that $\mathcal{F} = \{[a, b]\}^{**}$ in the notation of Example 6O), and if $\mathcal{P}_0$ is the partition of $[a, b]$ given by the eight endpoints of these four intervals, then $\ell^{\ddagger(\theta)}$ is a homeomorphism of $[a, b]$ into the rectangle R that is piecewise linear with respect to $\mathcal{P}_0$. Moreover, the subintervals of $\mathcal{P}_0$ on which the restrictions of $\ell^{\ddagger(\theta)}$ slant upward to the right are precisely the four intervals I_1, I_2, I_3 and I_4, and the restrictions of $\ell^{\ddagger(\theta)}$ to these four intervals are linear parametrizations of the (upward slanting to the right) diagonals of the four rectangles R_{00}, R_{01}, R_{10} and R_{11} obtained by removal of the central cross-shaped θth part of R (Ex. 6Q), while the restrictions of $\ell^{\ddagger(\theta)}$ to the other three subintervals of $\mathcal{P}_0$ are disjoint line segments that lie outside these four rectangles except for their endpoints. Finally, we observe that $\|\ell - \ell^{\ddagger(\theta)}\|_\infty < (7/9)\|x - y\|_2$, a fact that is disclosed by direct calculation. (If ℓ does not slant upward to the right, then $\ell^{\ddagger(\theta)} = \ell$, so $\|\ell - \ell^{\ddagger(\theta)}\|_\infty = 0$.) Figure 7 shows $\ell^{\ddagger(\theta)}$ for $\theta = 1/3$.

Turning now, once again, to a piecewise linear mapping π of $[a, b]$ into

$\mathbb{R}^2$ with respect to a partition $\mathcal{P} = \{a = u_0 < \ldots < u_n = b\}$, we define $N(\pi)$, just as before, to be the largest of the numbers $\|\pi(u_i) - \pi(u_{i-1})\|_2$ among all of those subintervals of $\mathcal{P}$ on which π slants upward to the right. Then it is clear that $N(\pi^{\ddagger(\theta)}) \leq (1/2)N(\pi)$ and $\|\pi - \pi^{\ddagger(\theta)}\|_\infty < (7/9)N(\pi)$. Moreover, π and $\pi^{\ddagger(\theta)}$ coincide on any subinterval of $\mathcal{P}$ on which π does not slant upward to the right, and if a rectangle contains one of the segments $\pi|[u_{i-1}, u_i]$, it also contains the subcurve $\pi^{\ddagger}|[u_{i-1}, u_i]$.

Suppose, finally, that $\{\theta_n\}_{n=0}^{\infty}$ is an arbitrarily prescribed sequence of proper ratios ($0 < \theta_n < 1$ for every n). Then we define a sequence of curves by mathematical induction, setting $\psi_0(u) = (u, u), 0 \leq u \leq 1$, and, assuming ψ_n already defined, setting $\psi_{n+1} = (\psi_n)^{\ddagger(\theta_n)}$. According to the above estimates we have $N(\psi_n) \leq \sqrt{2}/2^n$ and $\|\psi_n - \psi_{n+1}\|_\infty < (7/9)N(\psi_n)$ for each index n, and it follows that the sequence $\{\psi_n\}_{n=0}^{\infty}$ satisfies condition (C) with respect to $M = 7\sqrt{2}/9$ and $r = 1/2$. Hence $\{\psi_n\}$ converges uniformly to a curve ω on the parameter interval $[0, 1]$ by Propositions 8.6 and 8.7.

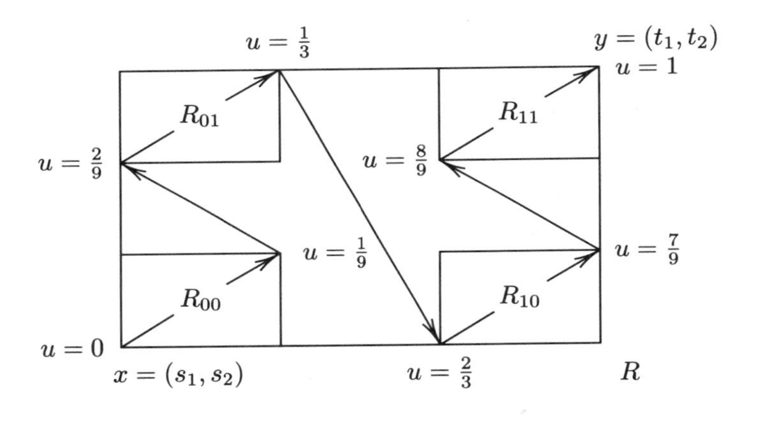

Figure 7

So far this construction parallels in every detail the construction of the space-filling curve in Example J. Let us turn now to the very critical ways in which the curve ω differs from the earlier one. To begin with, the curve ψ_0 is visibly simple, and ψ_1 is simple as well, as we have already noted. Indeed, a straightforward mathematical induction, which we omit, shows that for each nonnegative integer n the curve ψ_n is a simple arc in the unit square satisfying the following three conditions:

(1) The curve ψ_n is piecewise linear with respect to the partition $\mathcal{P}_{2n}$ of $[0, 1]$ consisting of the endpoints of the 4^n subintervals of $[0, 1]$ constituting the set $\mathcal{F}_{2n}$ used in the construction of the Cantor set in Example 6O,

(2) The restriction of ψ_n to a subinterval of $\mathcal{P}_{2n}$ slants upward to the right if and only if that subinterval is one of the intervals $I_{\varepsilon_1,\ldots,\varepsilon_{2n}}$ in $\mathcal{F}_{2n}$,

(3) The restriction of ψ_n to each subinterval $I_{\varepsilon_1,\ldots,\varepsilon_{2n}}$ is the linear parametrization on that interval of the (upward slanting to the right) diagonal of some one of the 4^n squares in the set $\mathcal{G}_n$ used in the construction of the planar Cantor set $P_{\{\theta_n\}}$ in Example 6Q, and these correspondences between $\mathcal{F}_{2n}$ and $\mathcal{G}_n$ are one-to-one and order preserving.

It follows from these three facts that if t is a number in $[0,1]$ such that $t \notin F_{2k}$ for some positive integer k, then the sequence $\{\psi_n(t)\}_{n=0}^{\infty}$ is constant for $n \geq k$, and hence that $\psi_n(t) = \omega(t), n \geq k$. Moreover, the point $\omega(t)$ does not belong to G_k, and therefore does not belong to any of the sets $G_n, n \geq k$. On the other hand, if t belongs to the Cantor set C, then $\psi_n(t) \in G_n, n \in \mathbb{N}_0$, and consequently $\omega(t) \in G_n, n \in \mathbb{N}_0$. But then $\omega(t) \in P_{\{\theta_n\}}$. From these observations it follows, in turn, that the curve ω is also a simple arc. Indeed, if t and t' are distinct points of $[0,1]$, and if neither t nor t' belongs to C, then both fall in $[0,1]\backslash F_{2k}$ for some sufficiently large k, and it is clear that in this case $\omega(t) \neq \omega(t')$. Next, if $t \notin C$ and $t' \in C$, then there is a positive integer k such that $\omega(t) \notin G_k$, and again $\omega(t) \neq \omega(t')$. Finally, if both t and t' are points of C, then for some positive integer k the numbers t and t' belong to two different intervals of the form $I_{\varepsilon_1,\ldots,\varepsilon_{2k}}$, whereupon it follows that the sequences $\{\psi_n(t)\}$ and $\{\psi_n(t')\}$ belong to distinct (and therefore disjoint) squares in $\mathcal{G}_n$ for all $n \geq k$. Thus ω is a homeomorphism of $[0,1]$ onto a simple arc in the unit square. Moreover, this discussion clearly shows that the restriction $\omega|C$ is also a homeomorphism of the Cantor set C into the planar Cantor set $P_{\{\theta_n\}}$.

As a matter of fact, $\omega|C$ maps C homeomorphically *onto* $P_{\{\theta_n\}}$. For if x_0 is a point of $P_{\{\theta_n\}}$, then for each positive integer n there is a unique square R_n in $\mathcal{G}_n$ that contains x_0, and the nested sequence $\{R_n\}$ corresponds to a similarly nested sequence $\{I_n\}$ of intervals, where $I_n \in \mathcal{F}_{2n}$ for each n. But then $\bigcap_{n=1}^{\infty} I_n$ is a singleton $\{t_0\}$, where $t_0 \in C$, and it is clear that $\omega(t_0) = x_0$. This fact is sometimes expressed by saying that the simple arc ω *threads* the Cantor set $P_{\{\theta_n\}}$.

PROBLEMS

A. Verify that the metric space obtained by equipping an arbitrary set with the discrete metric (Ex. 6C) is complete. Give an example of a discrete metric space (that is, one in which every point is isolated; see Problem 6O) that is not complete.

B. Let X be a metric space and let $\{x_n\}$ and $\{x'_n\}$ be equiconvergent sequences in X (Prob. 6M). Show that if either sequence is Cauchy, then both are. Conclude that if there exists a dense set M in X with the property that every Cauchy sequence in M converges in X, then X is complete.

C. Let Z be a nonempty set, let (X, ρ) be a metric space, and let X^Z denote, as usual, the collection of all mappings of Z into X. If ϕ and ψ denote two elements of X^Z, we define ϕ and ψ to be *boundedly equivalent* (notation: $\phi \overset{b}{\sim} \psi$) if

$$\sup_{z \in Z} \rho(\phi(z), \psi(z)) < +\infty.$$

(i) Verify that $\overset{b}{\sim}$ is, indeed, an equivalence relation on X^Z, and show that if we define

$$\sigma(\phi, \psi) = \sup_{z \in Z} \rho(\phi(z), \psi(z)) \tag{3}$$

whenever ϕ and ψ are boundedly equivalent elements of X^Z, then σ turns each equivalence class with respect to $\overset{b}{\sim}$ into a metric space.

(ii) Verify that the set $\mathcal{B}(Z; X)$ of bounded mappings of Z into X is one equivalence class in X^Z with respect to $\overset{b}{\sim}$, and that the metric σ defined in (3) on $\mathcal{B}(Z; X)$ is the metric of uniform convergence (Ex. 6H). (More generally, the metric σ on *any* equivalence class $[\phi_0]$ with respect to $\overset{b}{\sim}$ is called the *metric of uniform convergence* on $[\phi_0]$.)

(iii) Prove that each equivalence class $[\phi_0]$ in X^Z with respect to $\overset{b}{\sim}$ is complete in the metric of uniform convergence if the metric space (X, ρ) is complete.

(iv) Let (X, ρ) be a nonempty metric space, and for each point x of X let f_x denote the real-valued function defined on X by setting

$$f_x(y) = \rho(x, y), \quad y \in X.$$

Verify that, if x_0 is any one point of X, then all of the functions in the indexed family $\{f_x\}_{x \in X}$ are boundedly equivalent in $\mathbb{R}^X$ to f_{x_0}, and therefore all lie in the complete metric space obtained by equipping the equivalence class $[f_{x_0}]$ with the metric of uniform convergence. Show also that the mapping Φ of X into $\mathbb{R}^X$ defined by

$$\Phi(x) = f_x, \quad x \in X,$$

is an isometry of X into the complete space $[f_{x_0}]$. Finally, use these observations, along with the uniqueness assertion of Theorem 8.2, to give a new model for the completion $(\widehat{X}, \widehat{\rho})$ of the metric space (X, ρ).

D. Let (X, ρ) be a complete metric space.

(i) If ϕ is any continuous mapping of a subset A of X into a second metric space (Y, ρ'), and if we set

$$\widetilde{\rho}(x,y) = \rho(x,y) + \rho'(\phi(x), \phi(y)), \quad x, y \in A,$$

then $\widetilde{\rho}$ is a metric on A that is equivalent to the given (i.e., relative) metric on A.

(ii) Let U be an open subset of X, and set $f(x) = 1/d(x, X \backslash U), x \in U$. If f is used to define $\widetilde{\rho}$ as in (i) (so that $\widetilde{\rho}(x,y) = \rho(x,y) + |f(x) - f(y)|$ for $x, y \in U$), then U is complete with respect to the metric $\widetilde{\rho}$. Thus *every open set in a complete metric space can be remetrized by means of an equivalent metric with respect to which it becomes complete*; cf. Example E.

(iii) Let A be a G_δ in X, let $A = \bigcap_{n=1}^{\infty} U_n$, where each U_n is open, and for each index n set $f_n(x) = 1/d(x, X \backslash U_n), x \in U_n$. Then

$$\widehat{\rho}(x,y) = \rho(x,y) + \sum_{n=1}^{\infty} \frac{1}{2^n} \frac{|f_n(x) - f_n(y)|}{1 + |f_n(x) - f_n(y)|}, \quad x, y \in A,$$

defines a metric $\widehat{\rho}$ that is equivalent to the given metric on A, and A is complete with respect to $\widehat{\rho}$. Thus *every G_δ in a complete metric space can be remetrized by means of an equivalent metric with respect to which it becomes complete*. (Hint: In all parts of this problem it helps to recall Problem 6Y.)

E. A net $\{x_\lambda\}_{\lambda \in \Lambda}$ in a metric space (X, ρ) is said to be a *Cauchy* net, or to satisfy the *Cauchy criterion*, if for each positive number ε there exists an index λ_0 such that $\rho(x_\lambda, x_{\lambda'}) < \varepsilon$ whenever $\lambda, \lambda' \geq \lambda_0$.

(i) Prove that a net $\{x_\lambda\}$ in (X, ρ) is Cauchy if and only if the tail filter (base) associated with $\{x_\lambda\}$ (Prob. 7T) is Cauchy.

(ii) Use this fact to show that (X, ρ) is complete if and only if every Cauchy net in X is convergent.

F. For a real $n \times n$ matrix $A = (a_{ij})$ we define

$$N_1(A) = M_1 + \ldots + M_n$$

where, for each $i = 1, \ldots, n$, M_i denotes the largest of the absolute values of the entries in the ith row of $A (M_i = |a_{i1}| \vee \ldots \vee |a_{in}|)$. Verify that if T denotes the linear transformation on $\mathbb{R}^n$ having matrix A (Ex. 3Q), and if $\mathbb{R}^n$ is equipped with the norm $\| \quad \|_1$, then $\|T\| \leq N_1(A)$ (in the notation introduced in Example 7D). Using this information and following the line of argument employed in Example H, show that $1 - T$ is necessarily an isomorphism of $\mathbb{R}^n$ onto itself whenever $N_1(A) < 1$. Show too that this result neither subsumes nor is subsumed by the result of Example H. (Show, that is, that there are matrices A such that $N(A) < 1$ while $N_1(A) > 1$, and also matrices A such that $N_1(A) < 1$ while $N(A) > 1$.)

G. (Picard) Let $f(x, y)$ be a real-valued function defined and continuous on some open set U in the plane $\mathbb{R}^2$, and let (x_0, y_0) be a point of U.

(i) Show first that there exist positive numbers c_0 and d_0 such that the rectangle $R_0 = [x_0 - c_0, x_0 + c_0] \times [y_0 - d_0, y_0 + d_0]$ is contained in U and also such that $Ac_0 \leq d_o$ where $A = \max\{|f(x, y)| : (x, y) \in R_0\}$.

(ii) With c_0, d_0 as in (i), let c be a real number such that $0 < c \leq c_0$ and let $\mathcal{C}_c$ denote the set of all those continuous mappings g of $[x_0 - c, x_0 + c]$ into $[y_0 - d_0, y_0 + d_0]$ with the property that $g(x_0) = y_0$. Verify that $\mathcal{C}_c$ is a closed set in the space $\mathcal{C}_{\mathbb{R}}([x_0 - c, x_0 + c])$ (in the metric of uniform convergence). Show too that the function

$$h(x) = y_0 + \int_{x_0}^{x} f(t, g(t))dt, \quad |x - x_0| \leq c,$$

belongs to $\mathcal{C}_c$ whenever g does, and hence that setting

$$\Psi(g) = h, \quad g \in \mathcal{C}_c,$$

defines a mapping Ψ of $\mathcal{C}_c$ into itself.

(iii) Suppose now that f is not only continuous, but is also Lipschitzian as a function of y uniformly in x on R_0. Suppose, that is, that there exists a constant M such that $|f(x, y') - f(x, y'')| \leq M|y' - y''|$ for any two points (x, y') and (x, y'') in R_0. Prove that there exists a positive number c with the property that the initial value problem

$$\text{(I)} \qquad \begin{aligned} \frac{d}{dx}g(x) &= f(x, g(x)), \quad |x - x_0| < c, \\ g(x_0) &= y_0, \end{aligned}$$

possesses a unique solution g on $(x_0 - c, x_0 + c)$. (Hint: Show that if $c \leq c_0$, then a function g satisfies (I) if and only if it extends by continuity to a function on the closed interval $[x_0 - c, x_0 + c]$ that is a fixed point for Ψ. Choose c so as to make Ψ strongly contractive, and invoke Example G.)

H. Let $x = (s_1, s_2)$ and $y = (t_1, t_2)$ be points of $\mathbb{R}^2$, and let ℓ denote the linear parametrization on the parameter interval $[a, b]$ of the line segment joining x to y. We denote by a' and b' the points of trisection of the interval $[a, b]$, by s_1' and t_1' the points of trisection of the segment of real numbers joining s_1 to t_1, and likewise by s_2' and t_2' the points of trisection of the segment joining s_2 to t_2 in $\mathbb{R}$. (This somewhat awkward description is necessitated by the fact that, in the present construction, s_1 is allowed to be either less than or greater than t_1, and similarly, s_2 may be either less than or greater than t_2. Note that we have $s_i' = (2s_i + t_i)/3$ and $t_i' = (s_i + 2t_i)/3, i = 1, 2$, in any case.) We next define $\widetilde{\ell^{(x)}}$ to be the piecewise linear extension of the mapping of $\{a < a' < b' < b\}$ into $\mathbb{R}^2$ that takes a and b to x and y, respectively, and takes a' to (s_1', t_2) and b' to (t_1', s_2). Similarly, we denote by $\widetilde{\ell^{(y)}}$ the piecewise linear extension of the mapping of $\{a < a' < b' < b\}$ into $\mathbb{R}^2$ that carries these numbers to $x, (t_1, s_2'), (s_1, t_2'), y$, in that order. Then, with these preparations taken care of, we proceed to follow exactly the construction in Example J. That is, we define $\widetilde{\pi^{(x)}}$ and $\widetilde{\pi^{(y)}}$ for a curve π that is piecewise linear with respect to a partition $\{a = u_0 < \ldots < u_N = b\}$ of $[a, b]$ (so $\widetilde{\pi^{(x)}}|[u_{i-1}, u_i] = \widetilde{(\pi_i)^{(x)}}$ and $\widetilde{\pi^{(y)}}|[u_{i-1}, u_i] = \widetilde{(\pi_i)^{(y)}}$, where $\pi_i = \pi|[u_{i-1}, u_i], i = 1, \ldots, N$) and then set

$$\pi^{\S} = \pi^{\widetilde{(x)}\widetilde{(y)}}$$

(so that $\pi^{\S}|[u_{i-1}, u_i] = (\pi_i)^{\S}, i = 1, \ldots, N$).

(i) Let ℓ be as stated, and let R be the rectangle having ℓ as a diagonal. Verify that, if $\mathcal{P}$ denotes the partition of $[a, b]$ into *nine* subintervals of equal length (the result of trisecting $[a, b]$ twice), then $\ell^{\S}$ is a continuous mapping of $[a, b]$ into R that joins x to y, is piecewise linear with respect to $\mathcal{P}$, and satisfies the inequality $\|\ell^{\S} - \ell\|_\infty < \|x - y\|_2$. Show too that if $R^{\#}$ denotes the cellular partition of R into nine subrectangles effected by trisecting the sides of R, then $\ell^{\S}$ maps each subinterval of $\mathcal{P}$ linearly onto a diagonal of a cell in $R^{\#}$, thus establishing a one-to-one correspondence between $R^{\#}$ and the subintervals of $\mathcal{P}$. (Hint: Draw a sketch of $\ell^{\S}$.)

(ii) Set $\pi_0(u) = (u, u), 0 \le u \le 1$, and, supposing π_n already defined, set $\pi_{n+1} = (\pi_n)^{\S}$. Prove that the sequence $\{\pi_n\}_{n=0}^\infty$ thus inductively defined converges uniformly on the unit interval to a plane curve ψ, and that ψ is a Peano curve filling the unit square $[0, 1] \times [0, 1]$. (Hint: Let $\mathcal{P}_n$ be the partition of $[0, 1]$ into 3^n subintervals obtained by trisecting $[0, 1]$ n times, so that π_n is piecewise linear with respect to $\mathcal{P}_{2n}$ for each index n. Show first that $\{\pi_n\}$ satisfies condition (C) of Lemma 8.5 with $r = 1/3$ and $M = \sqrt{2}$. Show too that if $R^{(n)}$ denotes the cellular partition of the unit square $[0, 1] \times [0, 1]$ into 9^n subsquares obtained from the partition $\mathcal{P}_n$ of each of its sides, then for each index n there is a one-to-one correspondence κ_n between $R^{(n)}$ and the subintervals of $\mathcal{P}_{2n}$ such that π_n maps each subinterval I of $\mathcal{P}_{2n}$ linearly onto a diagonal of $\kappa_n(I)$, and such that a subinterval I'' of $\mathcal{P}_{2(n+1)}$ is contained in a subinterval I' of $\mathcal{P}_{2n}$ if and only if $\kappa_{n+1}(I'')$ is a subsquare of $\kappa_n(I')$. Note finally that ψ agrees with π_n on the points of $\mathcal{P}_{2n}$, and invoke Proposition 8.35.)

(iii) Show that, in fact (in the notation just introduced), each subinterval I of $\mathcal{P}_{2n}$ is mapped by $\psi|I$ continuously *onto* the corresponding subsquare $\kappa_n(I)$. Finally, verify that ψ is Lipschitz-Hölder continuous on $[0, 1]$

with respect to exponent $a = 1/2$ (and suitably chosen L.–H. constant M); cf. Problem 7K.

I. Let $x = (s_1, \ldots, s_m)$ and $y = (t_1, \ldots, t_m)$ be points in Euclidean space $\mathbb{R}^m$, let ℓ be the line segment (parametrized on the interval $[a, b]$) that joins x to y, and let Z denote the cell in $\mathbb{R}^m$ having ℓ for a diagonal. For each index $i = 1, \ldots, m$, let s_i' and t_i' denote the points that trisect the line segment joining s_i to t_i in $\mathbb{R}$ (so that $s_i' = (2s_i + t_i)/3, t_i' = (s_i + 2t_i)/3$), let x_i' be the point in $\mathbb{R}^m$ whose coordinates are the same as those of y except for the ith, where s_i' replaces t_i, and likewise let y_i' be the point whose coordinates are the same as those of x, except for the ith, where t_i' replaces s_i. Then, letting $\{a < a' < b' < b\}$ denote once again the partition of $[a, b]$ into three equal subintervals, we define $\ell^{\widetilde{(i)}}, i = 1, \ldots, m$, to be the piecewise linear extension to $[a, b]$ of the assignment $a \to x, a' \to x_i', b' \to y_i', b \to y$.

(i) Follow the example of the preceding problem to define

$$\pi^{\S(m)} = \pi^{\widetilde{(1)}\widetilde{(2)}\ldots\widetilde{(m)}}$$

for a piecewise linear curve π in $\mathbb{R}^m$ (so that, if π is piecewise linear with respect to a partition $\mathcal{P} = \{a = u_0 < \ldots < u_N = b\}$, then $\pi^{\S(m)}|[u_{i-1}, u_i] = (\pi|[u_{i-1}, u_i])^{\S(m)}, i = 1, \ldots, N$). Show that, for the line segment $\ell, \ell^{\S(m)}$ is a curve lying in the cell Z that is piecewise linear with respect to the partition $\mathcal{P}_m$ of $[a, b]$ into 3^m equal subintervals, and that $\|\ell^{\S(m)} - \ell\|_\infty < \|x - y\|_2$, while $\omega(\ell^{\S(m)}; \mathcal{P}_m) = \|x - y\|_2/3$. Show too that $\ell^{\S(m)}$ establishes a one-to-one correspondence between the subintervals of $\mathcal{P}_m$ and the 3^m cells in the cellular partition $Z^\#$ of Z obtained by trisecting each of its edges, and this in such a way that $\ell^{\S(m)}$ maps each subinterval of $\mathcal{P}_m$ linearly onto a diagonal of the corresponding cell in $Z^\#$. (Hint: We may suppose $m \geq 2$. If $i < m$ and if $\widehat{\ell}$ denotes the line segment in $\mathbb{R}^{m-1}$ that joins $\widehat{x} = (s_1, \ldots, s_{m-1})$ to $\widehat{y} = (t_1, \ldots, t_{m-1})$ and is parametrized on $[a, b]$, then the points assigned to $u = a, a', b', b$ by $\ell^{\widetilde{(i)}}$ are obtained from those assigned by $\widehat{\ell}^{\widetilde{(i)}}$ by adjoining the mth coordinates s_m, t_m, s_m, t_m (in that order).)

(ii) Let the sequence of curves $\{\pi_n\}_{n=0}^\infty$ in $\mathbb{R}^m$ be defined inductively by setting $\pi_0(u) = (u, \ldots, u), 0 \leq u \leq 1$, and $\pi_{n+1} = (\pi_n)^{\S(m)}, n \in \mathbb{N}_0$. Verify that $\{\pi_n\}$ converges uniformly on $[0, 1]$ to a curve ψ that is Lipschitz-Hölder continuous (Prob. 7K) with respect to exponent $a = 1/m$, and also that if, as above, $\mathcal{P}_n$ denotes for each index n in $\mathbb{N}_0$ the partition of $[0, 1]$ into 3^n equal subintervals, then ψ maps each subinterval of $\mathcal{P}_{mn}$ onto precisely one of the 3^{mn} cubes in the cellular partition of $W_0 = [0, 1] \times \ldots \times [0, 1]$ obtained from the partition $\mathcal{P}_n$ of each of the edges of W_0.

The curve constructed in Problem H is actually the original example of a space-filling curve given by Peano [21], who sets forth a direct definition of $\psi(t)$ in terms of the ternary expansion of t. This arithmetic definition, however, while entirely explicit, is far less perspicuous than the one employed in Example 7R. (The latter construction is due to Lebesgue [18; p. 44].) The sequential construction of Peano curves favored here is to be found in [20] and (in the m-dimensional case) in [19].

J. Let X be a perfect, separable and complete metric space, and let S be an arbitrary countable dense set in X. Prove that S is neither a G_δ nor of second category in X. Conclude that there does not exist any mapping of X into any metric space Y having S as its set of points of continuity.

K. (i) Prove that an *arbitrary* mapping of the metric space $\mathbb{Q}$ of rational numbers into any metric space is of Baire class one between those two spaces (Ex. Q).

(ii) As a matter of fact, any mapping of *any* countable metric space into a metric space is of Baire class one. (Hint: If x_0 is an arbitrary point in a countable metric space (X, ρ), then there exist radii r such that the set $\{x \in X : \rho(x, x_0) = r\}$ is empty.)

(iii) A real-valued [complex-valued] function f on a metric space X is a function of *Baire class one* if f is of Baire class one as a mapping of X into $\mathbb{R}[\mathbb{C}]$. (The continuous scalar-valued functions on X are the functions of *Baire class zero*.) Show that the collection $\mathcal{E}_1$ of all those subsets A of X such that χ_A is of Baire class one on X is *complemented* (i.e., $X \backslash A$ belongs to $\mathcal{E}_1$ whenever A does) and contains all open and closed sets. Conclude that on any metric space X there are functions of Baire class one that are not of Baire class zero unless X is discrete.

(iv) A scalar-valued function g on a metric space X is a function of *Baire class two* on X if there exists a sequence $\{f_n\}$ of functions of Baire class one such that $f_n(x) \to g(x)$ for every x in X. Show that the collection $\mathcal{E}_2$ of all those subsets A of X such that χ_A is of Baire class two on X is also complemented and contains all G_δs and F_δs in X. Thus in the metric space $\mathbb{R}$ the set $\mathbb{Q}$ of rational numbers belongs to $\mathcal{E}_2$ but not to $\mathcal{E}_1$. More generally, in any complete separable metric space X there are sets that belong to $\mathcal{E}_2$ but not to $\mathcal{E}_1$, unless X is countable. (Hint: An uncountable complete separable metric space has a nonempty Bendixson kernel; see Problem 6V.)

L. (Uniform Boundedness Theorem) Let $\mathcal{E}$ be Banach space, let $\mathcal{T}$ be a collection of bounded linear transformations of $\mathcal{E}$ into a normed space $\mathcal{F}$, and suppose that $\mathcal{T}$ is *pointwise bounded* on $\mathcal{E}$ in the sense that for each vector x in $\mathcal{E}$ the set $\{Tx : T \in \mathcal{T}\}$ is a bounded set in $\mathcal{F}$.

(i) Prove that for some sufficiently large number N there is a nonempty open set U in $\mathcal{E}$ such that $\|Tx\| \leq N$ for every vector x in U and every

T in $\mathcal{T}$. (Hint: For each positive integer n and for each T in $\mathcal{T}$, the set $\{x \in \mathcal{E} : \|Tx\| \leq n\}$ is closed, and so therefore is the set

$$F_n = \bigcap_{T \in \mathcal{T}} \{x \in \mathcal{E} : \|Tx\| \leq n\}.$$

Show that the sequence $\{F_n\}$ covers $\mathcal{E}$, and use Corollary 8.19.)

(ii) Conclude that $\mathcal{T}$ is actually *uniformly bounded* in the sense that there exists a positive constant M such that $\|T\| \leq M$ for every transformation T in $\mathcal{T}$. (Hint: With N and U as in (i) choose x_0 and $\varepsilon > 0$ such that $D_\varepsilon(x_0) \subset U$. Then for any vector x such that $\|x\| < \varepsilon$ we have $x = (x + x_0) - x_0$, and therefore $\|Tx\| \leq 2N$ for every T in $\mathcal{T}$.)

(iii) Let $\{T_n\}_{n=1}^\infty$ be a pointwise convergent sequence of bounded linear transformations of $\mathcal{E}$ into $\mathcal{F}$. Prove that the sequence $\{T_n\}$ is uniformly bounded and that $\lim_n T_n$ is also a bounded linear transformation.

M. Let X be a metric space and let $\{x_n\}_{n=1}^\infty$ be an arbitrary sequence of points in X. Prove that $\{x_n\}$ has either a Cauchy subsequence or a uniformly scattered subsequence. (Hint: Use the diagonal process.)

N. (i) Let $X_1, \ldots, X_n$ be complete metric spaces, and let $\Pi = X_1 \times \ldots \times X_n$ be equipped with a product metric ρ (Prob. 6H) with respect to which each projection π_i of Π onto $X_i, i = 1, \ldots, n$, is uniformly continuous. Show that (Π, ρ) is complete.

(ii) Does this result generalize to the product of an infinite sequence of complete metric spaces?

O. The central feature of the diagonal process is, as noted, the successive selection of ever finer subsequences from a given sequence. In our applications of the procedure up to now, the key element in the construction has been the fact that each subsequence in turn satisfied some more restrictive condition than its predecessor. There is another slightly different version of the diagonal process.

(i) Let $\{x_n\}_{n=1}^\infty$ be a sequence of objects, and let $\{p_k(\ )\}_{k=1}^\infty$ be a sequence of predicates, each of which is predicable of every term of $\{x_n\}$. Show that there exists a subsequence $\{x_{n_m}\}_{m=1}^\infty$ such that, for each index $k, p_k(x_{n_m})$ is either eventually true or eventually false. (Hint: For each k there is either a subsequence of $\{x_n\}$ on which $p_k(\ )$ is always true, or else a subsequence on which $p_k(\ )$ is always false. Use the diagonal process.)

(ii) For each real number x in $[0, 1]$ and each positive integer k let us say that $p_k(x)$ is true if x has a binary expansion $x = 0.\varepsilon_1 \ldots \varepsilon_n \ldots$ in which $\varepsilon_k = 0$. For a given sequence $\{x_n\}$ in $[0, 1]$ there is a subsequence $\{x_{n_m}\}$ such that each $p_k(\ )$ is either eventually true or eventually false of $\{x_{n_m}\}$. Prove that $\{x_{n_m}\}_{m=1}^\infty$ converges.

(iii) Prove that an arbitrary sequence $\{x_n\}_{n=1}^{\infty}$ of real numbers has a subsequence $\{x_{n_m}\}_{m=1}^{\infty}$ with the property that for every rational number r either $\{x_{n_m}\}$ is eventually $\le r$ or eventually $> r$. Show that $\{x_{n_m}\}$ is either convergent or tends to $\pm\infty$.

P. If $\{A_k\}_{k=1}^{\infty}$ and $\{E_n\}_{n=1}^{\infty}$ are sequences of subsets of a set X, then there is a subsequence $\{E_{n_m}\}$ of $\{E_n\}$ with the property that, for each index k, if A_k meets $\{E_{n_m}\}$ infinitely often, then it meets $\{E_{n_m}\}$ eventually.

(i) Conclude that an arbitrary sequence $\{E_n\}$ of sets in a separable metric space X possesses a subsequence $\{E_{n_m}\}$ such that $\mathrm{F}\limsup_m E_{n_m} = \mathrm{F}\liminf_m E_{n_m}$, and hence such that $\mathrm{F}\lim_m E_{n_m}$ exists. (Hint: Let $\{U_k\}_{k=1}^{\infty}$ be an enumeration of some countable base of open sets for X.)

(ii) Give an example of a sequence $\{F_n\}$ of nonempty closed subsets of $\mathbb{R}$ such that $\mathrm{F}\lim_n F_n = \varnothing$.

Q. If X, Y and Z denote arbitrary sets and ϕ is a mapping of $X \times Y$ into Z, then, for each fixed x in X, setting $\phi_x(y) = \phi(x,y), y \in Y$, defines a mapping $\phi_x : Y \to Z$. Similarly, if y is an arbitrary fixed element of Y, setting $\phi^y(x) = \phi(x,y), x \in X$, defines a mapping $\phi^y : X \to Z$. (These mappings are known as *partial mappings*.) Throughout the balance of this problem we suppose that $(X,\rho), (Y,\rho')$ and (Z,σ) are all metric spaces, and that ϕ is a given mapping of $X \times Y$ into Z. We also suppose that $X \times Y$ is equipped with a product metric ρ'' (Prob. 6H).

(i) Suppose first that all of the partial mappings ϕ_x are continuous on Y. If x', x'' are points of X, then the function

$$\sigma\left(\phi(x',y), \phi(x'',y)\right), \quad y \in Y,$$

is also continuous on Y. Consequently, if A is any nonempty subset of X, the function

$$g_A(y) = \sup\left\{\sigma(\phi(x',y), \phi(x'',y)) : x', x'' \in A\right\}$$

is a nonnegative, lower semicontinuous, extended real-valued function on Y. Verify that if $g_A(y) \le \eta$ for some positive number η and all y in some subset B of Y, then $\omega(\phi; (x,y)) \le 2\eta$ at every point (x,y) of $A^\circ \times B^\circ$. (Hint: By taking (x,y) sufficiently close to a point (x_0,y_0) with respect to the product metric ρ'' we can make x close to x_0 and y close to y_0 (why?). In particular, if $x_0 \in A^\circ$, we may arrange for x to belong to A° too, and also for $\phi_{x_0}(y)$ to be as close to $\phi_{x_0}(y_0)$ as desired. Use the fact that

$$\sigma\left(\phi(x,y), \phi(x_0,y_0)\right) \le \sigma\left(\phi(x,y), \phi(x_0,y)\right) + \sigma\left(\phi(x_0,y), \phi(x_0,y_0)\right).)$$

(ii) Suppose now that all of the partial mappings ϕ_x and ϕ^y are continuous, and also that the metric space (Y,ρ') is complete. For each point x of

X and positive integer n set $g_{x,n} = g_{U_n}$ where $U_n = D_{1/n}(x)$. Then for each x, $\{g_{x,n}\}_{n=1}^{\infty}$ is a decreasing sequence of nonnegative, lower semicontinuous, extended real-valued functions on Y with the property that (in the obvious sense) $\lim_n g_{x,n}(y) = 0$ at every point y of Y. Verify that if, for a positive number ε, we set $V_{x,n,\varepsilon} = \{y \in Y : g_{x,n}(y) > \varepsilon\}$, then $\{V_{x,n,\varepsilon}\}_{n=1}^{\infty}$ is a decreasing sequence of open sets in Y with empty intersection. Use this fact to conclude that if V is a nonempty open set in Y, x a point of X and $\varepsilon > 0$, then there is an index N and a nonempty open subset V_0 of V such that $\omega(\phi; (x,y)) \leq 2\varepsilon$ at every point (x,y) of $D_{1/N}(x) \times V_0$.

(iii) Complete the proof of the following result. *Let X and Y be complete metric spaces, and let ϕ be a mapping of the product $X \times Y$ into a metric space Z such that the partial mappings $\phi_x : Y \to Z, x \in X$, and $\phi^y : X \to Z, y \in Y$, are all continuous. If $X \times Y$ is equipped with a product metric, then the set of points of discontinuity of $\phi : X \times Y \to Z$ is an F_σ of first category in $X \times Y$ (and ϕ is accordingly continuous on a dense G_δ of the second category).* (Hint: Select a product metric on $X \times Y$ with respect to which it is complete; see Problem N.)

R. Let f be an upper semicontinuous extended real-valued function on a metric space X, and let K be a compact subset of X. Show that f assumes a maximum value on K. Conclude that an upper semicontinuous function is bounded above in $\mathbb{R}$ on compact sets if it does not assume the value $+\infty$, and give an example of an upper semicontinuous finite real-valued function on a compact space that is unbounded below (in $\mathbb{R}$). State and prove the dual assertions concerning lower semicontinuous functions.

S. (Dini) Let X be a compact metric space, and let $\{f_n\}_{n=1}^{\infty}$ be a sequence of continuous real-valued functions on X converging monotonely (either upward or downward) to a real-valued limit g. Prove that if the limit g is continuous, then the convergence of the sequence $\{f_n\}$ is necessarily uniform.

T. If ϕ is a continuous mapping of the Cantor set C into a metric space X, then the induced mapping $A \to \phi(A)$ of the subsets of C into the power class on X has the following properties. (Here and below we use the notation introduced in Example 6O; it will also be convenient to write $\mathcal{C}_n$ for the collection of sets $C_{\varepsilon_1,\ldots,\varepsilon_n}, \{\varepsilon_1, \ldots, \varepsilon_n\} \in S_n$, and to write $\mathcal{C} = \bigcup_n \mathcal{C}_n$.)

(a) For each set $C_{\varepsilon_1,\ldots,\varepsilon_n}$ in $\mathcal{C}$, the image $\phi(C_{\varepsilon_1,\ldots,\varepsilon_n})$ is closed and nonempty,

(b) If we set $m_n = \sup\{\text{diam}\ \phi(C_{\varepsilon_1,\ldots,\varepsilon_n}) : \{\varepsilon_1, \ldots, \varepsilon_n\} \in S_n\}$, then $\lim_n m_n = 0$,

(c) $\phi(C_{\varepsilon_1,\ldots,\varepsilon_{n+1}}) \subset \phi(C_{\varepsilon_1,\ldots,\varepsilon_n})$ for any $\{\varepsilon_1, \ldots, \varepsilon_n, \varepsilon_{n+1}\}$ in S_{n+1}.

(i) In the converse direction, suppose given, for some strictly increasing sequence $\{k_n\}_{n=1}^{\infty}$ of positive integers, a mapping Φ of $\mathcal{C}_0 = \bigcup_{n=1}^{\infty} \mathcal{C}_{k_n}$ into the power class on a *complete* metric space X, and suppose also that the mapping Φ satisfies the following three conditions.

(a′) For each $C_{\varepsilon_1,\ldots,\varepsilon_{k_n}}$ the set $\Phi(C_{\varepsilon_1,\ldots,\varepsilon_{k_n}})$ is closed and nonempty,

(b′) If $m_n = \sup\{\operatorname{diam} \Phi(C_{\varepsilon_1,\ldots,\varepsilon_{k_n}}) : \{\varepsilon_1,\ldots,\varepsilon_{k_n}\} \in S_{k_n}\}$, then $\lim_n m_n = 0$,

(c′) $\Phi(C_{\varepsilon_1,\ldots,\varepsilon_{k_{n+1}}}) \subset \Phi(C_{\varepsilon_1,\ldots,\varepsilon_{k_n}})$ for any $\{\varepsilon_1,\ldots,\varepsilon_{k_{n+1}}\}$ in $S_{k_{n+1}}$.

Prove that there exists a unique mapping ϕ of C into X such that $\Phi(C_0) = \phi(C_0)$ for every set C_0 in $\mathcal{C}_0$, and that the mapping ϕ is continuous. (Hint: For each point t of C there exists a uniquely determined sequence $\{\varepsilon_n\}_{n=1}^{\infty}$ of zeros and ones such that $t \in C_{\varepsilon_1,\ldots,\varepsilon_{k_n}}$ for every n. Consequently, ϕ must be given by

$$\{\phi(t)\} = \bigcap_{n=1}^{\infty} \Phi\left(C_{\varepsilon_1,\ldots,\varepsilon_{k_n}}\right).)$$

(ii) Prove that if X is an arbitrary nonempty compact metric space, then there exists a continuous mapping of the Cantor set C *onto* X. (Hint: Construct a mapping Φ as in (i) satisfying conditions (a′), (b′) and (c′), along with the following added conditions:

(d′) $\bigcup\{\Phi(C) : C \in \mathcal{C}_{k_1}\} = X$,

(e′) Each set $\Phi(C_{\varepsilon_1,\ldots,\varepsilon_{k_n}})$ is the union of all of the sets $\Phi(C_{\varepsilon_1,\ldots,\varepsilon_{k_{n+1}}})$ for which $\{\varepsilon_1,\ldots,\varepsilon_{k_{n+1}}\}$ is an extension of $\{\varepsilon_1,\ldots,\varepsilon_{k_n}\}$.

Use the fact that X is totally bounded (Th. 8.29).)

(iii) Prove also that if P is a nonempty perfect subset of a complete metric space X, then there exists a homeomorphism of the Cantor set C into P. (Hint: Take for $\{k_n\}$ the entire sequence $\mathbb{N}$, and construct a mapping Φ satisfying conditions (a′), (b′) and (c′), along with the following added condition.

(d″) If for any positive integer n, C' and C'' are distinct sets in $\mathcal{C}_n$, then $\Phi(C')$ and $\Phi(C'')$ are disjoint subsets of P.)

U. (Ascoli's Theorem) A collection Φ of mappings of a metric space (X,ρ) into a metric space (Y,ρ') is *equicontinuous* at a point x_0 of X if for every $\varepsilon > 0$ there is a $\delta > 0$ such that $\rho'(\phi(x),\phi(x_0)) < \varepsilon$ for every x in the ball $D_\delta(x_0)$ and every ϕ in Φ. Likewise, Φ is *equicontinuous on* X if it is equicontinuous at every point of X.

(i) Show that if $\{\phi_n\}_{n=1}^{\infty}$ is an equicontinuous sequence of mappings of X into Y, then the set F of those points x in X at which the sequence $\{\phi_n(x)\}_{n=1}^{\infty}$ is Cauchy in Y is a closed subset of X.

(ii) Show that if X is compact and $\{\phi_n\}_{n=1}^{\infty}$ is an equicontinuous sequence of mappings of X into Y that is pointwise Cauchy on X [pointwise convergent on X to some limit ψ], then $\{\phi_n\}$ is Cauchy in $\mathcal{C}(X;Y)$ in the metric of uniform convergence [uniformly convergent to ψ on X].

(iii) Prove that if X is a compact metric space and Y is any metric space, then a subset Φ of the space $\mathcal{C}(X;Y)$ of continuous mappings of X into Y is totally bounded in the metric of uniform convergence if and only if (a) Φ is equicontinuous on X, and (b) $\Phi(x) = \{\phi(x) : \phi \in \Phi\}$ is a totally bounded subset of Y for each x in X. (Hint: To go one way is easy; a finite ε-net in Φ is automatically equicontinuous and is carried onto an ε-net in $\Phi(x)$ under the point evaluation $\phi \to \phi(x)$. To go the other way, suppose Φ satisfies (a) and (b), and that $\{\phi_n\}$ is a sequence in Φ. Use Theorem 8.27 and the diagonal process to obtain a subsequence that is Cauchy at every point of some countable dense subset of X (recall Proposition 8.26).) This result, usually known as *Ascoli's theorem*, provides an effective criterion for the compactness of subsets of $\mathcal{C}(X;Y)$ in the metric of uniform convergence when X is compact and Y is complete.

(iv) As in (iii) let X be a compact metric space, and suppose further that the space Y is complete. Conclude that a subset Φ of $\mathcal{C}(X;Y)$ is compact in the metric of uniform convergence if and only if (a) Φ is equicontinuous on X, (b) the set $\Phi(x)$ is totally bounded in Y for each point x of X, and (c) Φ is closed in $\mathcal{C}(X;Y)$. Conclude, in particular, that a subset $\mathcal{F}$ of $\mathcal{C}_{\mathbb{C}}(X)$ is compact in the metric of uniform convergence if and only if it is closed in that metric and equicontinuous and pointwise bounded on X.

(v) Suppose both X and Y are compact metric spaces, and let a and M be positive constants, $a \leq 1$. Prove that the set $\Phi_{M,a}$ of all mappings of X into Y that are L.-H. continuous with respect to a and M (Prob. 7K) is a compact subset of $\mathcal{C}(X;Y)$.

V. Let $\{X_n\}_{n=0}^{\infty}$ be a sequence of nonempty sets, and let $\Pi = \prod_{n=0}^{\infty} X_n$ be equipped with the Baire metric ρ (recall Problem 6G). Under what conditions is (Π, ρ) complete? Under what conditions is (Π, ρ) compact?

W. Let X be a metric space and let $\mathcal{H}$ denote the collection of all closed, bounded, nonempty subsets of X equipped with the Hausdorff metric (Ex. 6X).

(i) Prove that $\mathcal{H}$ is compact when and only when X is compact. (Hint: Recall Problems P and 6X.)

(ii) Prove also that $\mathcal{H}$ is complete when and only when X is complete.

X. Let X and X' be metric spaces, and let $\mathcal{H}$ and $\mathcal{H}'$ denote the collections of all closed, bounded, nonempty subsets of X and X', respectively, equipped with the Hausdorff metric (Ex. 6X). Prove that if $\phi : X \to X'$ is closed and Lipschitzian with Lipschitz constant M, then the mapping Φ of 2^X into $2^{X'}$ induced by ϕ maps $\mathcal{H}$ into $\mathcal{H}'$, and $\Phi : \mathcal{H} \to \mathcal{H}'$ is also Lipschitzian with Lipschitz constant M. Show too that Φ maps $\mathcal{H}$ continuously into $\mathcal{H}'$ if ϕ is closed and uniformly continuous *provided* ϕ also has the property that it maps bounded subsets of X to bounded subsets of X'. Conclude that if X is compact and ϕ is an arbitrary continuous mapping of X into X', then Φ maps $\mathcal{H}$ continuously into $\mathcal{H}'$.

9 General topology

Many of the most important notions that arise in the theory of metric spaces do not depend on the numerical values of the metric at all and are totally unaffected if it is replaced by some equivalent metric. The properties of metric spaces that are so unaffected are, of course, the ones we have called *topological*, and the various concepts similarly unaffected are *topological concepts* (cf. Proposition 6.14). As it turns out, it is easy to define a context in which precisely these topological notions make sense, and it is also true that this idea of a *topological space* has had a major impact in analysis, and in other parts of mathematics as well. We close this summary of the fundamentals of mathematical analysis with a brief account of *general topology*, that is, the theory of topological spaces.

It should be acknowledged at the outset that this chapter will involve a good deal of repetition, particularly in the beginning. We shall be going over many—indeed most—of the ideas first encountered in Chapters 6–8, with a view to generalizing them into a metric-free context. In the course of this process it would be tedious to point out at every turn just which definition or argument from some earlier chapter is undergoing generalization, and we shall accordingly leave most such accounting chores to the reader. The following definition, based on the idea of an open set, is but one of several equivalent ones that could be employed (see Problems A and B).

Definition. A collection $\mathcal{T}$ of subsets of a set X is a *topology* on X if it satisfies the following three conditions:

(1) The whole set X and the empty set $\varnothing$ belong to $\mathcal{T}$,
(2) The intersection of any two sets in $\mathcal{T}$ (and hence of any nonempty finite collection of sets in $\mathcal{T}$) belongs to $\mathcal{T}$,
(3) The union of an arbitrary collection of sets in $\mathcal{T}$ belongs to $\mathcal{T}$.

A set X equipped with a topology on X is a *topological space*.

Notation and terminology. Condition (3) is often expressed by saying that a topology $\mathcal{T}$ is *closed with respect to the formation of (arbitrary) unions*; condition (2) says that a topology $\mathcal{T}$ is *closed with respect to the formation of finite intersections*. According to this definition, a topological space is, strictly speaking, a pair $(X, \mathcal{T})$, where X is a set—the *carrier* of the topological space—while the topology $\mathcal{T}$ is a (quite special) subset of 2^X, and we shall frequently use exactly this notation for topological spaces. However, in keeping with more or less universal practice, we often use the symbol X alone to denote the topological space $(X, \mathcal{T})$. Thus we shall refer simply to a "topological space X" or, when there is possible doubt as to which topology is in question—as when two or more topologies on the same carrier are under consideration simultaneously—to a "topological space X equipped with topology $\mathcal{T}$". (Since X can be recovered from a topology $\mathcal{T}$ on $X(X = \bigcup \mathcal{T})$, while $\mathcal{T}$ cannot possibly be recovered from X except in the most trivial cases, this notational practice is difficult to defend, but it is customary, and causes no confusion.)

If X is a topological space equipped with a topology $\mathcal{T}$, and if $U \in \mathcal{T}$, then we say that U is *open* in X, or is an *open subset* of X with respect to $\mathcal{T}$. Thus a topology $\mathcal{T}$ on a carrier X is identified at all times with the collection of all open subsets of X, and the phrases "U is open in X (with respect to $\mathcal{T}$)" and "U belongs to $\mathcal{T}$" have identical meanings.

Example A. If (X, ρ) is a metric space, then the collection $\mathcal{T}$ of open sets in X (with respect to ρ) is a topology on X (Prop. 6.12). This topology is called the *metric topology* on X, or the topology *induced by* ρ. The topology $\mathcal{T}$ induced by the metric ρ, and the topological space $(X, \mathcal{T})$ obtained by equipping X with $\mathcal{T}$, are said to be *metrized* by ρ. If $(X, \mathcal{T})$ is a topological space, and if there exists a metric ρ on X that metrizes $\mathcal{T}$, then X and $\mathcal{T}$ are *metrizable*. It is evident that the metric topologies induced by two different metrics on X coincide if and only if those metrics are equivalent (Prop. 6.14).

In keeping with the general policy enunciated above, in what follows we shall in most cases leave it to the reader to supply the proof (trivial in every individual case) that each "new" concept introduced actually retains its "old" meaning when the topology at hand happens to be a metric topology.

Example B. Every set X can be *topologized*, that is, equipped with a topology. Indeed it is clear that 2^X—the power class itself—is a topology on X, called the *discrete topology*. (In the discrete topology on a set X every subset is open; the discrete topology is metrized by the discrete metric (Ex. 6C) and is accordingly metrizable.) Another subset of 2^X that is a topology on X for an arbitrary set X is the smallest possible topology, viz., $\{\varnothing, X\}$. This topology is called the *indiscrete topology*. (The indiscrete topology on a set X is not metrizable unless X consists of at most one point, in which case the discrete and indiscrete topologies on X coincide and constitute the sole topology carried by X.)

Definition. A subset F of a topological space X is *closed* if its complement $X \backslash F$ is open.

The proof of the following proposition is just a matter of forming complements, and is therefore omitted.

Proposition 9.1. *In any topological space X the closed sets satisfy the following conditions:*

(1) *X and $\varnothing$ are closed,*
(2) *The union of any two (and hence of any finite number of) closed sets is closed,*
(3) *The intersection of an arbitrary nonempty collection of closed sets is closed.*

Proposition 9.2. *Let X be a topological space and let A be a subset of X. Then there exists a smallest closed set in X that contains A and a largest open set in X that is contained in A.*

PROOF. There exist closed subsets of X that contain A (since X itself is such a set). Hence the intersection of all such closed sets is a closed subset of X that contains A, and this is clearly the smallest closed subset of X that contains A. Dually, there exist open sets in X that are contained in A (since $\varnothing$ is such a set), and the union of all such open sets is clearly the largest open set in X that is contained in A. □

Definition. Let X be a topological space and let A be a subset of X. The *closure* A^- of A is the smallest closed subset of X that contains A, and the *interior* A° of A is the largest open subset of X that is contained in A. Likewise, the *boundary* of A is the set $\partial A = A^- \backslash A^\circ$. The set A is *dense* (in X) if $A^- = X$. More generally, A is *dense* in a subset B of X if $A^- \supset B$. A point a_0 of X is an *adherent point* of A if $a_0 \in A^-$, an *interior point* of A if $a_0 \in A^\circ$, and a *boundary point* of A if $a_0 \in \partial A$. A point a_0 of X is a *point of accumulation* of A if $a_0 \in (A \backslash \{a_0\})^-$. The set of all accumulation points of A is the *derived set* of A, denoted by A^*.

Proposition 9.3. *A subset A of a topological space X is closed [open] in X if and only if $A = A^-[A = A^\circ]$. A point a_0 of X is an adherent point of A if and only if every open set in X that contains a_0 meets A. Similarly, a_0 is an accumulation point of A if and only if every open set in X that contains a_0 also contains at least one point of A other than a_0.*

PROOF. If U is an open set in X that contains a_0 and is disjoint from A, then $X \backslash U$ is a closed set containing A, and therefore A^-, so $a_0 \notin A^-$. Conversely, if $a_0 \notin A^-$, then $X \backslash A^-$ is an open set containing a_0 that fails to meet A. Thus $a_0 \in A^-$ if and only if every open set in X that contains a_0 meets A. The rest of the proposition is either an immediate consequence of this fact or is obvious from the definitions. □

Proposition 9.4. *If A and B are subsets of a topological space X, and if $A \subset B$, then $A^- \subset B^-, A^\circ \subset B^\circ$ and $A^* \subset B^*$. Moreover the boundary ∂A is always closed, and for every subset A of X we have*

$$A^- = A \cup A^*.$$

Hence a subset A of X is closed if and only if $A \supset A^*$. Finally, a subset M of X is dense if and only if every nonempty open subset of X meets M.

PROOF. It is obvious from the various definitions that $A^- \subset B^-, A^\circ \subset B^\circ$ and $A^* \subset B^*$. Likewise, $\partial A = A^- \cap (X \backslash A^\circ)$ is closed for every A. If $x \in A^*$, then $x \in (A \backslash \{x\})^- \subset A^-$, so $A^* \subset A^-$, and therefore $A \cup A^* \subset A^-$. On the other hand, if $x \notin A \cup A^*$, then $A \backslash \{x\} = A$, so $(A \backslash \{x\})^- = A^-$, whence it follows that $x \notin A^-$. Thus $A^- = A \cup A^*$. If M is dense in X and if U is a nonempty open subset of X, then $X \backslash U$, being closed, cannot contain M, so $M \cap U \neq \varnothing$. On the other hand, if every nonempty open set in X has points in common with a set M, then $X \backslash M^- = \varnothing$, so $M^- = X$. □

In a metric space the derived set of an arbitrary set is closed (Prop. 6.8). The counterpart assertion was omitted from Proposition 9.4 for the good reason that it is not true in general in a topological space. Let X be a set consisting of at least two points, equip X with the indiscrete topology (Ex. B), let x_0 be a point of X, and set $A = \{x_0\}$.

Proposition 9.5. *For every topological space X the mapping $A \to A^-$, assigning to each subset A of X its closure, is a monotone increasing mapping of 2^X into itself possessing the following properties for all subsets A and B of X:*

(1) $\varnothing^- = \varnothing$,

(2) $A \subset A^-$,
(3) $(A^-)^- = A^-$,
(4) $(A \cup B)^- = A^- \cup B^-$.

PROOF. Clearly $A \to A^-$ is monotone increasing, and properties (1), (2) and (3) are obviously valid. To prove (4) we first note that $A^- \cup B^-$ is a closed set containing $A \cup B$, and hence that $(A \cup B)^- \subset A^- \cup B^-$. On the other hand, $(A \cup B)^-$ is a closed set containing both A and B. Hence $A^- \cup B^- \subset (A \cup B)^-$. □

Although topologies and filters are studied for different (but not unrelated) purposes, it is instructive to reflect on the similarities between them. Thus a topology on a set X is a subset of 2^X, just as is a filter in X. Moreover, as a collection of subsets of 2^X, the system of all topologies on X is naturally ordered in the inclusion ordering, just as is the system of all filters in X. Indeed, the same terminology is used. If $\mathcal{T}$ and $\mathcal{T}'$ are topologies on X such that $\mathcal{T} \subset \mathcal{T}'$, then $\mathcal{T}'$ is said to be *finer* than $\mathcal{T}$, or to *refine* $\mathcal{T}$, and $\mathcal{T}$ is said to be *coarser* than $\mathcal{T}'$. (Note that if $\mathcal{T}$ is refined by $\mathcal{T}'$, then every set that is open [closed] with respect to $\mathcal{T}$ is also open [closed] with respect to $\mathcal{T}'$; when more open sets are present, more closed sets are also automatically present.) Finally, continuing with this analogy, we observe that if $\mathcal{C}_0$ is an arbitrary nonempty collection of topologies on X, then it is clear from the definition that $\bigcap \mathcal{C}_0$ is again a topology on X (cf. Problem 7S), and from this fact we immediately deduce the following result.

Proposition 9.6. *If $\mathcal{G}$ is an arbitrary collection of subsets of a set X, then there is a coarsest topology $\mathcal{T}$ on X such that $\mathcal{G} \subset \mathcal{T}$. The system $\mathcal{G}$ is called* a set of generators *for $\mathcal{T}$, $\mathcal{T}$ is said to be* generated by $\mathcal{G}$, *and will be denoted by $\mathcal{T}(\mathcal{G})$ (cf. Problem C).*

PROOF. Since the power class 2^X is a topology on X, the collection $\mathcal{C}_0$ of topologies on X that contain $\mathcal{G}$ is never empty. Set $\mathcal{T}(\mathcal{G}) = \bigcap \mathcal{C}_0$. □

Example C. The finest topology on any set X is the discrete topology, generated by the collection of all singletons; the coarsest is the indiscrete topology, generated by the empty set. On any set X there is also a coarsest topology $\mathcal{T}_1$ in which each singleton $\{x\}$ is closed ($\mathcal{T}_1$ is the topology generated by the collection of all sets of the form $X \backslash \{x\}$). It is readily verified that the closed sets in X with respect to $\mathcal{T}_1$ are precisely X itself and the finite subsets of X. A topology with respect to which all singletons are closed sets is called a *T_1-topology*, and a topological space equipped with such a topology is a *T_1-space*.

It is a consequence of Proposition 9.6 that the collection of all topologies on an arbitrary set X is a complete lattice in the inclusion ordering. If

$\{\mathcal{T}_\gamma\}_{\gamma\in\Gamma}$ is an indexed family of topologies on X, then $\sup_\gamma \mathcal{T}_\gamma$ is the topology on X generated by $\bigcup_\gamma \mathcal{T}_\gamma$, while $\inf_\gamma \mathcal{T}_\gamma = \bigcap_\gamma \mathcal{T}_\gamma$. In connection with the concept of the topology generated by a collection of sets the following idea is of importance.

Definition. A collection $\mathcal{B}$ of open sets in a topological space $(X, \mathcal{T})$ is a *base* for the topology $\mathcal{T}$ on X (or a *topological base* for X) if every open set in X is the union of some subcollection of $\mathcal{B}$, i.e., if $\mathcal{T} = \mathcal{B}_u$ in the notation of Problem C.

Proposition 9.7. *If X is a given topological space, then a collection $\mathcal{B}$ of open sets in X is a topological base for X if and only if for each point x of X and each open set U in X such that $x \in U$ there exists a set V in $\mathcal{B}$ such that $x \in V \subset U$. If X is an arbitrary set and $\mathcal{B}$ a collection of subsets of X, then $\mathcal{B}$ is a base for a topology on X (viz., the topology generated by $\mathcal{B}$) if and only if the following two criteria are satisfied:*

(1) *$\mathcal{B}$ covers X,*
(2) *If V and V' are sets in $\mathcal{B}$ and if $x \in V \cap V'$, then there exists a set V'' in $\mathcal{B}$ such that $x \in V'' \subset V \cap V'$ (alternatively: every intersection $V \cap V'$ of two sets in $\mathcal{B}$ is a union of sets in $\mathcal{B}$).*

PROOF. If $\mathcal{B}$ is a base for a given topology on X, and if U is open with respect to that topology, then U is the union of some subcollection $\mathcal{B}_U$ of $\mathcal{B}$. Hence if $x \in U$, then $x \in V$ for some V in $\mathcal{B}_U$, and therefore $x \in V \subset U$. On the other hand, if $\mathcal{B}$ satisfies the stated conditions, and if U is an open set in X, then

$$U = \bigcup\{V \in \mathcal{B} : V \subset U\},$$

so $\mathcal{B}$ is a base for the topology.

Suppose next that X is an abstract set and $\mathcal{B}$ is a collection of subsets of X. If $\mathcal{B}$ is a base for some topology $\mathcal{T}$ on X, then $\mathcal{T}$ must consist of the unions of all possible subcollections of $\mathcal{B}$, and must therefore coincide with the topology generated by $\mathcal{B}$. Moreover (1) must hold since the whole space X is open, and if V and V' are any two sets in $\mathcal{B}$, then V and V', and with them $V \cap V'$, also belong to $\mathcal{T}$, so $V \cap V'$ is the union of some subcollection of $\mathcal{B}$.

Suppose, finally, that a collection $\mathcal{B}$ of s bsets of a set X satisfies both (1) and (2), and let $\mathcal{T}$ denote the collection $\mathcal{B}_u$ of unions of all subcollections of $\mathcal{B}$. Then it is clear that $\varnothing \in \mathcal{T}$ and that $\mathcal{T}$ is closed with respect to the formation of unions (Prob. C). Moreover, $X \in \mathcal{T}$ by (1). All that remains therefore to complete the proof that $\mathcal{T}$ is a topology on X (for which $\mathcal{B}$ is obviously a base) is to verify that the intersection of any two sets in $\mathcal{T}$ is again in $\mathcal{T}$. Accordingly, let U and U' belong to $\mathcal{T}$, and let $\mathcal{B}_U$ and $\mathcal{B}_{U'}$ be subcollections of $\mathcal{B}$ such that $U = \bigcup \mathcal{B}_U$ and $U' = \bigcup \mathcal{B}_{U'}$.

If $x \in U \cap U'$, then there exist sets V and V' in $\mathcal{B}_U$ and $\mathcal{B}_{U'}$, respectively, such that $x \in V \cap V'$. But then there also exists a set V'' in $\mathcal{B}$ such that $x \in V'' \subset V \cap V'$. Thus $U \cap U' = \bigcup\{V \in \mathcal{B} : V \subset U \cap U'\} \in \mathcal{T}$. □

There is a weaker notion than that of topological base that is also sometimes useful.

Definition. A collection $\mathcal{S}$ of open sets in a topological space X is a *subbase* for the topology on X (or a *topological subbase* for X) if the collection $\mathcal{S}_d$ of all finite intersections of sets belonging to $\mathcal{S}$ is a topological base for X.

The following result is a more or less obvious consequence of the definition and Proposition 9.7 (cf. Problem C).

Proposition 9.8. *If X is a given topological space, then a collection $\mathcal{S}$ of open sets in X is a topological subbase for X if and only if for each point x of X and each open set U in X such that $x \in U$ there exist sets $W_1, \dots, W_n$ in $\mathcal{S}$ such that $x \in W_1 \cap \dots \cap W_n \subset U$. If X is an arbitrary set and $\mathcal{S}$ a collection of subsets of X, then $\mathcal{S}$ is a subbase for a topology on X (viz., the topology generated by $\mathcal{S}$) if and only if $\mathcal{S}$ covers X.*

Example D. Let X be a simply ordered set, and for each element x of X write $A_x = \{z \in X : z < x\}$ for the initial segment in X determined by x (cf. Lemma 5.1). If $\mathcal{L}$ denotes the system of all such initial segments, then $\mathcal{B}_\ell = \mathcal{L} \cup \{X\}$ is a base for a topology $\mathcal{T}_\ell$, called the *left-ray topology* on X. (To see this, all that is necessary is to verify the provisions of Proposition 9.7. The reason for explicitly including X in $\mathcal{B}_\ell$ is that if X happens to have a greatest element, the collection $\mathcal{L}$ does not cover X; if X is unbounded above, the system $\mathcal{L}$ itself is a base for $\mathcal{T}_\ell$.) Dually, the collection $\mathcal{R}$ of all sets of the form $B_x = \{z \in X : z > x\}$, together with X itself (the latter being required only if X possesses a least element), is a base $\mathcal{B}_r$ for a topology $\mathcal{T}_r$ on X, called the *right-ray topology*.

It is obvious that all of the sets belonging to $\mathcal{T}_\ell$ are initial segments, and likewise that all of the sets in $\mathcal{T}_r$ are terminal segments, that is, complements of initial segments. Hence the infimum $\mathcal{T}_\ell \wedge \mathcal{T}_r = \mathcal{T}_\ell \cap \mathcal{T}_r$ consists exclusively of sets possessing both of these properties, and is therefore the indiscrete topology on X (Ex. B). Rather more interesting is the supremum $\mathcal{T}_o = \mathcal{T}_\ell \vee \mathcal{T}_r$. A subbase for $\mathcal{T}_o$ is given by $\mathcal{B}_\ell \cup \mathcal{B}_r$, and a base for $\mathcal{T}_o$ is accordingly given by the union $\mathcal{B}_\ell \cup \mathcal{B}_r \cup \mathcal{O}$, where $\mathcal{O}$ denotes the collection of all *open intervals* in X, i.e., sets of the form $(a, b) = \{x \in X : a < x < b\}$, where a and b denote arbitrary elements of X (cf. Problem D). The topology $\mathcal{T}_o$ is known as the *order topology* on X. (Note that if X has no least element, then every set in $\mathcal{L}$ is a union of open intervals. Dually, if X has no greatest element, then every set in $\mathcal{R}$ is a union of open intervals. Thus if X has neither a least nor a greatest element, then $\mathcal{O}$ alone constitutes a base for the order topology.)

Example E. Of particular interest in this connection are the cases $X = \mathbb{R}$ and $X = \mathbb{R}^\natural$. Since $\mathbb{R}$ is unbounded both above and below as a simply ordered set, a base for the order topology $\mathcal{T}_o$ on $\mathbb{R}$ is given by the system $\mathcal{O}$ of open intervals (a, b), and it is at once clear that $\mathcal{T}_o$ coincides with the *usual* metric topology on $\mathbb{R}$. Moreover, because of the completeness of the ordering of $\mathbb{R}$ (see Chapter 2), the system $\mathcal{B}_\ell$ of all open rays to the left in $\mathbb{R}$ (including $\mathbb{R}$ itself) is not just a base for the left-ray topology on $\mathbb{R}$ but actually constitutes the entire topology except for $\varnothing$: $\mathcal{T}_\ell = \mathcal{B}_\ell \cup \{\varnothing\}$. Dually, of course, the right-ray topology on $\mathbb{R}$ consists of $\varnothing$ along with the collection $\mathcal{B}_r$ of all open rays to the right.

In the extended real number system $\mathbb{R}^\natural$ matters are slightly different. Here the system $\mathcal{L}$ of all initial segments (i.e., the empty set together with those sets of the form $A_x = (-\infty, x) \cup \{-\infty\}, x \in \mathbb{R}^\natural \backslash \{-\infty\}$) is closed with respect to the formation of unions, but fails to cover $\mathbb{R}^\natural$. Thus in this case the left-ray topology on $\mathbb{R}^\natural$ is simply $\mathcal{T}_\ell = \mathcal{B}_\ell$. Likewise the right-ray topology on $\mathbb{R}^\natural$ coincides with $\mathcal{B}_r$, while the order topology $\mathcal{T}_o$ on $\mathbb{R}^\natural$ has $\mathcal{L} \cup \mathcal{O} \cup \mathcal{R}$ for a base. It is this order topology that we take as the *usual* topology on $\mathbb{R}^\natural$, i.e., the topology with which it is assumed to be equipped whenever it is regarded as a topological space, unless some other topology is expressly stipulated.

Definition. Let X be a topological space and let A be a subset of X. A subset B of A is said to be *open relative to* A if B is of the form $B = A \cap U$ where U is open in X. It is readily verified that the collection of all such relatively open subsets of A is, in fact, a topology on A, called the *relative topology* on A. When a subset A of X is equipped with this relative topology, it is known as a *subspace* of X. Whenever a subset of a topological space is regarded as a topological space in its own right, it is this relative topology that is understood to be in use unless the contrary is expressly stipulated.

Proposition 9.9. *If A is a subspace of a topological space X, then the closed sets in A are precisely the sets of the form $A \cap F$ where F is closed in X. Moreover, the closure $B^{-(A)}$ of a subset B of A in the subspace A coincides with $A \cap B^-$ (where B^- denotes the closure of B in X).*

PROOF. That sets of the specified form are closed in the relative topology on A follows at once from the identity $(X \backslash F) \cap A = A \backslash (A \cap F)$, valid for arbitrary subsets A and F of X. Suppose, on the other hand, that B is a closed set in the subspace A. Then there exists an open set U in X such that $A \backslash B = A \cap U$, whence it follows at once that

$$B = A \cap (X \backslash U).$$

Thus B has the specified form. Finally, if B is an arbitrary subset of A, and if C is a relatively closed subset of A containing B, then $C = A \cap F$, where F is closed in X, and it is clear that F contains B. But then $F \supset B^-$, and it follows that the relative closure $B^{-(A)}$ of B in A is just $A \cap B^-$. □

Corollary 9.10. *Let X be a topological space. The open subsets of an open subspace of X are open sets in X. Dually, the closed subsets of a closed subspace of X are closed sets in X.*

Example F. If (X, ρ) is a metric space and A is a subset of X, then it is readily verified that the metric topology induced on A by the relative metric $\rho|(A \times A)$ coincides with the relative topology induced on A by the metric topology on X. Thus the notion of subspace is unambiguously defined when X is a metric space.

One key construct in the theory of metric spaces, viz., the notion of the *open ball* of given center and radius, has thus far been accorded no general topological counterpart. The idea is a tricky one. An open ball with center x_0 in a metric space is a "neighborhood" of x_0 in that it contains all of the points of that space that are "close" to x_0, but it also possesses a very special shape. Unfortunately, the second of these two rather vague ideas is entirely meaningless in the present context, and must simply be foregone. Hence one is led, perhaps a little reluctantly, to accept the following substitute notion, which certainly captures the first and more important of the ideas of what a neighborhood should be.

Definition. Let X be a topological space, and let x_0 be a point of X. Then a subset V of X is a *neighborhood* of x_0 if there exists an open set U in X such that $x_0 \in U \subset V$; in other words, if $x_0 \in V^\circ$.

As it happens, this somewhat unwelcome, but altogether necessary, relaxation of the idea of what constitutes a neighborhood of a point in a topological space brings with it an unexpected bonus.

Proposition 9.11. *For each point x of a topological space X the collection $\mathcal{V}_x$ of all neighborhoods of x in X is a filter in X, called the* neighborhood filter *at x.*

PROOF. If x belongs to X, then since every set V in $\mathcal{V}_x$ contains the point x, it is clear that $\mathcal{V}_x$ is a nonempty collection of nonempty sets, and it is equally obvious that any set containing a neighborhood of x is itself a neighborhood of x. What remains is to verify that $\mathcal{V}_x$ is closed with respect to the formation of finite intersections. Let $V_1, \ldots, V_n$ be neighborhoods of x. Then $x \in V_i^\circ$, $i = 1, \ldots, n$, so

$$x \in V_1^\circ \cap \cdots \cap V_n^\circ \subset V_1 \cap \cdots \cap V_n,$$

and it follows that $V_1 \cap \cdots \cap V_n \in \mathcal{V}_x$. □

It is of importance to observe that the various neighborhood filters $\mathcal{V}_x$ determine the topology on X completely (in this context see also Problem B). All parts of the following proposition are obvious consequences of the various definitions and Proposition 9.3.

Proposition 9.12. *A subset U of a topological space X is open in X if and only if U is a neighborhood of every point of U, i.e., if and only if $U \in \mathcal{V}_x$ for every point x of U. Likewise, a point a_0 of X is an adherent point of a subset A of X if and only if every neighborhood of a_0 meets A.*

Example G. If $\mathcal{T}_1$ and $\mathcal{T}_2$ are two topologies on the same set X, and if $\mathcal{T}_2$ refines $\mathcal{T}_1$, then every set V that is a neighborhood of a point x of X with respect to $\mathcal{T}_1$ is also a neighborhood of x with respect to $\mathcal{T}_2$. Thus the neighborhood filter with respect to $\mathcal{T}_2$ refines the neighborhood filter with respect to $\mathcal{T}_1$ at each point x of X. Conversely, if this condition is satisfied, then $\mathcal{T}_2$ refines $\mathcal{T}_1$ by Proposition 9.12. In the same vein we observe that if A is a subset of a topological space X and if V is a neighborhood (in X) of a point x of A, then $x \in V^\circ \cap A$, and the latter set is open relative to A. Thus $V \cap A$ is a neighborhood of x relative to A. On the other hand, if W is an arbitrary neighborhood of x relative to A, then there exists an open subset U of X such that $x \in U$ and such that $U \cap A \subset W$. But then $U \cup W$ is a neighborhood of x in X, and $W = (U \cup W) \cap A$. Thus the neighborhood filter at x relative to A is the trace $\{V \cap A : V \in \mathcal{V}_x\}$ on A of the neighborhood filter $\mathcal{V}_x$ at x in X.

In many situations an important role is played by determinative systems of neighborhoods that are special in one way or another.

Definition. A system $\mathcal{W}_x$ of neighborhoods of a point x of a topological space X is a *neighborhood base* at x if $\mathcal{W}_x$ is a base for the filter $\mathcal{V}_x$ of all neighborhoods of x.

The following result is also an immediate consequence of the various definitions.

Proposition 9.13. *A collection $\mathcal{W}_x$ of neighborhoods of a point x in a topological space X is a neighborhood base at x if and only if, whenever U is an open set in X such that $x \in U$, there exists a set W in $\mathcal{W}_x$ such that $W \subset U$.*

Example H. Let (X, ρ) be a metric space, let M be a set of positive real numbers with the property that $\inf M = 0$, and let x_0 be a point of X. Then the collection of open balls $\{D_r(x_0)\}_{r \in M}$ is a neighborhood base at x_0 in the metric topology induced by ρ. The same is true of the set of all closed balls $\{x \in X : \rho(x, x_0) \leq r\}$, $r \in M$. In particular, if we take for X Euclidean space $\mathbb{R}^n$ equipped with the metric ρ_∞, this shows that the open [closed] cubes with center x_0 and edge $2r$, $r \in M$, form a neighborhood base at each point x_0 of $\mathbb{R}^n$ (with respect to the usual metric topology on $\mathbb{R}^n$). A neighborhood base at an arbitrary point y_0 of a set Y with respect to the discrete topology on Y is provided by the singleton $\{y_0\}$.

Using neighborhoods one defines convergence in a topological space in a wholly natural way.

Definition. Let X be a topological space and let $\{x_\lambda\}_{\lambda \in \Lambda}$ be a net in X. Then the net $\{x_\lambda\}$ *converges* to a point a_0 of X or is *has limit* a_0 (notation: $\lim_\lambda x_\lambda = a_0$ or $x_\lambda \to a_0$) if for every neighborhood V of a_0 there exists an index λ_0 such that $x_\lambda \in V$ for all $\lambda \geq \lambda_0$. Similarly, a filter base $\mathcal{B}$ in X *converges* to a point a_0 (notation: $\lim \mathcal{B} = a_0$) if for every neighborhood V of a_0 there exists a set E in $\mathcal{B}$ such that $E \subset V$.

We observe that these definitions include the definition of a convergent *sequence* in a topological space, as well as that of a convergent *filter*, and also that they are generalizations of the corresponding notions in a metric space (that is, a net or a filter base in a metric space X converges to a point a_0 of X in the sense of Chapters 6 and 7 if and only if it converges to a_0 in the sense just defined with respect to the metric topology on X). Note that a filter $\mathcal{F}$ in a topological space X converges to a point a_0 of X if and only if $\mathcal{F}$ refines the neighborhood filter $\mathcal{V}_{a_0}$. Using the notion of convergence for nets it is easy to formulate the appropriate generalization of the familiar sequential criterion for being an adherent point of a set.

Proposition 9.14. *Let X be a topological space, let A be a subset of X, and let a be a point of X. Then a is an adherent point of A if and only if there exists a net $\{x_\lambda\}$ in A such that* $\lim_\lambda x_\lambda = a$.

PROOF. If such a net exists, then it is clear that every neighborhood of a contains points of A, and hence that $a \in A^-$. On the other hand, for each point a of X, and for an arbitrary neighborhood base $\mathcal{W}_a$ at a in X, the system $\mathcal{W}_a$ is itself a directed set (in the inverse inclusion ordering; Example 1Q), and if $a \in A^-$, then every set W in $\mathcal{W}_a$ contains a point x_W of A. Thus $\{x_W\}_{W \in \mathcal{W}_a}$ is a net in A that converges to a. □

That nets, as opposed to sequences, are really needed in the preceding proposition is shown by the following elementary example.

Example I. Let X be an uncountable set, and let x_0 be any one fixed point of X. For each point y of the set $A = X \backslash \{x_0\}$ we define $\mathcal{W}_y = \{\{y\}\}$, and we define $\mathcal{W}_{x_0}$ as follows: A subset W of X belongs to $\mathcal{W}_{x_0}$ if and only if W contains x_0 and $X \backslash W$ is countable. It is more or less obvious that each $\mathcal{W}_x$, $x \in X$, is a filter base in X, and it is readily verified (Prob. B) that there exists a unique topology $\mathcal{T}$ on X with respect to which each $\mathcal{W}_x$ is, in fact, a neighborhood base at x. Moreover, given this fact, it is clear that A is dense in X in the topology $\mathcal{T}$, but there is no *sequence* in A that converges to x_0 with respect to $\mathcal{T}$. Indeed, if $\{y_n\}$ is an arbitrary sequence in A, then the set B consisting of all the points in $\{y_n\}$ is countable, so $W = X \backslash B$ belongs to $\mathcal{W}_{x_0}$. Thus W is a neighborhood of x_0 containing no point of $\{y_n\}$.

On the other hand, there are topological spaces other than the metrizable ones in which sequences *do* suffice to characterize the closure of an arbitrary set, and thus to determine the entire topology.

Definition. A topological space X is said to satisfy the *first axiom of countability* if there is a countable neighborhood base at every point of X. Likewise, just as in the case of metric spaces, X is said to satisfy the *second axiom of countability* if there exists a countable base for the topology on X. Obviously the second axiom of countability implies the first. (Since every metric space satisfies the first axiom of countability (Ex. H), it is evident why this concept had to wait till now to be introduced. A nonmetrizable space satisfying the first axiom of countability will be found in Example J below.)

Proposition 9.15. *If a topological space X satisfies the first axiom of countability, and if A is a subset of X, then a point a_0 of X is an adherent point of A if and only if there exists a sequence in A that converges to a_0.*

PROOF. The sufficiency of the condition has already been established (Prop. 9.14). To prove its necessity suppose $a_0 \in A^-$, and let $\{W_n\}_{n=1}^{\infty}$ be a countable neighborhood base at a_0 arranged, somehow, into a sequence. If for each positive integer n we set

$$V_n = W_1 \cap \cdots \cap W_n,$$

then $\{V_n\}_{n=1}^{\infty}$ is a nested neighborhood base at a_0, and if for each n we choose x_n such that $x_n \in A \cap V_n$, then it is easily seen that the sequence $\{x_n\}_{n=1}^{\infty}$ converges to a_0. □

It is entirely possible for a topological space to satisfy the first axiom of countability without satisfying the second, even in the presence of a countable dense set (cf. Theorem 6.18).

Example J. Let X denote the set $\mathbb{R}$ of real numbers, deprived for the moment of its natural topology and thought of simply as a set. Let M be a countable set of positive real numbers with the property that $\inf M = 0$, and for each t in X and r in M set

$$W_{t,r} = \{t\} \cup \{q \in \mathbb{Q} : |q - t| < r\}.$$

Then $\mathcal{W}_t = \{W_{t,r} : r \in M\}$ is a filter base in X, and it is not difficult to see that the filter bases $\mathcal{W}_t$, $t \in X$, are neighborhood bases at the various points of X with respect to a uniquely determined topology $\mathcal{T}$ on X (cf. Problem B). As regards this topology, it is obvious that $(X, \mathcal{T})$ satisfies the first axiom of countability (since M is countable), and also that X is separable with respect to $\mathcal{T}$ in the sense that a countable subset of X (viz., $\mathbb{Q}$) is dense in X. But X does not satisfy the second axiom of countability with respect to $\mathcal{T}$. Indeed, if $\mathcal{B}$ is an arbitrary topological base for $(X, \mathcal{T})$, and if r_0 is any one fixed half-length in M, then, for each irrational number $t, \mathcal{B}$ must contain a set V_t such that $t \in V_t \subset W_{t,r_0}$, and any two such sets are distinct. Thus the cardinal number of $\mathcal{B}$ is at least $\aleph$. This example shows that the counterpart of Theorem 6.18 fails to hold in a general topological space. Moreover, the relative topology on the set of irrational numbers in $(X, \mathcal{T})$ is simply the discrete topology, so it is also possible for a topological space in which a countable set is dense to possess subspaces that do not have this property (cf. Corollary 6.19).

Up to this point we have dealt in this chapter almost exclusively with topics introduced in Chapter 6. How do matters stand as regards Chapter 7? The root idea of that chapter, viz., the notion of continuity, generalizes at once. (Indeed, continuity has already been observed to be a topological concept; recall Propositions 7.1 and 7.2.)

Definition. Let X and Y be topological spaces and let ϕ be a mapping of X into Y. Then ϕ is *continuous* at a point x_0 of X if for every neighborhood W of $\phi(x_0)$ in Y there exists a neighborhood V of x_0 in X such that $\phi(V) \subset W$ (or, equivalently, if, for every neighborhood W of $\phi(x_0)$ in $Y, \phi^{-1}(W)$ is a neighborhood of x_0). If ϕ is continuous at a point x_0 of X, then x_0 is a *point of continuity* of ϕ; otherwise x_0 is a *point of discontinuity* of ϕ and ϕ is *discontinuous* at x_0. If ϕ is continuous at every point of X, then ϕ is *continuous* on X.

The sequential criterion for continuity given in Proposition 7.1 fails in general, as was to be expected, but we do have the following two results. (The proof of Proposition 9.17 is substantially the same as that of Proposition 7.1, and is therefore omitted.)

Proposition 9.16. *Let ϕ be a mapping of a topological space X into a topological space Y, and let x_0 be a point of X. Then ϕ is continuous at x_0 if and only if the net $\{\phi(x_\lambda)\}$ converges in Y to $\phi(x_0)$ whenever $\{x_\lambda\}$ is a net in X that converges to x_0.*

PROOF. That the stated criterion is necessary is more or less obvious: let ϕ be continuous at x_0 and let $\{x_\lambda\}_{\lambda\in\Lambda}$ be net in X that converges to x_0. If W is an arbitrary neighborhood of $\phi(x_0)$ in Y, then $\phi^{-1}(W)$ is a neighborhood of x_0 in X, so, for some index $\lambda_0, x_\lambda \in \phi^{-1}(W)$ for all $\lambda \geq \lambda_0$, and therefore $\phi(x_\lambda) \in W$ for $\lambda \geq \lambda_0$.

The sufficiency of the criterion is somewhat more interesting. Suppose that ϕ is discontinuous at a point x_0 in X. Then for a suitably chosen ("sufficiently small") neighborhood W_0 of $\phi(x_0)$ there is no neighborhood V of x_0 such that $\phi(V) \subset W_0$. Consequently for each neighborhood V in $\mathcal{V}_{x_0}$ there is a point x_V in V such that $\phi(x_V) \notin W_0$. But then $\{x_V\}_{V\in\mathcal{V}_{x_0}}$ is a net in X that obviously converges to x_0, while, equally obviously, the net $\{\phi(x_V)\}$ does not converge to $\phi(x_0)$. Thus the stated condition is sufficient. □

Proposition 9.17. *Let ϕ be a mapping of a topological space X into a topological space Y, and suppose X satisfies the first axiom of countability. Then ϕ is continuous at a point x_0 of X if and only if the sequence $\{\phi(x_n)\}$ converges in Y to $\phi(x_0)$ whenever $\{x_n\}$ is a sequence in X that converges to x_0.*

The following elementary results are frequently useful.

Proposition 9.18. *Let ϕ be a mapping of a topological space $(X, \mathcal{T})$ into a topological space $(Y, \mathcal{T}')$. If $\mathcal{T}''$ is another topology on X that refines $\mathcal{T}$, then $\phi : (X, \mathcal{T}'') \to (Y, \mathcal{T}')$ is continuous [at a point x of X] whenever $\phi : (X, \mathcal{T}) \to (X, \mathcal{T}')$ is continuous [at x]. Dually, if $\mathcal{T}''$ is another topology on Y, and if $\mathcal{T}''$ is refined by $\mathcal{T}'$, then $\phi : (X, \mathcal{T}) \to (Y, \mathcal{T}'')$ is continuous [at a point x of X] whenever $\phi : (X, \mathcal{T}) \to (Y, \mathcal{T}')$ is continuous [at x].*

PROOF. If $\mathcal{T}_1$ and $\mathcal{T}_2$ are topologies on the same set Z such that $\mathcal{T}_1$ is refined by $\mathcal{T}_2$, then the neighborhood filter at each point of Z with respect to $\mathcal{T}_1$ is also refined by the neighborhood filter at that point with respect to $\mathcal{T}_2$

(Ex. G). Thus the validity of the proposition is an immediate consequence of the very definition of continuity. □

Proposition 9.19. *Let $\mathcal{T}$ be a topology on a set X, let $\{\mathcal{T}_\gamma\}_{\gamma\in\Gamma}$ be an indexed family of topologies on a set Y, and set $\mathcal{T}_0 = \sup_\gamma \mathcal{T}_\gamma$. Then a mapping $\phi : (X, \mathcal{T}) \to (Y, \mathcal{T}_0)$ is continuous [at a point x of X] if and only if all of the mappings $\phi : (X, \mathcal{T}) \to (Y, \mathcal{T}_\gamma), \gamma \in \Gamma$, are continuous [at x].*

PROOF. If $\phi : (X, \mathcal{T}) \to (Y, \mathcal{T}_0)$ is continuous at some point x of X, then so is $\phi : (X, \mathcal{T}) \to (Y, \mathcal{T}_\gamma)$ for every index γ by what has just been shown. Suppose, on the other hand, that this latter condition is satisfied, and set $y = \phi(x)$. For any neighborhood W of y with respect to $\mathcal{T}_0$ there exists a finite set of indices $\{\gamma_1, \ldots, \gamma_n\}$ and, for each index $i = 1, \ldots, n$, an open set U_i in $\mathcal{T}_{\gamma_i}$ such that

$$y \in U_1 \cap \ldots \cap U_n \subset W$$

(see Problem D). Moreover, for each index i there exists a neighborhood of V_i of x such that $\phi(V_i) \subset U_i$. But then $\phi(V_1 \cap \ldots \cap V_n) \subset W$, and since $V_1 \cap \ldots \cap V_n$ is a neighborhood of x in X, this shows that $\phi : (X, \mathcal{T}) \to (Y, \mathcal{T}_0)$ is continuous at x. □

For mappings defined only on some subset of a topological space continuity is defined in terms of the relative topology. The following formal definition clarifies this point. (Since the relative metric on a subset of a metric space induces the relative topology on that subset (Ex. F), this is in accord with the notion of relative continuity introduced in Chapter 7.)

Definition. Let X and Y be topological spaces, and let ϕ be a mapping of some subset A of X into Y. Then ϕ is *continuous* at a point x_0 of A *relative to* A if the restriction $\phi|A$ (regarded as a mapping on the subspace A) is continuous at x_0. If ϕ is continuous at a point x_0 of A relative to A, then x_0 is a *point of continuity* of ϕ *relative to* A; otherwise, ϕ is *discontinuous* at x_0 *relative to* A, and x_0 is a *point of discontinuity* of ϕ *relative to* A. Similarly, ϕ is *continuous* on A (*relative to* A) if $\phi|A : A \to Y$ is continuous on the subspace A.

The following result generalizes Proposition 7.3. (If the subspace A in Proposition 9.20 satisfies the first axiom of countability, then the nets in its statement may be replaced by sequences.)

Proposition 9.20. *Let X and Y be topological spaces, and let ϕ be a mapping of a subset A of X into Y. Then ϕ is continuous at a point*

x_0 of A relative to A if and only if for each neighborhood W of $\phi(x_0)$ in Y there is a neighborhood V of x_0 in X such that $\phi(V \cap A) \subset W$. Equivalently, ϕ is continuous at x_0 relative to A if and only if $\{\phi(x_\lambda)\}$ converges to $\phi(x_0)$ whenever $\{x_\lambda\}$ is a net in A that converges to x_0.

PROOF. The validity of the latter of these two criteria is an immediate consequence of Proposition 9.16 and the above definition. As for the former, its validity follows at once from the fact that the neighborhoods of x_0 in the subspace A are precisely the sets of the form $V \cap A$ where V is a neighborhood of x_0 in X; see Example G. □

The proof of the following theorem may be obtained from that of Theorem 7.4 by systematically substituting neighborhoods for open balls and is therefore omitted.

Theorem 9.21. *Let X and Y be topological spaces and let ϕ be a mapping of X into Y. Then the following are equivalent:*

(1) ϕ is continuous,
(2) For every open set U in Y the inverse image $\phi^{-1}(U)$ is open in X,
(3) For every closed set F in Y the inverse image $\phi^{-1}(F)$ is closed in X.

The following notions also make sense in an arbitrary topological space.

Definition. A mapping ϕ of a topological space X into a topological space Y is *open* [*closed*] if $\phi(A)$ is open [closed] in Y whenever A is open [closed] in X. A one-to-one mapping ϕ of a topological space X onto a topological space Y is a *homeomorphism* if both ϕ and ϕ^{-1} are continuous.

The following result consists essentially of paraphrases of Theorem 9.21 in the special context of one-to-one mappings, and no proof is needed or given.

Corollary 9.22. *A one-to-one mapping ϕ of a topological space X into a topological space Y is continuous if and only if the inverse mapping ϕ^{-1} (regarded as a mapping of the range of ϕ onto X) is open, or, equivalently, closed. A one-to-one mapping ϕ of X onto Y is a homeomorphism if and only if ϕ is both continuous and open, or what comes to the same thing, if and only if the mapping of the power class 2^X onto 2^Y induced by ϕ carries the topology on X onto the topology on Y.*

Example K. If X and Y are simply ordered sets and if ϕ is an order isomorphism of X onto Y, then ϕ is automatically a homeomorphism of X onto Y as well, if X and Y are equipped either with their order topologies, or with their left- or right-ray topologies (Ex. D). Thus, in particular, the extended real number system $\mathbb{R}^\natural$ is homeomorphic to an arbitrary nondegenerate closed interval $[a, b]$ (in its relative topology; see Problem E). This remark shows, of course, that the space $\mathbb{R}^\natural$ is metrizable; there is just no one "natural" metric to assign to it.

Similarly, if ψ is an order anti-isomorphism between simply ordered sets X and Y (i.e., an order isomorphism between X and Y^*; Example 1P), then ψ is a homeomorphism between $(X, \mathcal{T}_0)$ and $(Y, \mathcal{T}_0)$ that interchanges the left- and right-ray topologies. Thus, in particular, $t \to -t$ is such a homeomorphism of $\mathbb{R}[\mathbb{R}^\natural]$ onto itself.

The statements of the following three results are obtained by systematically substituting the phrase "topological space" for "metric space" in Proposition 7.9, Corollaries 7.10 and 7.11, and Proposition 7.12. None of them needs to be proved anew.

Proposition 9.23. *Let X be a topological space, let ϕ be a mapping of X into a topological space Y, and let U be an open set in X. Then ϕ is continuous at a point x_0 of U relative to U if and only if ϕ is continuous at x_0 (relative to X).*

Corollary 9.24. *If two mappings ϕ and ψ of a topological space X into a topological space Y coincide on some open subset U of X, then ϕ and ψ are continuous (and discontinuous) at precisely the same points of U. A mapping ϕ of a topological space X into a topological space Y is continuous if and only if there exists an open covering of X consisting of sets U such that ϕ is continuous on U relative to U.*

Proposition 9.25. *Let X, Y and Z be topological spaces, let ϕ be a mapping of X into Y, and let ψ be a mapping of Y into Z, so that $\psi \circ \phi$ is a mapping of X into Z. If ϕ is continuous at a point x_0 of X and ψ is continuous at the point $y_0 = \phi(x_0)$, then $\psi \circ \phi$ is also continuous at x_0. In particular, if ϕ and ψ are both continuous, then $\psi \circ \phi$ is also continuous.*

Example L. Let X be a set consisting of at least three points, and let X be equipped with the indiscrete topology (Ex. B). Then any mapping of X into itself is continuous according to Theorem 9.21. In particular, the identity mapping on X is continuous, but its level sets (Ex. 1H), being the various singletons in X, are not closed. Likewise, the mapping of X onto itself that interchanges some one pair of points of X and leaves all other points alone agrees with the identity mapping at at least one point. But then these two continuous mappings agree on a dense subset of X without being identical. Thus the counterparts of Corollaries 7.6 and 7.7 fail to hold for general topological spaces.

The theory of semicontinuous functions actually assumes a more natural appearance in the context of the topological space $\mathbb{R}^\natural$ of extended real numbers.

Definition. An extended real-valued function f on a topological space X is *upper semicontinuous* at a point x_0 of X if for each extended real number u such that $f(x_0) < u$ there exists a neighborhood V of x_0 such that $x \in V$ implies $f(x) < u$ (equivalently, if for any such u the set $\{x \in X : f(x) < u\}$ is a neighborhood of x_0). Dually, f is *lower semicontinuous* at x_0 if for each extended real number s such that $f(x_0) > s$ there exists a neighborhood V of x_0 such that $x \in V$ implies $f(x) > s$. Finally, f is *upper [lower] semicontinuous on* X if it is upper [lower] semicontinuous at every point of X.

Proposition 7.15 now assumes the following somewhat tidier form.

Proposition 9.26. *Let f be an extended real-valued function defined on a topological space X, and let x_0 be a point of X. Then f is upper semicontinuous at x_0 if and only if f is continuous at x_0 as a mapping of X into $\mathbb{R}^\natural$ equipped with the left-ray topology $\mathcal{T}_\ell$ (Ex. E). Dually, f is lower semicontinuous at x_0 if and only if f is continuous at x_0 as a mapping of X into $\mathbb{R}^\natural$ equipped with the right-ray topology $\mathcal{T}_r$. Consequently, f is upper [lower] semicontinuous if and only if it is continuous as a mapping of X into $\mathbb{R}^\natural$ equipped with the topology $\mathcal{T}_\ell[\mathcal{T}_r]$. Finally, f is continuous [at x_0] if and only if it is both upper and lower semicontinuous [at x_0].*

PROOF. It is easily seen that Proposition 9.26 follows at once from its first assertion (recall Proposition 9.19 and Example K). To verify the latter, suppose first that f is upper semicontinuous at x_0, and let W be a neighborhood of $f(x_0)$ with respect to $\mathcal{T}_\ell$. If $f(x_0) = +\infty$, then $W = \mathbb{R}^\natural$, and $f^{-1}(W) = X$. Otherwise, W contains an initial segment A_u for some u such that $f(x_0) < u$, and it is a consequence of the definition that there exists a neighborhood V of x_0 in X such that $f(V) \subset A_u \subset W$. Thus, in any case, $f^{-1}(W)$ is a neighborhood of x_0, so $f : X \to (\mathbb{R}^\natural, \mathcal{T}_\ell)$ is continuous at x_0. On the other hand, if $f : X \to (\mathbb{R}^\natural, \mathcal{T}_\ell)$ is continuous at x_0 and if u is an extended real number such that $f(x_0) < u$, then there exists a neighborhood V of x_0 such that $f(V) \subset A_u$, which shows that f is upper semicontinuous at x_0. □

The proof of the following proposition may readily be supplied by the reader (see Example 6L). (As before, nets can be replaced by sequences whenever X satisfies the first axiom of countability.)

Proposition 9.27. *Let f be an extended real-valued function defined on* *a topological space X. Then f is upper semicontinuous at a point x_0 of X if and only if*

$$\limsup_{\lambda} f(x_\lambda) \leq f(x_0)$$

for every net $\{x_\lambda\}_{\lambda\in\Lambda}$ in X such that $x_\lambda \to x_0$. Dually, f is lower semicontinuous at x_0 if and only if

$$\liminf_{\lambda} f(x_\lambda) \geq f(x_0)$$

for every net $\{x_\lambda\}$ in X such that $x_\lambda \to x_0$. Likewise, f is upper [lower] semicontinuous on X if and only if the set $\{x \in X : f(x) < t\}$ [the set $\{x \in X : f(x) > t\}$] is open in X for every finite real number t.

Corollary 9.28. *Let f be an extended real-valued function on a topological space X. If f is upper semicontinuous, then for each extended real number t the set $\{x \in X : f(x) \leq t\}$ is a G_δ in X (i.e., the intersection of countably many open sets). Dually, if f is lower semicontinuous, then the set $\{x \in X : f(x) \geq t\}$ is a G_δ in X for each extended real number t.*

Corollary 9.29. *The infimum (taken pointwise in $\mathbb{R}^\natural$) of an arbitrary collection of upper semicontinuous functions on a topological space X is again upper semicontinuous on X. Dually, the supremum of an arbitrary collection of lower semicontinuous functions is lower semicontinuous.*

The remarkable result set forth in Proposition 7.20 is false in general topological spaces, as might be expected in view of its sequential nature.

Example M. Consider once again the topological space X of Example I. The singleton $\{x_0\}$ is a closed set in that space, so the characteristic function $h = \chi_{\{x_0\}}$ is upper semicontinuous (cf. Problem 7P). On the other hand, if f is an arbitrary continuous real-valued function on X such that $f \geq h$, and if n is an arbitrary positive integer, then there exists a neighborhood V_n of x_0 such that $f(x) > 1 - 1/n$ whenever $x \in V_n$. Hence $f(x) \geq 1$ whenever $x \in V = \bigcap_{n=1}^\infty V_n$, and V is the complement of a countable subset of X. Similarly, of course, if $\{f_n\}_{n=1}^\infty$ is a sequence of continuous real-valued functions on X such that $f_n \geq h$ for all n, then $\inf_n f_n \geq 1$ on the complement of some countable subset of X. Clearly, then, no such sequence can converge downward to h.

Thus far our discussion of general topology has dealt almost exclusively with the similarities and differences between this theory and the theory of metric spaces. There are also topics of interest in general topology that have no counterpart in the theory of metric spaces (the first axiom of countability is an instance of this). We turn next to the notion, or rather notions, of *separation* in a topological space, another topic having, as we shall see, no counterpart in the theory of metric spaces. We have had occasion more than once to invoke the fact that if x and y are distinct points of a metric space (X, ρ), then for sufficiently small radius $\varepsilon > 0$, the open balls $D_\varepsilon(x)$ and $D_\varepsilon(y)$ are disjoint. This elementary but extremely important property is used to classify abstract topological spaces.

Definition. A topological space X is said to be a *Hausdorff space*—or, simply, to be *Hausdorff*—if for any two distinct points x and y of X there exist neighborhoods V_x and V_y of x and y, respectively, such that $V_x \cap V_y = \varnothing$.

Thus every metrizable topological space is Hausdorff. (The converse is false, of course, as many examples attest. That the converse of the following elementary proposition is also false may be seen by considering the space $(X, \mathcal{T}_1)$ of Example C.)

Proposition 9.30. *Every Hausdorff space is a T_1-space.*

PROOF. If $y \neq x$ in a Hausdorff space X, then there is a neighborhood V_y of y that does not contain x. Thus $X \backslash \{x\}$ is open (Prop. 9.12). □

If (X, ρ) is a metric space, then the two singletons $\{x\}$ and $\{y\}$ in the formulation of the definition of a Hausdorff space can be replaced by any two disjoint closed sets. Indeed, let E and F be closed nonempty subsets of X such that $E \cap F = \varnothing$. If $d(E, F) = d_0 > 0$, it is a triviality to construct disjoint open sets U and V such that $E \subset U$ and $F \subset V$; in fact, $U = D_{d_0/2}(E)$ and $V = D_{d_0/2}(F)$ do the trick. If E and F are merely closed and disjoint, the construction is harder but it can still be done. To see this, let us write $r_x = d(x, F)/2$ for each point x of E and, likewise, $s_y = d(y, E)/2$ for each point y of F. Then

$$U = \bigcup_{x \in E} D_{r_x}(x) \quad \text{and} \quad V = \bigcup_{y \in F} D_{s_y}(y)$$

are clearly open subsets of X containing E and F, respectively, and U and V are, in fact, disjoint. For if $x_0 \in U \cap V$, then there must exist x in E and y in F such that $x_0 \in D_{r_x}(x) \cap D_{s_y}(y)$, and, therefore, such that $\rho(x, y) < [d(x, F) + d(y, E)]/2$, clearly an impossibility. This stronger property of metric spaces is also used to classify topological spaces.

Definition. A Hausdorff space (or, what comes to the same thing, a T_1-space) is said to be *normal* if it has the property that, whenever E and F are disjoint closed sets in X, there exist disjoint open subsets U and V of X such that $E \subset U$ and $F \subset V$. (Such open sets U and V are said to *separate* E and F.)

Thus every metrizable topological space is normal, and, once again, the converse is false, though less obviously so this time; see Example X. The following criterion is nothing more that a modest reworking of the definition, but it is sometimes useful.

Proposition 9.31. *A T_1-space X is normal if and only if for every closed set F and open set U in X such that $F \subset U$ there exists an open set V in X such that $F \subset V$ and $V^- \subset U$.*

PROOF. If X is normal and F and U are as stated, set $E = X \backslash U$. Then E and F are disjoint and closed, so there exist disjoint open sets W and V such that $E \subset W$ and $F \subset V$. But then $V^- \subset X \backslash W$, so $V^- \subset U$. On the other hand, if the stated condition is satisfied, and if E and F are disjoint closed sets in X, then $U = X \backslash E$ is open and contains F. Let V be an open set such that $F \subset V$ and $V^- \subset U$. Then $X \backslash V^-$ is open and contains $X \backslash U = E$. □

The following result is of crucial importance in the study of metrizability and in other contexts as well. (In this connection the reader should recall Example 7G and Problem 7E.)

Theorem 9.32 (Urysohn's Lemma). *If E and F are disjoint closed subsets of a normal topological space X, then there exists a continuous function $f : X \to [0, 1]$ such that f is identically zero on E and identically one on F, i.e., such that $\chi_F \leq f \leq \chi_{X \backslash E}$.*

PROOF. Let $\mathcal{D}$ denote the collection of all dyadic fractions (Th. 2.12). The first step in the construction of the function f is the definition of a mapping $t \to V_t$ assigning to each number t in $\mathcal{D}$ an open set V_t in X in such a way that

(a) $V_t = \varnothing$ for $t < 0$ and $V_t = X$ for $t > 1$,
(b) $E \subset V_t$ for $t \geq 0$ and $V_t \cap F = \varnothing$ for $t \leq 1$,
(c) $V_t^- \subset V_{t'}$, for all t, t' in $\mathcal{D}$ such that $t < t'$.

Suppose for the moment that such a mapping has been defined. Then for each point x of X the set $R_x = \{t \in \mathcal{D} : x \in V_t\}$ is, by (a), a nonempty subset of $\mathcal{D}$ such that $0 \leq \inf R_x \leq 1$. Thus setting

$$f(x) = \inf R_x, \quad x \in X,$$

defines a mapping f of X into $[0, 1]$. Moreover, it is clear from (b) that $f(x) = 0$ whenever $x \in E$ and that $f(x) = 1$ whenever $x \in F$. Finally, the function f is continuous because of (c); the proof is the same as that given in Example 7G and may therefore be omitted.

Thus everything comes down to defining a mapping of $\mathcal{D}$ into the topology $\mathcal{T}$ on X satisfying (a), (b) and (c). Moreover, the dyadic fractions not in [0,1] take care of themselves by virtue of (a). Thus what is really needed is a mapping of the set $\widetilde{\mathcal{D}}$ of all dyadic fractions in [0,1] into $\mathcal{T}$ satisfying (b) and (c) for all t, t' in $\widetilde{\mathcal{D}}$. The definition is by mathematical induction and uses Proposition 9.31 repeatedly. For each nonnegative integer n let $\mathcal{D}_n$ denote the set of those numbers in $\widetilde{\mathcal{D}}$ of the form $k/2^n$, $k = 0, 1, \ldots, 2^n$, and consider, to begin with, the set $\mathcal{D}_0 = \{0, 1\}$. Here the definition presents no problem: We set $V_1 = X \backslash F$ and choose for V_0 any open set V in X such that $E \subset V$ and $V^- \subset V_1$. Consider next $\mathcal{D}_1 = \{0, 1/2, 1\}$. We require that the mapping of $\mathcal{D}_1$ into $\mathcal{T}$ extend the one already defined on $\mathcal{D}_0$, so V_t is defined for $t = 0$ and $t = 1$, and $V_0^- \subset V_1$. Hence for $V_{1/2}$ we may select any open set V in X such that $V_0^- \subset V$ and $V^- \subset V_1$.

The inductive step presents no new difficulty. Suppose nested mappings of $\mathcal{D}_k$ into $\mathcal{T}$ have already been defined for $k = 0, \ldots, n$ such that (c) is satisfied for all t, t' in $\mathcal{D}_n$. (We can forget about (b) since all of these mappings extend the one defined above on $\mathcal{D}_0$.) If $t = k/2^{n+1} \in \mathcal{D}_{n+1}$, and if $k = 2j$ is even, then $V_t = V_{j/2^n}$ is already defined. On the other hand, if k is odd, then $V' = V_{(k-1)/2^n}$ and $V'' = V_{(k+1)/2^n}$ are defined, and we have $V'^- \subset V''$. Thus we may choose for V_t any open set V in X such that $V'^- \subset V$ and $V^- \subset V''$. In this way we obtain a mapping of $\mathcal{D}_{n+1}$ into $\mathcal{T}$ that extends the one already defined on $\mathcal{D}_n$, and it is readily verified that (c) holds for this extension whenever t and t' are both in $\mathcal{D}_{n+1}$.

Finally, we form the supremum of this inductively defined sequence of mappings, thus obtaining a mapping of $\widetilde{\mathcal{D}}$ into $\mathcal{T}$ that clearly satisfies (b). But it also satisfies (c). Indeed, for any two numbers t and t' in $\widetilde{\mathcal{D}}$ there is a positive integer n such that t and t' both belong to $\mathcal{D}_n$, so (c) holds in general. □

Definition. A function f as in Urysohn's lemma is known as a *Urysohn function* for the pair of sets E and F, and is said to *separate* E and F.

The method employed in Chapter 7 to prove the Tietze extension theorem (Ex. 7S), based as it was on the Hahn interpolation theorem (Prop. 7.21), is unavailable in a general topological space. Nevertheless, the result remains valid in a normal space by virtue of Urysohn's lemma. In the following discussion we shall write $\|g\|_X$ for the sup norm

$$\|g\|_X = \sup_{x \in X} |g(x)|$$

of a bounded scalar-valued function g defined on a space X, and also, if A is a subset of X,

$$\|g\|_A = \sup_{x \in A} |g(x)|$$

for a bounded scalar-valued function g defined on either A or X. The following result contains the needed consequence of Urysohn's lemma.

Lemma 9.33. *Let X be a normal topological space, let F be a closed nonempty subset of X, and let a be a positive real number. For any continuous real-valued function g on F such that $\|g\|_F \le a$ there is a continuous real-valued function ϕ on X such that $\|\phi\|_X \le a/3$ while $\|g - \phi\|_F \le 2a/3$.*

PROOF. Set $E_+ = \{x \in F : g(x) \ge a/3\}$ and $E_- = \{x \in F : g(x) \le -a/3\}$. The sets E_+ and E_- are closed and disjoint in X, and it is an immediate consequence of Urysohn's lemma that there exists a continuous mapping ϕ of X into $[-a/3, +a/3]$ such that $f = \pm a/3$ on $E_\pm$. But then, of course, $\|\phi\|_X \le a/3$, and it is easily seen that $\|g - \phi\|_F \le 2a/3$ as well. □

Theorem 9.34 (Tietze Extension Theorem; Version II). *Let X be a normal topological space, let F be a nonempty closed set in X, and let f_0 be a continuous real-valued function defined and bounded on F. Then there exists a continuous real-valued function f on X that has the same upper and lower bounds on X as f_0 has on F, and that agrees with f_0 on F.*

PROOF. Let

$$a = \inf_{x \in F} f_0(x) \quad \text{and} \quad b = \sup_{x \in F} f_0(x).$$

If $a = b$, then f_0 is constant on F, and there is nothing to prove, as we may set $\phi = a = b$ on X. Otherwise, there is a linear function $h(x) = Ax + B$ carrying the interval $[a, b]$ onto $[-1, +1]$, and therewith f_0 onto $\widetilde{f}_0 = h \circ f_0$ with $\|\widetilde{f}_0\|_F = 1$, and it clearly suffices to prove the theorem for $\widetilde{f}_0$ in place of f_0 (for if $\widetilde{\phi}$ is a suitable extension of $\widetilde{f}_0$, then $\phi = h^{-1} \circ \widetilde{\phi}$ provides a suitable extension of f_0).

According to the lemma, there exists a continuous real-valued function ϕ on X satisfying the conditions

$$\|\phi\|_X \le 1/3 \quad \text{and} \quad \|f_0 - \phi\|_F \le 2/3.$$

We set $\phi_1 = \phi$ on X and $f_1 = f_0 - \phi_1$ on F, and apply the lemma once again to obtain a function ϕ_2 such that

$$\|\phi_2\|_X \le (1/3)(2/3) \quad \text{and} \quad \|f_1 - \phi_2\|_F \le (2/3)^2,$$

and observe that $f_1 - \phi_2 = f_0 - (\phi_1 + \phi_2)$. Continuing in this way by mathematical induction, we obtain a sequence $\{\phi_n\}_{n=1}^\infty$ of continuous real-valued functions on X such that

$$\|\phi_n\|_X \le (1/3)(2/3)^{n-1} \quad \text{and} \quad \|f_0 - (\phi_1 + \ldots + \phi_n)\|_F \le (2/3)^n, \; n \in \mathbb{N}.$$

The series $\sum_{n=1}^{\infty} \phi_n$ is then uniformly convergent on X to a sum f such that

$$\|f\|_X \le \sum_{n=1}^{\infty} \|\phi_n\|_X \le (1/3) \sum_{n=0}^{\infty} (2/3)^n = 1$$

and such that

$$\|f - (\phi_1 + \ldots + \phi_n)\|_X \le \sum_{k=n+1}^{\infty} \|\phi_k\|_X = (2/3)^n,$$

and therefore such that

$$\|f_0 - f\|_F \le 2(2/3)^n$$

for each index n. Hence $f = f_0$ on F. □

Definition. A subset C of a topological space X that is both closed and open is called *closed-open.* A partition $\{M, N\}$ of X into the union of two nonempty disjoint closed-open sets M and N is a *disconnection* of X. If there exists a disconnection of X, then X is *disconnected.* If X is not disconnected, then it is *connected.* A subset A of X is *connected* [*disconnected*] if it is connected [disconnected] as a subspace of X.

From the fact that open and closed sets are complements of one another it is apparent that a disconnection of an arbitrary topological space X may equally well be described as a partition of X into two disjoint nonempty open sets, or two disjoint nonempty closed sets. Thus (Cor. 9.10) a disconnection of an open subset U in X is a partition of U into two disjoint nonempty sets that are open in X, while a disconnection of a closed set F in X is a partition of F into two disjoint nonempty sets that are closed in X. It is desirable to have a description of a disconnection of an arbitrary subspace.

Definition. Two subsets M and N of a topological space X satisfy the *Hausdorff-Lennes condition* if $M^- \cap N = M \cap N^- = \varnothing$.

Proposition 9.35. *If A is an arbitrary subset of a topological space X, and if $\{M, N\}$ is an arbitrary partition of A into two nonempty subsets, then $\{M, N\}$ is a disconnection of A if and only if M and N satisfy the Hausdorff-Lennes condition.*

PROOF. By Proposition 9.9 the closure of M relative to A is $M^- \cap A = M \cup (M^- \cap N)$. Thus M is closed relative to A if and only if $M^- \cap N = \varnothing$. Similarly, N is closed relative to A if and only if $M \cap N^- = \varnothing$. □

The fact that the definition of connectedness is stated in terms of a negation makes the concept somewhat awkward to work with. Here is a positive characterization of connectedness that summarizes the foregoing remarks.

Proposition 9.36. *A topological space X is connected if and only if the only closed-open subsets of X are $\varnothing$ and X itself or, equivalently, if and only if C closed-open in X implies either $C = \varnothing$ or $C = X$. A subspace A of X is connected if and only if, whenever $A = M \cup N$, where M and N are two subsets satisfying the Hausdorff-Lennes condition, either $M = \varnothing$ or $N = \varnothing$.*

Corollary 9.37. *If $\{M, N\}$ is a pair of subsets of a topological space X satisfying the Hausdorff-Lennes condition, and if C is a connected subset of X such that $C \subset M \cup N$, then either $C \subset M$ or $C \subset N$.*

PROOF. The pair $\{M \cap C, N \cap C\}$ also satisfies the Hausdorff-Lennes condition. □

Proposition 9.38. *If C is a connected subset of a topological space X, and if A is a set such that $C \subset A \subset C^-$, then A is also connected. In particular, C^- is connected.*

PROOF. Let $A = M \cup N$ where $\{M, N\}$ is a pair of sets satisfying the Hausdorff-Lennes condition. By Corollary 9.37, either $C \subset M$ or $C \subset N$. Suppose without loss of generality that $C \subset M$. Then $A \subset C^- \subset M^-$, so $N = A \cap N = \varnothing$. □

Proposition 9.39. *If $\mathcal{C}$ is an arbitrary chained collection of connected subsets of a topological space X (Ex. 1J), then $\bigcup \mathcal{C}$ is also connected.*

PROOF. Let $A = \bigcup \mathcal{C}$, and let $A = M \cup N$, where M and N are sets satisfying the Hausdorff-Lennes condition. If C is any one of the sets in $\mathcal{C}$, then either $C \subset M$ or $C \cap M = \varnothing$. Suppose $M \neq \varnothing$. Then there exists a set C in $\mathcal{C}$ such that $C \cap M \neq \varnothing$, and therefore such that $C \subset M$. But if D is any set in $\mathcal{C}$, then there exist sets $C_0, \ldots, C_p$ in $\mathcal{C}$ such that $C_0 = C, C_p = D$, and $C_i \cap C_{i-1} \neq \varnothing$, $i = 1, \ldots, p$. Then $C_0 \subset M$ and if $C_{i-1} \subset M$, then $C_i \cap M \supset C_i \cap C_{i-1} \neq \varnothing$, so $C_i \subset M$. Thus $C_i \subset M$ for all $i = 0, \ldots, p$ by induction. In particular, $C_p = D \subset M$, and it follows that $A = M$ (and hence that $N = \varnothing$). □

Corollary 9.40. *For any topological space X and any point a of X there exists a unique largest connected subset $[a]$ of X containing a.*

PROOF. The collection of all connected subsets of X containing the point a is obviously chained, and its union is connected, contains a (for the singleton $\{a\}$ is surely connected), and is clearly the largest such set. □

Definition. For each point a of a topological space X the largest connected subset $[a]$ of X containing a is the (*connected*) *component* of a (in X).

Proposition 9.41. *The component $[a]$ of a point a in a topological space X is a closed subset of X. Moreover, if x is some point of X lying in $[a]$, then the component $[x]$ of x coincides with $[a]$. Hence the collection of all connected components of the various points of X constitutes a partition of X into closed equivalence classes. (For this reason the connected components of the various points of X are also referred to as the* (connected) components of X.)

PROOF. The closure $[a]^-$ is connected along with $[a]$ (Prop. 9.38), and since $[a]$ is the largest connected subset of X containing a, this shows that $[a]^- = [a]$. As for the rest, if $x \in [a]$, then $[x] \cap [a] \neq \varnothing$ since both sets contain x. Thus the collection $\{[a], [x]\}$ is chained, so $[a] \cup [x]$ is also connected. But then $[a] \cup [x] = [a] = [x]$. Thus the relation $x \sim a$, defined to mean $x \in [a]$, is an equivalence relation on X. □

Example N. Let C be a nonempty connected set of real numbers. For any real number t the two rays $(-\infty, t)$ and $(t, +\infty)$ satisfy the Hausdorff-Lennes condition. Hence if C contains numbers s and u such that $s < t < u$, then C must also contain t. But then, if $a = \inf C$ and $b = \sup C$ and if $a < t < b$, then $t \in C$, and it follows that C consists of the interval (a, b) along with, perhaps, one or both of its endpoints; briefly, C is an interval. (The cases $a = -\infty$ and\or $b = +\infty$ are not excluded.) On the other hand, all intervals are, in fact, connected in $\mathbb{R}$. Indeed, $\mathbb{R}$ itself cannot be partitioned into two nonempty open sets, since these would, in turn, be disjoint unions of nonempty open intervals and rays, and this would yield a representation of $\mathbb{R}$ as the union of a pairwise disjoint collection of nonempty open intervals and rays other than the unique such expression: $\mathbb{R} = \mathbb{R}$ (cf. Example 6S). For exactly the same reasons all open rays and open intervals are connected in $\mathbb{R}$, whence it follows that all intervals and rays are connected (Prop. 9.38).

Example O. Example N shows that the connected components of an open subset U of $\mathbb{R}$ are precisely the open intervals and/or rays that are the constituent intervals of U. For other sets of real numbers matters can be much more complicated. Consider, for instance, the Cantor set C. If s and u are any two points of C such that $s < u$ and if n is a positive integer such that $1/3^n < u - s$, then it is impossible for s and u to belong to any one interval of length $1/3^n$. Hence s and u must belong to different intervals in the set $\mathcal{F}_n$. (The reader is referred, once again, to Example 6O for details of the construction of C.) Hence there exists a real number t such that $s < t < u$ and $t \notin F_n = \bigcup \mathcal{F}_n$. But then if A is any subset of C containing both s and u, then $\{A \cap (-\infty, t), A \cap (t, +\infty)\}$ is a disconnection of A. Thus there is no connected subset of C containing two distinct points, and we see that each connected component of C is a singleton. A topological space with this property is said to be *totally disconnected*.

Of the topics discussed in Chapter 8 the notion of compactness is of paramount interest in general topology. This idea generalizes without difficulty, the only problem being to decide which of the several criteria for compactness to turn into a definition. The covering property of Theorem 8.31 is the one that leads to important consequences.

Definition. A topological space X is *compact* if every open covering of X contains a finite subcovering. A subset K of a topological space X is *compact* if it is compact as a subspace.

Example P. Let α be a positive ordinal number, and let the ordinal number segment $W(\alpha)$ be equipped with its order topology (Ex. D). If η is a positive number belonging to $W(\alpha)$ and U is an open set containing η, then, according to the definition of the order topology, there exists an ordinal number $\xi < \eta$ such that $\{\zeta : \xi < \zeta \leq \eta\} \subset U$. Hence such sets constitute an open neighborhood base at η in $W(\alpha)$. (In particular then, a point η of $W(\alpha)$ is an isolated point, i.e., an open singleton, if and only if it is a number of type I (cf. Problem 5G).)

Suppose now that $\mathcal{U}$ is an open covering of $W(\alpha)$, and consider the set Ξ of those numbers ξ in $W(\alpha)$ with the property that the set $\widehat{W}(\xi) = W(\xi) \cup \{\xi\}$ of all ordinal numbers less than or equal to ξ is covered by some finite subcollection of $\mathcal{U}$. Clearly 0 belongs to Ξ. But also if $0 < \eta < \alpha$ and if $W(\eta) \subset \Xi$, there is a set U_0 in $\mathcal{U}$ such that $\eta \in U_0$ and hence a number ξ such that $\xi < \eta$ and such that $\widehat{W}(\eta) \backslash \widehat{W}(\xi) = \{\zeta : \xi < \zeta \leq \eta\} \subset U_0$. But then $\xi \in \Xi$, so there are sets $U_1, \ldots, U_k$ in $\mathcal{U}$ that cover $\widehat{W}(\xi)$, and it follows that $\{U_0, U_1, \ldots, U_k\}$ covers $\widehat{W}(\eta)$. Thus $\eta \in \Xi$, and therefore $\Xi = W(\alpha)$ by transfinite induction. Thus we see that $W(\alpha)$ *is compact in its order topology if and only if* α *is a number of type* I. (If $\alpha = \lambda$ is a limit number, then $\mathcal{U} = \{W(\eta) : \eta < \lambda\}$ is an open covering of $W(\lambda)$ containing no finite subcovering.)

Regarding compact sets in a topological space, we have the following elementary result (cf. Corollary 8.32).

Proposition 9.42. *A subset K of a topological space X is compact if and only if every covering of K by means of open subsets of X contains a finite subcovering.*

PROOF. If $\mathcal{U}$ is a covering of K consisting of open subsets of X, then $\widetilde{\mathcal{U}} = \{U \cap K : U \in \mathcal{U}\}$ is an open covering of the subspace K. Conversely, every covering of K consisting of relatively open subsets of K can be obtained in this fashion. □

Proposition 9.43. *A closed subset of a compact topological space is compact. A compact subset of a Hausdorff space is closed. Consequently, a subset of a compact Hausdorff space is compact if and only if it is closed.*

PROOF. Let X be a compact topological space, and let F be closed in X. If $\mathcal{U}$ is a covering of F consisting of open subsets of X, then $\mathcal{U} \cup \{X \backslash F\}$ is an open covering of X. This covering contains a finite covering, from which we may simply discard the set $X \backslash F$ to obtain a finite covering of F consisting of sets in $\mathcal{U}$. Thus F is compact by the foregoing proposition.

Suppose next that X is a Hausdorff space and that K is a compact subset of X. If $a \in X \backslash K$, then for each point x in K there exist open neighborhoods U_x of x and V_x of a such that $U_x \cap V_x = \varnothing$. The open sets $\{U_x\}_{x \in K}$ cover K, so there exist points $x_1, \ldots, x_n$ of K such that

$$K \subset U = U_{x_1} \cup \cdots \cup U_{x_n}.$$

Let $V = V_{x_1} \cap \cdots \cap V_{x_n}$. Then V is an open neighborhood of a and $U \cap V = \varnothing$. In particular, V is a neighborhood of a that does not meet K, and it follows that $X \backslash K$ is open. □

The central idea of the preceding proof yields further useful information.

Proposition 9.44. *Every compact Hausdorff space is normal.*

PROOF. Let X be a compact Hausdorff space and let E and F be disjoint closed sets in X. Both E and F are compact by Proposition 9.43. Moreover, as was shown in the proof of that proposition, for each point y of $X \backslash E$ there exist disjoint open sets U_y and V_y such that $E \subset U_y$ and $y \in V_y$. The sets $\{V_y\}_{y \in F}$ cover F, so there exist points $y_1, \ldots, y_m$ in F such that

$$F \subset V = V_{y_1} \cup \cdots \cup V_{y_m}.$$

Let $U = U_{y_1} \cap \cdots \cap U_{y_m}$. Then U is an open set containing E, and U and V separate E and F. □

It is a simple matter to generalize the material of Proposition 8.35 and Theorem 8.37.

Proposition 9.45. *Let ϕ be a continuous mapping of a topological space X into a topological space Y. If K is a compact subset of X, then $\phi(K)$ is compact in Y. Hence if the entire space X is compact, and if Y is Hausdorff, then ϕ is a closed mapping.*

PROOF. Let $\mathcal{U}$ be a covering of $\phi(K)$ consisting of open subsets of Y. Then $\phi^{-1}(\mathcal{U}) = \{\phi^{-1}(U) : U \in \mathcal{U}\}$ is a collection of open sets in X, and it is clear that $K \subset \bigcup \phi^{-1}(\mathcal{U})$. Since K is compact there are sets $U_1, \ldots, U_n$ in $\mathcal{U}$ such that $K \subset \phi^{-1}(U_1) \cup \cdots \cup \phi^{-1}(U_n)$, and it follows that $\phi(K) \subset U_1 \cup \cdots \cup U_n$. Thus $\phi(K)$ is compact along with K. To complete the proof, suppose that X is compact and Y is Hausdorff. If F is a closed set in X, then F is compact by Proposition 9.43. But then $\phi(F)$ is also compact, and therefore closed in Y, by the same proposition. □

Corollary 9.46. *If f is a continuous real-valued function on a topological space X, and if K is a compact subset of X, then f assumes both a maximum and a minimum value on K.*

Theorem 9.47. *Let X be a compact topological space and let ϕ be a continuous one-to-one mapping of X into a Hausdorff space Y. Then ϕ is a homeomorphism of X onto $\phi(X)$. In particular, if both X and Y are compact Hausdorff spaces, then any continuous one-to-one mapping of X onto Y is a homeomorphism.*

PROOF. Clearly the subspace $\phi(X)$ is Hausdorff along with Y. Hence the mapping ϕ is closed, and therefore also open, as a mapping of X onto $\phi(X)$. □

Example Q. Let $\mathcal{T}$ and $\mathcal{T}'$ be topologies on a set X such that $(X, \mathcal{T})$ is Hausdorff, while $(X, \mathcal{T}')$ is compact, and suppose that $\mathcal{T}'$ refines $\mathcal{T}$. Then the identity mapping $\iota : (X, \mathcal{T}') \to (X, \mathcal{T})$ is continuous and one-to-one, whence it follows that ι is a homeomorphism. Thus $\mathcal{T} = \mathcal{T}'$ and $(X, \mathcal{T}) = (X, \mathcal{T}')$ is a compact Hausdorff space. In particular, if $\mathcal{T}$ and $\mathcal{T}'$ are any two comparable topologies on X each of which turns X into a compact Hausdorff space, then $\mathcal{T} = \mathcal{T}'$.

Just as was the case in Chapter 8, there are a number of different characterizations of compactness in general topological spaces. Before stating the next result, we introduce a notion that turns out to be useful in other contexts as well.

Definition. An *ultrafilter* in a (nonempty) set X is a filter in X that is a maximal element in the collection of all filters in X. That is, $\mathcal{U}$ is an ultrafilter in X if (a) $\mathcal{U}$ is a filter in X and (b) there is no filter in X that properly refines $\mathcal{U}$.

Proposition 9.48. *The following conditions are equivalent for an arbitrary topological space X:*

(1) *X is compact,*
(2) *Every collection of closed sets in X possessing the finite intersection property (Prob. 7S) has nonempty intersection,*
(3) *Every collection of sets in X possessing the finite intersection property has an adherent point,*
(4) *Every filter in X has an adherent point,*
(5) *Every ultrafilter in X is convergent.*

PROOF. Suppose first that X is compact and that $\mathcal{F}$ is a collection of closed subsets of X possessing the finite intersection property. Then $\mathcal{U} = \{X \backslash F : F \in \mathcal{F}\}$ is a collection of open sets in X with the property that no finite subset of $\mathcal{U}$ covers X. But then $\mathcal{U}$ itself cannot cover X, so $\bigcap \mathcal{F} \neq \varnothing$. Thus (1) implies (2). To see that (2) implies (3) suppose (2) holds and let $\mathcal{C}$ be a collection of subsets of X possessing the finite intersection property. Then $\mathcal{F} = \{E^- : E \in \mathcal{C}\}$ also possesses the finite intersection property, so $\bigcap \mathcal{F} = \bigcap_{E \in \mathcal{C}} E^- \neq \varnothing$, and any point of this set is an adherent point of $\mathcal{C}$.

It is obvious that (3) implies (4), since a filter possesses the finite intersection property by definition. Suppose next that (4) holds, and let $\mathcal{U}$ be an ultrafilter in X. Then $\mathcal{U}$ has an adherent point a in X, and if V is a neighborhood of a, then V meets every set in $\mathcal{U}$. But then V is itself an element of $\mathcal{U}$ (see Problem M). Thus $\mathcal{U}$ refines the neighborhood filter $\mathcal{V}_a$, and $\lim \mathcal{U} = a$.

Finally, suppose (5) holds, and let $\mathcal{U}$ be an open covering of X with the property that no finite subset of $\mathcal{U}$ covers X. Then $\mathcal{F} = \{X \backslash U : U \in \mathcal{U}\}$ is a collection of closed sets in X possessing the finite intersection property, and there exists an ultrafilter $\mathcal{U}_0$ in X such that $\mathcal{F} \subset \mathcal{U}_0$. (This follows from Zorn's lemma; see Problem M once again.) If $a = \lim \mathcal{U}_0$, and if V is a neighborhood of a, then $V \in \mathcal{U}_0$, so V meets every set in $\mathcal{F}$. Since this is true of every neighborhood of a, it follows that a belongs to every set in $\mathcal{F}$. But then $a \notin \bigcup \mathcal{U}$, contrary to hypothesis. Thus we have reached a contradiction, and the proof of the proposition is complete. □

Special topologies are sometimes introduced to make certain mappings continuous. (The ray topologies on $\mathbb{R}^\natural$ introduced in Example E may be viewed as instances of this.) Here is a fairly general version of this kind of construction.

Definition. Let X be a set and for each γ in an index set Γ let ϕ_γ be a mapping of X into a topological space Y_γ. Then the coarsest topology on X making all of the mappings ϕ_γ continuous is the topology *inversely induced* on X by the family $\{\phi_\gamma\}_{\gamma\in\Gamma}$. (That such a coarsest topology always exists is an immediate consequence of Proposition 9.6.)

Constructions of this sort have appeared before, albeit in other garb.

Example R. If ϕ is a single mapping of a set X into a topological space $(Y, \mathcal{T})$, then the topology $\mathcal{T}_\phi$ *inversely induced* on X by ϕ, that is, the coarsest topology on X with respect to which ϕ is continuous, is clearly just $\phi^{-1}(\mathcal{T})$ $(= \{\phi^{-1}(U) : U \in \mathcal{T}\})$. If ϕ is a constant mapping, then $\mathcal{T}_\phi$ is the indiscrete topology on X (Ex. B). If ϕ is one-to-one and if the topology on Y is metrized by a metric ρ, then $\mathcal{T}_\phi$ is metrized by the induced metric σ, where

$$\sigma(x, x') = \rho(\phi(x), \phi(x')), \quad x, x' \in X$$

(cf. Example 8E). If $X = A$ is a subset of Y and ϕ is the inclusion mapping ι_A (Ex. 1E), then $\mathcal{T}_\phi$ is simply the relative topology on A. (Still another instance of an inversely induced topology will be found in Problem W below.)

Likewise, if we are given some indexed family $\{\phi_\gamma\}$ of mappings of X into topological spaces Y_γ, and if for each index γ we write $\mathcal{T}_\gamma$ for the topology inversely induced on X by ϕ_γ, then the topology induced by the family $\{\phi_\gamma\}$ is clearly just $\sup_\gamma \mathcal{T}_\gamma$. Moreover, this latter construction itself acquires a slightly different form in the present context. Let $\{\mathcal{T}_\gamma\}_{\gamma\in\Gamma}$ be an indexed family of topologies on a set X, and for each index γ let X_γ denote the topological space consisting of X equipped with the topology $\mathcal{T}_\gamma$. If, for each index γ, ι_γ denotes the identity mapping on X regarded as a mapping of X onto X_γ, then the topology inversely induced on X by the family $\{\iota_\gamma\}_{\gamma\in\Gamma}$ is the supremum $\sup_\gamma \mathcal{T}_\gamma$.

The following result is frequently useful in dealing with inversely induced topologies.

Proposition 9.49. *Let $\{\phi_\gamma\}_{\gamma\in\Gamma}$ be an indexed family of mappings of a set X into an indexed family $\{Y_\gamma\}$ of topological spaces, as above, and let $\mathcal{T}$ denote the topology inversely induced on X by $\{\phi_\gamma\}$. Then a base for $\mathcal{T}$ is given by the collection of all sets of the form*

$$\phi_{\gamma_1}^{-1}(U_1) \cap \cdots \cap \phi_{\gamma_n}^{-1}(U_n), \tag{1}$$

where n denotes an arbitrary positive integer, $\{\gamma_1, \ldots, \gamma_n\}$ is an arbitrary set of n elements of the index sets Γ, and U_i is an open set in Y_{γ_i}, $i = 1, \ldots, n$. Moreover, if, for each index $\gamma, \mathcal{B}_\gamma$ is any one fixed topological base for Y_γ, then the sets U_i in (1) may be required to belong to $\mathcal{B}_{\gamma_i}$

for all $i = 1, \ldots, n$. *If* $\{x_\lambda\}_{\lambda \in \Lambda}$ *is a net in* X, *then* $\{x_\lambda\}$ *converges to a point* x_0 *of* X *with respect to* $\mathcal{T}$ *if and only if the net* $\{\phi_\gamma(x_\lambda)\}_{\lambda \in \Lambda}$ *converges to* $\phi_\gamma(x_0)$ *in* Y_γ *for each index* γ. *Finally, if* ψ *is a mapping of some topological space* Z *into* X, *then* ψ *is continuous as a mapping of* Z *into* $(X, \mathcal{T})$ *if and only if each* $\phi_\gamma \circ \psi : Z \to Y_\gamma$ *is continuous.*

PROOF. In view of Example R, the first two assertions of this proposition are merely paraphrases of the relevant parts of Problem D. Moreover, if a net $\{x_\lambda\}$ converges in X to a limit x_0 with respect to $\mathcal{T}$, then the net $\{\phi_\gamma(x_\lambda)\}$ also converges in Y_γ to $\phi_\gamma(x_0)$ since ϕ_γ is continuous with respect to $\mathcal{T}$ for each index γ (Prop. 9.16). Suppose, on the other hand, that $\{x_\lambda\}$ is a net in X and that there is a point x_0 of X such that $\{\phi_\gamma(x_\lambda)\}_{\lambda \in \Lambda}$ converges to $\phi_\gamma(x_0)$ in Y_γ for each index γ. If V is an arbitrary neighborhood of x_0 with respect to $\mathcal{T}$, then there are indices $\gamma_1, \ldots, \gamma_n$ and open sets U_i in Y_{γ_i} such that $x_0 \in \phi_{\gamma_i}^{-1}(U_i)$, $i = 1, \ldots, n$, and such that

$$\phi_{\gamma_1}^{-1}(U_1) \cap \cdots \cap \phi_{\gamma_n}^{-1}(U_n) \subset V.$$

But then $\phi_{\gamma_i}(x_0) \in U_i$, $i = 1, \ldots, n$. Hence there exist indices λ_i in Λ such that $\phi_{\gamma_i}(x_\lambda) \in U_i$ for $\lambda \geq \lambda_i$, $i = 1, \ldots, n$. Since Λ is directed, there is also a single index λ_0 in Λ such that $\lambda_i \leq \lambda_0$, $i = 1, \ldots, n$, and it follows that $x_\lambda \in V$ for $\lambda \geq \lambda_0$. Thus $\{x_\lambda\}$ converges to x_0 with respect to $\mathcal{T}$.

Finally, if ψ is a continuous mapping of Z into $(X, \mathcal{T})$ then it is obvious that each of the compositions $\phi_\gamma \circ \psi$ is continuous since each ϕ_γ is continuous on $(X, \mathcal{T})$. To complete the proof, suppose, conversely, that ψ is a mapping of Z into X having the stated property, and let $\{z_\lambda\}_{\lambda \in \Lambda}$ be a net in Z converging to a limit z_0. Then each net $\{\phi_\gamma(\psi(z_\lambda))\}_{\lambda \in \Lambda}$ converges to $\phi_\gamma(\psi(z_0))$ in Y_γ by Proposition 9.16. But then, by what has just been shown, $\{\psi(z_\lambda)\}$ converges in X to $\psi(z_0)$ with respect to $\mathcal{T}$. Thus ψ is continuous by virtue of the same proposition. □

The most important application of the idea of an inversely induced topology is concerned with Cartesian products. (In this connection the reader may wish to refer to Chapter 1.)

Definition. Let $\{Y_\gamma\}_{\gamma \in \Gamma}$ be an indexed family of topological spaces, and let $X = \prod_\gamma Y_\gamma$ be the Cartesian product of the sets Y_γ, $\gamma \in \Gamma$. Then the *product topology* on X is the topology inversely induced on X by the family $\{\pi_\gamma\}$ of projections. In the event that the index set Γ is $\{1, \ldots, n\}$, it is customary to write $Y_1 \times \cdots \times Y_n$ for the product of the topological spaces $\{Y_i\}_{i=1}^n$ equipped with the product topology. Whenever a product of topological spaces is regarded as a topological space, it is the product topology that is understood to be in use unless the contrary is expressly stipulated.

The following summary of facts concerning product topologies is little more than a paraphrase of Proposition 9.49. (That the projection π_γ of X onto Y_γ is an open mapping follows from the observation that if U is a nonempty open set of the form (2), then $\pi_\gamma(U) = U_\gamma$.)

Proposition 9.50. *Let $\{Y_\gamma\}_{\gamma\in\Gamma}$ be an indexed family of topological spaces and let $X = \prod_\gamma Y_\gamma$, as above. Then a base for the product topology on X is provided by the set of all products of the form*

$$\prod_\gamma U_\gamma \tag{2}$$

where U_γ is an open subset of Y_γ for each index γ, and where $U_\gamma = Y_\gamma$ except for a finite number of indices. Moreover, if $\mathcal{B}_\gamma$ is a specified topological base for Y_γ, $\gamma \in \Gamma$, then the sets U_γ in (2) may be required to belong to $\mathcal{B}_\gamma$ for each index γ for which $U_\gamma \neq Y_\gamma$. A net in X converges with respect to the product topology if and only if it converges coordinatewise. That is, if $\{x_\lambda\}_{\lambda\in\Lambda}$ is a net in X, and if $x_\lambda = \{y_\gamma^{(\lambda)}\}_{\gamma\in\Gamma}$ for each λ, then $\{x_\lambda\}$ converges to a limit $x_0 = \{y_\gamma^{(0)}\}$ in the product topology if and only if $\lim_\lambda y_\gamma^{(\lambda)} = y_\gamma^{(0)}$ for each index γ. The projections π_γ are continuous and open mappings of X onto the various factors Y_γ. A mapping ψ of a topological space Z into X is continuous if and only if each composition $\pi_\gamma \circ \psi$ is continuous.

It follows, of course, that if $X = Y_1 \times \ldots \times Y_n$ is a finite product of topological spaces, and if $\mathcal{B}_i$ is a topological base in Y_i for each index i, then the products $U_1 \times \ldots \times U_n$, where $U_i \in \mathcal{B}_i$, $i = 1, \ldots, n$, constitute a base for the product topology on X. We also have the following result.

Corollary 9.51. *If $\{Y_\gamma\}$ is a countable indexed family of topological spaces, and if each space Y_γ satisfies the second axiom of countability, then $X = \prod_\gamma Y_\gamma$ also satisfies the second axiom of countability.*

PROOF. Let $\mathcal{B}_\gamma$ be a countable base for the topology on Y_γ, $\gamma \in \Gamma$. For any one finite set $\{\gamma_1, \ldots, \gamma_n\}$ of indices there are but countably many products of the form (2) where $U_{\gamma_i} \in \mathcal{B}_{\gamma_i}$, $i = 1, \ldots, n$, and $U_\gamma = Y_\gamma$ for $\gamma \notin \{\gamma_1, \ldots, \gamma_n\}$. Moreover, there are but countably many finite subsets of the index set Γ. Thus the base described in Proposition 9.50 is countable (Cor. 4.5). □

Proposition 9.50 also permits us to give a useful description of the neighborhoods in a product space.

Proposition 9.52. *Let $\{Y_\gamma\}_{\gamma\in\Gamma}$ be an indexed family of topological spaces, let $X = \prod_\gamma Y_\gamma$, and for each index γ let $\mathcal{W}_\gamma$ be a neighbor-*

hood base at a point y_γ of Y_γ. Then a neighborhood base at the point $x = \{y_\gamma\}$ of X in the product topology is given by the collection of products of the form

$$W = \prod_\gamma W_\gamma$$

where $W_\gamma = Y_\gamma$ for all but a finite number of indices γ and $W_\gamma \in \mathcal{W}_\gamma$ whenever $W_\gamma \neq Y_\gamma$.

Corollary 9.53. *If $\{Y_\gamma\}_{\gamma\in\Gamma}$ is a countable indexed family of topological spaces, and if each space Y_γ satisfies the first axiom of countability, then $X = \prod_\gamma Y_\gamma$ also satisfies the first axiom of countability. On the other hand, if the index set Γ is uncountable, and if the factors Y_γ are all nonempty Hausdorff spaces, then X does not satisfy the first axiom of countability unless all but countably many of the factors Y_γ are singletons.*

PROOF. The first assertion follows from Proposition 9.52 exactly as Corollary 9.51 followed from Proposition 9.50. To prove the final assertion of the corollary, let $x_0 = \{y_\gamma^{(0)}\}_{\gamma\in\Gamma}$ be a point of X (nonempty by virtue of the axiom of products), and let V be a neighborhood of x_0 in X. Then by Proposition 9.52 there is a finite subset Γ_0 of Γ with the property that if $x = \{y_\gamma\}$ is a point of X such that $y_\gamma = y_\gamma^{(0)}$ for each γ in Γ_0, then $x \in V$. It follows at once that if $\{V_n\}$ is any countable collection of neighborhoods of x_0, then there exists a countable subset Γ_1 of Γ with the property that if $x = \{y_\gamma\}$ is an arbitrary point of X such that $y_\gamma = y_\gamma^{(0)}$ for every γ in Γ_1, then $x \in \bigcap_n V_n$. But if the set Γ_2 of indices for which Y_γ is a Hausdorff space with more than one element is uncountable, then there is an index γ_0 in $\Gamma_2\backslash\Gamma_1$ and a point $\widetilde{y}_{\gamma_0}$ of Y_{γ_0} different from $y_{\gamma_0}^{(0)}$, and the set of all $x = \{y_\gamma\}$ such that $y_{\gamma_0} \neq \widetilde{y}_{\gamma_0}$ is a neighborhood of x_0 that does not contain $\bigcap_n V_n$. Thus there does not exist a countable neighborhood base at x_0. □

Example S. Let $\{Y_\gamma\}_{\gamma\in\Gamma}$ be an indexed family of nonempty topological spaces, and let $X = \prod_{\gamma\in\Gamma} Y_\gamma$. If, for a fixed index γ_0, we select a point $y_\gamma^{(0)}$ in each factor $Y_\gamma, \gamma \neq \gamma_0$, then the subset

$$S = \{x = \{y_\gamma\} \in X : y_\gamma = y_\gamma^{(0)}, \gamma \neq \gamma_0\}$$

is a *slice* in X *parallel* to the factor Y_{γ_0}. It is easily seen that if X is equipped with the product topology, then $\phi = \pi_{\gamma_0}|S$ becomes a homeomorphism of S onto Y_{γ_0}. (If U is a basic open set in X of the form (2), then $\phi(U)$ is either $\varnothing$ or U_{γ_0}, which shows that ϕ is open.) Thus a product of nonempty spaces contains (in general many) homeomorphic copies of each of its factors.

Example T. Let $\{Y_\gamma\}_{\gamma\in\Gamma}$ be a countable indexed family of metric spaces, and let $X = \prod_\gamma Y_\gamma$. Let ρ be a product metric on X (Prob. 6H) and let $\mathcal{T}$ denote the product topology on X, where each Y_γ is equipped with its metric topology. A sequence $\{x_n\}$ in X converges to a limit in X with respect to both ρ and $\mathcal{T}$ if and only if it converges to that limit coordinatewise (Prop. 9.50). But $(X, \mathcal{T})$ also satisfies the first axiom of countability according to the foregoing corollary, and it follows that the same sets are closed with respect to $\mathcal{T}$ that are closed with respect to ρ (Prop. 9.15). In other words, *the product metric metrizes the product topology on* X.

Example U. Let f and g be continuous complex-valued functions on a topological space X. Then the mapping $\phi : X \to \mathbb{C} \times \mathbb{C}$ defined by setting $\phi(x) = (f(x), g(x))$, $x \in X$, is continuous, and it follows, exactly as in Chapter 7, that $f + g, f - g$ and fg are also continuous on X. Likewise, f/g is continuous on the (open) subset of X on which it is defined. If $p(\lambda_1, \ldots, \lambda_n)$ is an arbitrary complex polynomial in n indeterminates, and if $f_1, \ldots, f_n$ are any n continuous complex-valued functions on X, then the function $x \to p(f_1(x), \ldots, f_n(x))$ is continuous on X. Similarly, if $r(\lambda_1, \ldots, \lambda_n)$ is a complex rational function in n indeterminates, then $x \to r(f_1(x), \ldots, f_n(x))$ is a continuous function on the (open) subset of X on which it is defined. (This is the counterpart in the theory of general topology of Example 7M; it appears this late in the present chapter only because we have chosen, once again, to treat these ideas in terms of products.)

Example V. Let X be a set, let $\{\phi_\gamma\}_{\gamma\in\Gamma}$ be an indexed family of mappings $\phi_\gamma : X \to Y_\gamma$ of X into topological spaces Y_γ, and suppose that for any pair of distinct points x_1, x_2 of X there is some mapping ϕ_γ such that $\phi_\gamma(x_1) \neq \phi_\gamma(x_2)$. (Such a family of mappings is said to be *separating* on X.) If we form the product $\Pi = \prod_\gamma Y_\gamma$ and define

$$F(x) = \{\phi_\gamma(x)\}_{\gamma\in\Gamma}, \quad x \in X,$$

then F is a one-to-one mapping of X into Π. Moreover, if x_0 is a point of X and $\{x_\lambda\}$ a net in X, then, according to Proposition 9.49, $\lim_\lambda x_\lambda = x_0$ in the topology $\mathcal{T}$ inversely induced by the family $\{\phi_\gamma\}$ if and only if $\lim_\lambda \phi_\gamma(x_\lambda) = \phi_\gamma(x_0)$ for every index γ. On the other hand, the net $\{F(x_\lambda)\}$ tends to $F(x_0)$ in Π if and only if the *very same condition* is satisfied (Prop. 9.50). Hence, by Proposition 9.16, F is a homeomorphism of $(X, \mathcal{T})$ onto the subspace $F(X)$ of Π. (A mapping such as F is called a *topological embedding* of $(X, \mathcal{T})$ in Π.)

The formation of products also behaves well with respect to the formation of subspaces.

Proposition 9.54. *Let $\{Y_\gamma\}_{\gamma\in\Gamma}$ be an indexed family of topological spaces, and for each index γ let A_γ be a subset of Y_γ. Then the relative topology on the subspace $A = \prod_\gamma A_\gamma$ in the product $X = \prod_\gamma Y_\gamma$ coincides with the product topology on A that results from equipping each A_γ with its relative topology in Y_γ. Moreover, we have $A^- = \prod_\gamma A_\gamma^-$.*

PROOF. Both parts of the proposition are immediate consequences of the elementary fact that if, for each index γ, A_γ and B_γ are arbitrary subsets of Y_γ, then

$$\left(\prod_\gamma A_\gamma\right) \cap \left(\prod_\gamma B_\gamma\right) = \prod_\gamma (A_\gamma \cap B_\gamma). \qquad \square$$

The following two results combine to form one of the most powerful tools of modern analysis. The former of the two is essentially trivial, while the latter is among the deepest results of general topology.

Proposition 9.55. *The product X of an arbitrary indexed family $\{Y_\gamma\}_{\gamma\in\Gamma}$ of nonempty topological spaces is a Hausdorff space if and only if all of the factors $Y_\gamma, \gamma \in \Gamma$, are Hausdorff.*

PROOF. Let $x_1 = \{y_\gamma^{(1)}\}_{\gamma\in\Gamma}$ and $x_2 = \{y_\gamma^{(2)}\}_{\gamma\in\Gamma}$ be distinct points of X, and let γ_0 be an index such that $y_{\gamma_0}^{(1)} \neq y_{\gamma_0}^{(2)}$. If Y_{γ_0} is Hausdorff, there exist disjoint open sets $U_{\gamma_0}^{(1)}$ and $U_{\gamma_0}^{(2)}$ in Y_{γ_0} such that $y_{\gamma_0}^{(i)} \in U_{\gamma_0}^{(i)}$, $i = 1, 2$, and if we set

$$U_i = \prod_{\gamma\in\Gamma} U_\gamma^{(i)}, \quad i = 1, 2,$$

where $U_\gamma^{(1)} = U_\gamma^{(2)} = Y_\gamma$ for all $\gamma \neq \gamma_0$, then U_1 and U_2 are disjoint open sets in X such that $x_i \in U_i$, $i = 1, 2$. Thus a product of Hausdorff spaces is Hausdorff. The converse follows at once from the fact that the product of nonempty topological spaces contains homeomorphic copies of all of its factors (Ex. S). $\square$

Proposition 9.56. *The product X of an arbitrary indexed family $\{Y_\gamma\}_{\gamma\in\Gamma}$ of nonempty topological spaces is compact if and only if all of the factors $Y_\gamma, \gamma \in \Gamma$, are compact.*

PROOF. Since X is nonempty, $\pi_\gamma(X) = Y_\gamma$ for each index γ. Thus if X is compact, then all of the factors Y_γ must also be compact by Proposition 9.45.

To go the other way, suppose all of the spaces Y_γ are compact, and let $\mathcal{U}$ be an ultrafilter in X. For each index γ in Γ the collection of sets

$\pi_\gamma(\mathcal{U}) = \{\pi_\gamma(E) : E \in \mathcal{U}\}$ is a filter base in Y_γ (Prob. 7S). Hence $\pi_\gamma(\mathcal{U})$ has an adherent point $y_\gamma^{(0)}$ (Prop. 9.48), and if V_γ is a neighborhood of $y_\gamma^{(0)}$ in Y_γ, then V_γ meets every set in $\pi_\gamma(\mathcal{U})$, so $\pi_\gamma^{-1}(V_\gamma)$ meets every set in $\mathcal{U}$. But then $\pi_\gamma^{-1}(V_\gamma)$ belongs to $\mathcal{U}$ (Prob. M).

Consider now the point $x_0 = \{y_\gamma^{(0)}\}_{\gamma\in\Gamma}$. If V is an arbitrary neighborhood of x_0 in X, then there exist indices $\gamma_1, \ldots, \gamma_n$ and corresponding neighborhoods V_{γ_i} of $y_{\gamma_i}^{(0)}$, $i = 1, \ldots, n$, such that

$$x_0 \in \pi_{\gamma_1}^{-1}(V_{\gamma_1}) \cap \cdots \cap \pi_{\gamma_n}^{-1}(V_{\gamma_n}) \subset V,$$

and it follows by what has just been shown that V belongs to $\mathcal{U}$. Thus $\mathcal{U}$ refines $\mathcal{V}_{x_0}$, and therefore $x_0 = \lim \mathcal{U}$. This shows that the ultrafilter $\mathcal{U}$ converges in X, and hence that X is compact, by Proposition 9.48 once again. □

Corollary 9.57 (Tikhonov's Theorem). *The product of an arbitrary indexed family of compact Hausdorff spaces is a compact Hausdorff space.*

Example W. Let Γ be an index set, Y a topological space, and let $Y_\gamma = Y$ for each index γ. Then the product $\prod_\gamma Y_\gamma$ coincides as a set with the set Y^Γ of all mappings of Γ into Y. Thus the set of all mappings of an arbitrary set Γ into a topological space Y may be equipped with the product topology. In this topology on Y^Γ a net $\{\phi_\lambda\}$ converges to a limit if and only if it converges *pointwise* to that limit. For this reason, the product topology on Y^Γ is also known as the *topology of pointwise convergence.* A base for the topology of pointwise convergence on Y^Γ is given by the collection of all sets of the form

$$\{\phi \in Y^\Gamma : \phi(\gamma_i) \in U_i, \quad i = 1, \ldots, n\},$$

where n is an arbitrary positive integer, $\{\gamma_1, \ldots, \gamma_n\}$ an arbitrary set of n indices, and U_i is an open subset of Y for each $i = 1, \ldots, n$. If Y is a compact Hausdorff space, then, according to the foregoing result, so is the space Y^Γ in the topology of pointwise convergence.

Example X. Let Ω be the first uncountable ordinal number (Ex. 5I), let the ordinal number segment $W(\Omega)$ be equipped with its order topology, and, for each ζ in $W(\Omega)$, let S_ζ denote the *big square*

$$S_\zeta = \{(\xi, \eta) \in \Pi_0 : \xi, \eta \geq \zeta\}$$

in the product $\Pi_0 = W(\Omega) \times W(\Omega)$. Suppose that A is some subset of Π_0 that contains no such big square. Starting with any point (ξ_1, η_1) in $\Pi_0 \backslash A$, we set $\zeta_1 = \xi_1 \vee \eta_1$ and select a second point (ξ_2, η_2) in $S_{\zeta_1} \backslash A$. Then, continuing in this way, we obtain by mathematical induction a sequence $\{(\xi_n, \eta_n)\}_{n=1}^{\infty}$ in $\Pi_0 \backslash A$ with the property that

$$\xi_{n+1} \wedge \eta_{n+1} \geq \xi_n \vee \eta_n \tag{3}$$

for each index n. It follows, of course, that both of the sequences $\{\xi_n\}$ and $\{\eta_n\}$ are monotone increasing and, since they are bounded above in $W(\Omega)$ (Ex. 5L), each sequence converges to its supremum. But then

$$\lim_n \xi_n = \lim_n \eta_n,$$

because of (3). Hence, denoting this common limit by α, we see that

$$\lim_n (\xi_n, \eta_n) = (\alpha, \alpha)$$

in Π_0, and therefore $(\alpha, \alpha) \in (\Pi_0 \backslash A)^-$. Thus if a subset A of Π_0 fails to contains a big square, then $(\Pi_0 \backslash A)^-$ meets the diagonal Δ. But then (contrapositively), *if U is any open set in Π_0 that contains Δ, then U must contain a big square* (for $\Pi_0 \backslash U$ is a closed set that does not meet Δ).

Next let us consider the product $\Pi_1 = \widehat{W}(\Omega) \times W(\Omega)$, where $\widehat{W}(\Omega)$ denotes the ordinal number segment $W(\Omega+1) = W(\Omega) \cup \{\Omega\}$. The product topology on Π_0 coincides with its relative topology in Π_1 (Prop. 9.54) and the diagonal Δ in Π_0 is readily seen to be closed in Π_1, as is the set $F = \{(\Omega, \xi) : \xi < \Omega\} = \{\Omega\} \times W(\Omega)$. But if U is any open set in Π_1 containing Δ, then U must contain a big square in Π_0, and the closure U^- of U in Π_1 therefore contains an entire tail of F. Thus Π_1 is *not a normal topological space*, even though it is the product of two completely normal spaces, one of which is, in fact, a compact Hausdorff space (see Example P and Problem L).

Finally, let $\Pi_2 = \widehat{W}(\Omega) \times \widehat{W}(\Omega)$. The product Π_2 is also a compact Hausdorff space by Tikhonov's theorem, and is therefore normal (Prop. 9.44). But Π_2 contains the (nonnormal) product Π_1 as a subspace, and is therefore not completely normal.

(Another favorite, and entirely similar, example is based on the product $\widetilde{\Pi} = \widehat{W}(\Omega) \times \widehat{W}(\omega)$, where $\widehat{W}(\omega) = W(\omega + 1) = W(\omega) \cup \{\omega\}$ in its order topology. The product $\widetilde{\Pi}$ is a compact Hausdorff space, but the subspace $P = \widetilde{\Pi} \backslash \{(\Omega, \omega)\}$—known as the *Tikhonov plank*—is not normal. Indeed, the disjoint sets $E = \{(\xi, \omega) : \xi < \Omega\}$ and $F = \{\Omega, n) : n < \omega\}$ are readily seen to be closed in P, but if U is open and $F \subset U$, then for each nonnegative integer n there is an ordinal number $\xi_n < \Omega$ such that $\{(\zeta, n) : \xi_n < \zeta \leq \Omega\} \subset U$. But then, taking $\eta < \Omega$ such that $\xi_n \leq \eta$ for every n, we see that the product $\{\zeta : \eta < \zeta \leq \Omega\} \times W(\omega)$ is contained in U, so that $U^- \cap E$ contains $\{(\zeta, \omega) : \eta < \zeta < \Omega\}$.)

We close this introductory account of the theory of topological spaces with a brief glance at the notion of the limit of a mapping at a point. If a_0

is an adherent point of a subset A of a topological space X, then, just as before (Ex. 7T), the trace $\mathcal{D} = \{V \cap A : V \in \mathcal{V}_{a_0}\}$ on A of the neighborhood filter $\mathcal{V}_{a_0}$ at a_0 is a filter base in A (and in X), and if ϕ is some mapping of A into a topological space Y, then $\phi(\mathcal{D})$ is also a filter base in Y (cf. Problem 7S). Thus the criterion set forth in Proposition 7.24 may be used to define the idea of limit in general.

Definition. Let A be a subset of a topological space X, let ϕ be a mapping of A into a topological space Y, and let $a_0 \in A^-$. Then a point y_0 of Y is a *limit of* $\phi(x)$ *as* x *approaches* (or *tends to*) a_0 *through* A (notation: $y_0 = \lim_{\substack{x \to a_0 \\ x \in A}} \phi(x)$) if $\lim \phi(\mathcal{D}) = y_0$, where, as above, $\mathcal{D}$ denotes the filter base $\{V \cap A : V \in \mathcal{V}_{a_0}\}$. (When $A = X$ we write simply $y_0 = \lim_{x \to a_0} \phi(x)$.)

In the spirit of Proposition 7.23 we have the following result. (It is readily seen that this criterion may also be stated in terms of filter bases in place of nets. Moreover, as always, sequences may also be used in place of nets when X satisfies the first axiom of countability.)

Proposition 9.58. *Let X and Y be topological spaces, let ϕ be a mapping of a subset A of X into Y, and let a_0 be a point of X belonging to A^-. Then $y_0 = \lim_{\substack{x \to a_0 \\ x \in A}} \phi(x)$ if and only if $\phi(x_\lambda) \to y_0$ for every net $\{x_\lambda\}$ in A such that $x_\lambda \to a_0$.*

PROOF. Suppose first that $\lim_{\substack{x \to a_0 \\ x \in A}} \phi(x) = y_0$, let $\{x_\lambda\}$ be a net in A converging to a_0, and let V be a neighborhood of y_0 in Y. By definition there exists a neighborhood W of a_0 in X such that $\phi(W \cap A) \subset V$. Moreover, there is an index λ_0 such that $x_\lambda \in W$ for $\lambda \geq \lambda_0$. But then $\phi(x_\lambda) \in V$, $\lambda \geq \lambda_0$, and this shows that $\lim_\lambda \phi(x_\lambda) = y_0$. Thus the stated criterion is necessary. To see that it is sufficient, suppose it is not the case that $\lim_{\substack{x \to a_0 \\ x \in A}} \phi(x) = y_0$. Then there exists a neighborhood V of y_0 such that for each neighborhood W of a_0 there is at least one point x_W of $W \cap A$ such that $\phi(x_W) \notin V$, and it is obvious that the net $\{x_W\}_{W \in \mathcal{V}_{a_0}}$ (indexed by the neighborhood filter $\mathcal{V}_{a_0}$ itself) converges to a_0 in A while the net $\{\phi(x_W)\}$ fails to converge to y_0 in Y. □

The analog of Proposition 7.22 does not hold in this general context without some extra qualification (if X is equipped with the indiscrete topology, then the identity mapping on X has every point of X as its limit at each point of X), but we do have the following result (the proof of which is exactly like that of Proposition 7.22, and is therefore omitted).

Proposition 9.59. *If ϕ is a mapping of a subset A of a topological space X into a* Hausdorff *space Y, and if $a_0 \in A^-$, then either there is no point*

y_0 of Y such that $y_0 = \lim_{\substack{x \to a_0 \\ x \in A}} \phi(x)$ (in which case the limit of $\phi(x)$ as x tends to a_0 though A fails to exist) *or there is exactly one such limit (so that the limit of $\phi(x)$ as x tends to a_0 through A is unique when it exists).*

A similar generalization of Proposition 7.25 goes through with the same restriction on the codomain.

Proposition 9.60. *Let ϕ be a mapping of a subset A of a topological space X into a Hausdorff space Y, and let a_0 be a point of A^-. Then the limit of $\phi(x)$ as x tends to a_0 through A exists if and only if there exists a (necessarily unique) point y_0 of Y with the property that if ϕ is extended to a mapping $\widehat{\phi}$ on $A \cup \{a_0\}$ by defining $\widehat{\phi}(a_0) = y_0$, the extended mapping $\widehat{\phi}$ is continuous at a_0 relative to $A \cup \{a_0\}$. Moreover, if such a point y_0 exists, then $y_0 = \lim_{\substack{x \to a_0 \\ x \in A}} \phi(x)$. Finally, if $a_0 \in A$, then the limit of $\phi(x)$ as x tends to a_0 through A exists if and only if ϕ is continuous at a_0 (relative to A), and in this case the limit must coincide with $\phi(a_0)$.*

PROOF. Everything is clear except for the uniqueness of the point y_0, and that follows at once from Problem F. □

While the assumption that the codomain is a Hausdorff space suffices to ensure the uniqueness and one-point continuity characterization of limits in general topological spaces, some further restriction is needed in order to obtain the existence of extensions by continuity (Prop. 7.26).

Proposition 9.61. *Let ϕ be a mapping of a subset A of a topological space X into a* regular *topological space Y (see Problem G). Let $\widetilde{A}$ denote the subset of A^- consisting of those points a at which the limit of $\phi(x)$ exists as x tends to a through A, and for each point a of $\widetilde{A}$ set*

$$\widetilde{\phi}(a) = \lim_{\substack{x \to a \\ x \in A}} \phi(x).$$

Then $\widetilde{\phi} : \widetilde{A} \to Y$ is continuous. (In particular, if ϕ is itself continuous on A, so that $A \subset \widetilde{A}$, then $\widetilde{\phi}$ is a continuous extension of ϕ, and we say that $\widetilde{\phi}$ results from extending ϕ by continuity.)

PROOF. Let a_0 be a point of $\widetilde{A}$, let $y_0 = \widetilde{\phi}(a_0)$, and let V be a neighborhood of y_0 in Y. According to Problem G there is a *closed* neighborhood W of y_0 such that $W \subset V$, and by the definition of $\widetilde{\phi}$ there is an *open* neighborhood U of a_0 with the property that $\phi(U \cap A) \subset W$. Let a be a point of

$\widetilde{A}$ that is contained in U. Then, of course, U is a neighborhood of a such that $\phi(U \cap A) \subset W$, and it follows that $\widetilde{\phi}(a) \in W^- = W$. Thus $\widetilde{\phi}(U \cap A) \subset W \subset V$, which shows that $\widetilde{\phi}$ is continuous at a_0. □

That the regularity of the codomain suffices to yield this result is not simply fortuitous. In fact, it can be shown that the validity of Proposition 9.61 for mappings taking their values in a topological space Y is yet another characterization of the regularity of Y; see [3].

For (extended) real-valued functions the notion of limit splits into two dual notions, just as before.

Definition. Let A be a subset of a topological space X, and let f be an extended real-valued function defined on A. For each point a_0 of the closure A^- we define the *limit superior* (or *upper limit*) of $f(x)$ *as* x *tends to* a_0 *through* A (notation: $\limsup_{\substack{x \to a \\ x \in A}} f(x)$) as follows: For each neighborhood V of a_0 we first set $M(f; a_0, V) = \sup\{f(x); x \in V \cap A\}$, and then define

$$\limsup_{\substack{x \to a_0 \\ x \in A}} f(x) = \inf_V M(f; a_0, V),$$

where the infimum is taken over the entire neighborhood filter at a_0. Dually, we define $m(f; a_0, V) = \inf\{f(x) : x \in V \cap A\}$ for each neighborhood V of a_0, and then define the *limit inferior* (or *lower limit*) of $f(x)$ *as* x *tends to* a_0 *through* A (notation: $\liminf_{\substack{x \to a \\ x \in A}} f(x)$) by setting

$$\liminf_{\substack{x \to a_0 \\ x \in A}} f(x) = \sup_V m(f; a_0, V),$$

where, once again, the supremum is taken over $\mathcal{V}_{a_0}$.

The proof of the following summary of properties exactly parallels that of Proposition 7.27, and is therefore omitted.

Proposition 9.62. *Let A be a subset of a topological space X, let f be an extended real-valued function defined on A, and for each point a_0 of A^- set*

$$m(a_0) = \liminf_{\substack{x \to a_0 \\ x \in A}} f(x), \quad M(a_0) = \limsup_{\substack{x \to a_0 \\ x \in A}} f(x).$$

Then $m \leq f \leq M$ on A, while $m \leq M$ holds everywhere on A^-. Moreover, M is an upper semicontinuous function on A^- with the property that f is upper semicontinuous at a point x_0 of A (relative to A) if and only if $f(x_0) = M(x_0)$, and also with the property that if h is an arbitrary upper semicontinuous function on A^- such that $f \leq h$ on A, then

$M \leq h$ as well. Dually, m is a lower semicontinuous function on A^- with the property that f is lower semicontinuous at a point x_0 of A if and only if $f(x_0) = m(x_0)$, and also with the property that if k is an arbitrary lower semicontinuous function on A^- such that $k \leq f$ on A, then $k \leq m$ as well. (Because of these extremal properties, M and m are called the upper *and* lower envelopes *of f, respectively.)*

PROBLEMS

A. Let X be a set and suppose given a collection $\mathcal{F}$ of subsets of X satisfying the conditions of Proposition 9.1. Verify that there exists a unique topology on X with respect to which $\mathcal{F}$ is precisely the collection of all closed sets. Show too, in the same vein, that if $A \to A^-$ is a mapping of 2^X into itself satisfying conditions (1)–(4) of Proposition 9.5, then there exists a unique topology on X with respect to which A^- is the closure of A for each subset A of X.

B. Let X be a set and suppose given, for each point x of X, a filter $\mathcal{F}_x$ in X satisfying condition (a) every set V in $\mathcal{F}_x$ contains the point x.

(i) Define a subset U of X to be *open* if and only if U belongs to $\mathcal{F}_x$ whenever x belongs to U (in other words, if and only if $x \in U$ implies $U \in \mathcal{F}_x$). Prove that the collection $\mathcal{T}$ of open sets thus obtained is, in fact, a topology on X.

(ii) Verify that the filter $\mathcal{F}_x$ refines the neighborhood filter $\mathcal{V}_x$ with respect to the topology $\mathcal{T}$ at every point x of X. Give an example showing that it is possible for $\mathcal{F}_x$ to be distinct from $\mathcal{V}_x$.

(iii) Prove that each set V in $\mathcal{F}_x$ is a neighborhood of x with respect to $\mathcal{T}$ (and hence that $\mathcal{F}_x = \mathcal{V}_x$) for each x in X if and only if the given assignment $x \to \mathcal{F}_x$ also satisfies condition (b) for each point x of X and each V in $\mathcal{F}_x$ there exists a set U in $\mathcal{F}_x$ such that V belongs to $\mathcal{F}_y$ for each y in U. (Hint: For given V in $\mathcal{F}_x$ set $W = \{z \in X : V \in \mathcal{F}_z\}$, and prove $W \in \mathcal{T}$.)

C. Let X be a set and let $\mathcal{C}$ be a collection of subsets of X.

(i) The collection $\mathcal{C}_u$ of all unions of subcollections of $\mathcal{C}$ always possesses roughly half of the properties of a topology on X, in that $\varnothing \in \mathcal{C}_u$ and $\mathcal{C}_u$ is closed with respect to the formation of arbitrary unions.

(ii) The collection $\mathcal{C}_d$ of finite intersections of sets in $\mathcal{C}$, that is, the set of all intersections of nonempty finite subcollections of $\mathcal{C}$, contains $\mathcal{C}$ and is closed with respect to the formation of finite intersections.

(iii) If $\mathcal{C}'$ is another collection of subsets of X, and if we denote by $\mathcal{C}''$ the collection of all sets $C \cap C'$, where $C \in \mathcal{C}$ and $C' \in \mathcal{C}'$, then

$$\left(\bigcup \mathcal{C}\right) \cap \left(\bigcup \mathcal{C}'\right) = \bigcup \mathcal{C}''.$$

(iv) Conclude that $(\mathcal{C}_d)_u$ is a topology on X if and only if $\mathcal{C}$ covers X, while the topology on X generated by $\mathcal{C}$ is given in general by

$$\mathcal{T}(\mathcal{C}) = (\mathcal{C}_d)_u \cup \{X\}.$$

D. Let $\{\mathcal{T}_\gamma\}_{\gamma \in \Gamma}$ be a nonempty indexed family of topologies on a set X, and set $\mathcal{T}_0 = \sup_\gamma \mathcal{T}_\gamma$.

(i) The union $\mathcal{U} = \bigcup_\gamma \mathcal{T}_\gamma$ is a set of generators for $\mathcal{T}_0$, so the set $\mathcal{U}_d$ of finite intersections of sets in $\mathcal{U}$ is a base for $\mathcal{T}_0$. Verify that $\mathcal{U}_d$ coincides with the collection of all sets $U_1 \cap \ldots \cap U_k$ where $U_i \in \mathcal{T}_{\gamma_i}, i = i, \ldots, k$, and $\{\gamma_1, \ldots, \gamma_k\}$ is a finite set of indices. More generally, if, for each index γ, $\mathcal{B}_\gamma$ is some base for $\mathcal{T}_\gamma$, then the collection of all sets $V_1 \cap \ldots \cap V_k$ where $V_i \in \mathcal{B}_{\gamma_i}, i = 1, \ldots, k$, and $\{\gamma_1, \ldots, \gamma_k\}$ is a finite set of indices, is a base for $\mathcal{T}_0$.

(ii) The union $\mathcal{U}$ itself is a subbase for $\mathcal{T}_0$. More generally, if $\mathcal{S}_\gamma$ is some subbase for $\mathcal{T}_\gamma$ for each index γ, then $\bigcup_\gamma \mathcal{S}_\gamma$ is a subbase for $\mathcal{T}_0$.

E. If X is a simply ordered set and E is a subset of X, then X has its order topology (Ex. D) and E acquires its relative topology $\mathcal{T}_{rel}$ as a subspace of X. But E is also a simply ordered set and consequently has an order topology $\mathcal{T}_o$ of its own. Show that these topologies are always comparable ($\mathcal{T}_o \subset \mathcal{T}_{rel}$), and give an example of a set E of real numbers on which $\mathcal{T}_o$ and $\mathcal{T}_{rel}$ are different. Show, however, that if $X = \mathbb{R}$ or $\mathbb{R}^\natural$, and if I is an interval in X, then $\mathcal{T}_o = \mathcal{T}_{rel}$ on I. Show too that if F is closed in $\mathbb{R}$ (or $\mathbb{R}^\natural$), then $\mathcal{T}_o$ and $\mathcal{T}_{rel}$ agree on F. In the same vein, show that if α and β are ordinal numbers with $\alpha < \beta$, then $W(\alpha)$, in its order topology, is a subspace of $W(\beta)$ in *its* order topology.

F. Let $(X, \mathcal{T})$ be a topological space.

(i) Prove that if ϕ and ϕ' are continuous mappings of X into a Hausdorff space Y, then $\{x \in X : \phi(x) = \phi'(x)\}$ is closed in X. (In particular, if ϕ and ϕ' agree on a dense set in X, then $\phi = \phi'$.)

(ii) Show, in the other direction, that if $(X, \mathcal{T})$ is a T_1-space, then there exist a T_1-space Y and continuous mappings ϕ and ϕ' of X into Y such that $\{x \in X : \phi(x) = \phi'(x)\}$ is an arbitrarily prescribed subset Z of X. (Hint: Take for Y the set X itself equipped with its coarsest T_1-topology $\mathcal{T}_1$ (Ex. C). Every one–to–one mapping of $(X, \mathcal{T})$ into $(X, \mathcal{T}_1)$ is continuous.)

(iii) The space $(X, \mathcal{T})$ is Hausdorff if and only if every net in X converges to *at most one* limit. (Hint: If X is not Hausdorff, then there exist two distinct points x and y of X such that $\mathcal{V}_x \cup \mathcal{V}_y$ possesses the finite intersection property.)

G. (i) A Hausdorff space X (or, what comes to the same thing, a T_1-space X) is said to be *regular* if it has the property that if F is a closed set in X and x is a point of X not belonging to F, then there exist disjoint open subsets U and V of X such that $x \in U$ and $F \subset V$. Verify that a Hausdorff space X is regular if and only if for each point x of X and each neighborhood V of x there exists a neighborhood W of x such that $W^- \subset V$. Conclude that X is regular if and only if there exists a neighborhood base consisting exclusively of closed sets at every point x of X.

(ii) A Hausdorff space X (or, what comes to the same thing, a T_1-space X) is said to be *completely regular* if for each point x of X and neighborhood V of x there exists a continuous mapping f of X into the closed unit interval such that $f(x) = 1$ and $f(y) = 0$ on $X \backslash V$ (i.e., such that $\chi_{\{x\}} \leq f \leq \chi_{X \backslash V}$). Show that every normal topological space is completely regular, and also that every completely regular topological space is regular.

The degrees of separation that a general topological space can enjoy that have been recognized so far are five in number. The weakest of these (in order of implication) is the property of being a T_1-space, in which all finite sets are closed. The next degree of separation in this ranking is the property of being a Hausdorff space—sometimes called a *T_2-space*. Then come, in order of specialization, the regular topological spaces (also known as *T_3-spaces*), the completely regular spaces (sometimes called *T_π-spaces*), and the normal topological spaces (also known as *T_4-spaces*). While it is clear that a topological space possessing any one of these properties automatically possesses all those of lower rank as well, it is by no means evident that none of these properties implies any one of the higher ranking ones, but that is, in fact, the case. To verify this assertion one needs, for each $i = 1, 2, 3, \pi, 4$, an example of a T_i-space that fails to be a T_j-space for any $j > i$. We note in this connection that the topological space $(X, \mathcal{T}_1)$ of Example C is a T_1-space that is not Hausdorff (provided the carrier X is not finite). Likewise, a Hausdorff space that is not regular is readily constructed as follows. Let D be a set of positive real numbers such that $D^* = \{0\}$, and for each nonzero point t of $\mathbb{R}$ let $\widehat{\mathcal{V}}_t$ be the neighborhood filter of t in the ordinary metric topology on $\mathbb{R}$, while for $\widehat{\mathcal{V}}_0$ we take the filter in $\mathbb{R}$ generated by the filter base consisting of all sets of the form $(-\varepsilon, +\varepsilon) \backslash D, \varepsilon > 0$. Then it is obvious that D is a closed set with respect to the unique topology $\mathcal{T}$ on $\mathbb{R}$ for which $\widehat{V}_t$ is the neighborhood filter at t for each real t (Prob. B), and it is also clear that $\mathcal{T}$ refines the usual topology on $\mathbb{R}$ (Ex. G), and hence that $(X, \mathcal{T})$ is a Hausdorff space. But $(X, \mathcal{T})$ is not regular. Indeed, if U is a set in $\mathcal{T}$ containing 0, then there exists a positive number ε such that $(-\varepsilon, +\varepsilon)\backslash D \subset U$, and an element t of D such that $t < \varepsilon$. But if V is a set in $\mathcal{T}$ containing D, then $t \in V$, so V contains some open interval about t, and therefore must meet U.

In this same connection we remark that a completely regular topological space that is not normal appears in Example X (see Problem H below). Thus the only really difficult part of showing that all of these five degrees of separation are actually distinct is the construction of a regular topological space that is not completely regular. Such a construction is given by A. Tikhonov [22], but it is quite complicated.

H. (i) Show that the relation of being a subspace is transitive, in that a subspace of a subspace of a topological space X is also a subspace of X.

(ii) Verify that if a topological space X is a T_1-space or a Hausdorff space, and if A is a subspace of X, then A is also a T_1-space, respectively, a Hausdorff space. In the same vein, prove that if X is a [completely] regular space, then every subspace of X is also [completely] regular. Show, finally, that if X is a normal topological space and if F is a closed subspace of X, then F is also normal. (An arbitrary—i.e., not necessarily closed—subspace of a normal space need not be normal; recall Example X.)

I. Let X be a normal topological space and let $\{U_1, \ldots, U_N\}$ be a finite open covering of X.

(i) Show that there exists a corresponding closed covering $\{F_1, \ldots, F_N\}$ of X such that $F_i \subset U_i, i = 1, \ldots, N$. (Hint: Suppose the stated fact holds for some N, and that $\{U_1, \ldots, U_N, U_{N+1}\}$ is an open covering of X. Then $F = X\backslash(U_1 \cup \cdots \cup U_N)$ is a closed set contained in U_{N+1}, so there exists an open set V such that $F \subset V$ and $V^- \subset V_{N+1}$, and the closed subspace $X\backslash V$ is normal. Use mathematical induction.)

(ii) Show too that there exists a corresponding family $\{g_1, \ldots, g_N\}$ of continuous nonnegative real-valued functions on X such that $g_i(x) = 0$ on $X\backslash U_i, i = 1, \ldots, N$, and such that

$$g_1 + \cdots + g_N \equiv 1$$

on X. (Such a family of functions is called a *partition of unity subordinate to* the covering $\{U_1, \ldots, U_N\}$.) (Hint: Let $\{F_1, \ldots, F_N\}$ be a closed covering of X as in (i), and for each i let f_i be a Urysohn function on X that is equal to one on F_i and vanishes on $X\backslash U_i$. Then

$g = f_1 + \cdots + f_N$ is a continuous function on X such that $g(x) \geq 1$ for every x in X.)

J. A topological space X is said to be *completely normal* if every subspace of X is normal.

(i) Prove that X is completely normal if and only if for every pair $\{M, N\}$ of subsets of X satisfying the Hausdorff-Lennes condition there exist disjoint open sets V and W such that $M \subset V$ and $N \subset W$. (Hint: If M and N satisfy the Hausdorff-Lennes condition, then $U = X \backslash (M^- \cap N^-)$ is an open subspace of X in which $E = M^- \cap U$ and $F = N^- \cap U$ are closed and disjoint.)

(ii) Verify that every metrizable space is completely normal.

Thus complete normality and metrizability are yet two more in the scale of separation axioms that may be imposed on a topological space. (Indeed, a completely normal space is sometimes called a *T_5-space.*) Clearly a completely normal space is normal. A normal space that is not completely normal appears, as noted above, in Example X. Completely normal spaces that are not metrizable figure in Problem L below.

K. The analysis of the structure of open sets in $\mathbb{R}$ given in Example 6S can be extended to open sets in the order topology on an arbitrary simply ordered set if the notion of interval is suitably generalized. A subset C of a simply ordered set X is said to be *convex* in X if $a, b \in C$ and $a < x < b$ imply $x \in C$. Equivalently, C is convex if and only if for every y in $X \backslash C$ it is the case that either every element of C is less than $y (C < y)$ or every element of C is greater than $y (C > y)$.

(i) All intervals and rays in X are automatically convex. (Hence there is a base for the order topology on X consisting exclusively of convex sets.) Give an example of a convex subset of a simply ordered set X that is neither an interval nor a ray in X.

(ii) Show that if $\mathcal{C}$ is a chained collection of convex subsets of X (Ex. 1J), then $\bigcup \mathcal{C}$ is also convex. Conclude that if A is an arbitrary subset of X and a is a point of A, then there exists a (unique) largest convex subset of X that contains a and is contained in A. Show too that if U is an open subset of X in the order topology, then the largest convex subset of U containing a point a of U is also open in X.

(iii) Prove that each open set U in the order topology on X is the union of a uniquely determined, simply ordered collection $\mathcal{C}$ of pairwise disjoint nonempty convex open subsets of X, each of which is maximal with respect to being convex and contained in U. (The sets belonging to $\mathcal{C}$ might be called simply the *constituents* of U, the term "constituent interval" being inappropriate in this context.)

L. Let X be a simply ordered set equipped with its order topology, let M and N be two subsets of X satisfying the Hausdorff-Lennes condition, and let $\mathcal{C}$ and $\mathcal{D}$ denote the collections of (nonempty) constituents (Prob. K) of the open sets $X \backslash N^-$ and $X \backslash M^-$, respectively.

(i) Prove that for any one set C belonging to $\mathcal{C}$ there is an open set U_C such that

$$M \cap C \subset U_C \subset C \quad \text{and} \quad U_C^- \cap N = \varnothing.$$

(Hint: The intersection $C^- \cap N$ can contain at most two points. If it is empty, we may set $U_C = C$. If $C^- \cap N = \{y\}$, and if, say, $C < y$, then y has a convex open neighborhood V such that $V \cap M = \varnothing$, and V must meet C, so V contains a point $z < y$, and we may set $U_C = C \cap A_z$, where $A_z = \{x \in X : x < z\}$. Consider all cases.)

(ii) Show that each set D in $\mathcal{D}$ meets at most two of the constituents C of $X \backslash N^-$ such that $C \cap M \neq \varnothing$. (Hint: Suppose C and C' are distinct elements of $\mathcal{C}$, and suppose x and x' are points of M belonging to C and C', respectively, where, say, $x < x'$. Then for each set D in $\mathcal{D}$ we have either $D < x$, $D > x'$, or $x < D < x'$. Show first that if $D < x$, then $D \cap C' = \varnothing$.)

(iii) Verify that there exist open sets U and V separating M and N, and conclude that X is completely normal. (Hint: If C is a constituent of $X \backslash N^-$ that is disjoint from M we may set $U_C = \varnothing$ in (i). Define $U = \bigcup_C U_C$, and show that $U^- \cap N = \varnothing$.)

(iv) In particular, then, every ordinal number segment $W(\alpha)$ is completely normal in its order topology. But for $\alpha > \Omega$ these spaces fail to satisfy the first axiom of countability (for there is no countable base in the neighborhood filter $\mathcal{V}_\Omega$) and are therefore not metrizable. Similarly, $W(\Omega)$ itself is completely normal without being metrizable (see Problem N below).

M. (i) Prove that a collection $\mathcal{U}$ of subsets of a nonempty set X is an ultrafilter in X if and only if $\mathcal{U}$ possesses the finite intersection property (Prob. 7S) and is maximal (in the inclusion ordering on 2^X) with respect to that property. Use this fact to prove that every collection of subsets of X possessing the finite intersection property is contained in an ultrafilter in X. Conclude, in particular, that every filter in X is refined by an ultrafilter in X. (Hint: Use Zorn's lemma.)

(ii) Let X be a nonempty set and let $\mathcal{U}$ be an ultrafilter in X. Show that if A is a subset of X that meets every set in $\mathcal{U}$, then A is itself in $\mathcal{U}$. Show, in the same vein, that if $E \in \mathcal{U}$ and if $\{A_1, \ldots, A_p\}$ is an arbitrary finite covering of E consisting of subsets of X, then some one of the sets A_i belongs to $\mathcal{U}$. (Hint: Suppose, for each $i = 1, \ldots, p$, there is a set E_i in $\mathcal{U}$ such that $E_i \cap A_i = \varnothing$.)

N. The property of compactness can be viewed as the conjunction of two other properties in an interesting way. A topological space X is said to be a

Lindelöf space if every open covering of X contains a countable subcovering. Likewise, X is said to be *countably compact* if every countable open covering of X contains a finite subcovering.

(i) A topological space X is compact if and only if it is both countably compact and a Lindelöf space.

(ii) Every topological space satisfying the second axiom of countability is a Lindelöf space. (Hint: Cf. Problem 6S.)

(iii) Every regular Lindelöf space is normal. (Hint: If E and F are disjoint closed subsets of a regular Lindelöf space X, then E and F are themselves Lindelöf spaces in their relative topologies, and it follows that there exist sequences $\{U_n\}_{n=1}^{\infty}$ and $\{V_n\}_{n=1}^{\infty}$ of open sets in X such that $\{U_n\}$ covers E, while $F \cap U_n^- = \varnothing$ for all n, and likewise such that $\{V_n\}$ covers F while $E \cap V_n^- = \varnothing$ for all n. If $\widetilde{U}_n = U_n \backslash (V_1^- \cup \ldots \cup V_n^-)$ and $\widetilde{V}_n = V_n \backslash (U_1^- \cup \ldots \cup U_n^-)$ for each n, then $\widetilde{U}_p \cap \widetilde{V}_q = \varnothing$ for any two indices p and q.)

(iv) A topological space X is countably compact if and only if every decreasing sequence $\{F_n\}$ of nonempty closed sets in X has nonempty intersection. Equivalently, X is countably compact if and only if every countable filter base in X possesses an adherent point. Conclude that every countably compact metric space is compact.

(v) Let $W = W(\Omega)$ where Ω denotes the first uncountable ordinal number (Ex. 5I), and let W be equipped with its order topology. Show that the intersection of a countably infinite collection of closed cofinal subsets of W is again (closed and) cofinal (recall Example 5L), and use this fact to prove that W is countably compact. Conclude that W is not metrizable.

O. A topological space X will be said to have property (COB) if there exists a topological base for X consisting exclusively of closed-open sets.

(i) Verify that every T_1-space X with property (COB) is totally disconnected (Ex. O). Can the assumption that X is a T_1-space be dropped?

(ii) Let X be a compact Hausdorff space, let x be a point of X, and let D denote the intersection of all the closed-open subsets of X that contain x. Show that D coincides with the connected component of x. (Hint: To show that D is connected, suppose $D = D_1 \cup D_2$ where D_1 and D_2 are closed and disjoint. Then there exist disjoint open sets U_1 and U_2 such that $D_i \subset U_i$, $i = 1, 2$ (Prop. 9.44) and then a closed-open set E such that $D \subset E \subset U_1 \cup U_2$.) Use this fact to show that if C is an arbitrary connected component of X, and if U is an open set in X such that $C \subset U$, then there exists a closed-open set E in X such that $C \subset E \subset U$.

(iii) Conclude, finally, that a totally disconnected compact Hausdorff space has property (COB).

P. A collection $\mathcal{F}$ of Urysohn functions on a topological space X is *hyperseparating* if for each open set U in X and point x of U there is a function f in $\mathcal{F}$ that separates $\{x\}$ and $X \backslash U$. (Thus a T_1-space X is completely regular if and only if there exists a hyperseparating collection of Urysohn functions on X.)

(i) Let $\mathcal{F}$ be a hyperseparating collection of Urysohn functions on a T_1-space X, for each f in $\mathcal{F}$ set $Y_f = [0,1]$, and let $\Pi = \prod_{f \in \mathcal{F}} Y_f$. (Such a product of copies of $[0,1]$ is a *Tikhonov cube*.) Prove that

$$F(x) = \{f(x)\}_{f \in \mathcal{F}}, \quad x \in X,$$

defines a homeomorphism F of X into Π. (Hint: Show that the topology inversely induced on X by $\mathcal{F}$ coincides with the given topology on X, and recall Example V.)

(ii) For an arbitrary topological space X the following conditions are equivalent:

(1) X is completely regular,
(2) X is homeomorphic to a subspace of a Tikhonov cube,
(3) X is homeomorphic to a subspace of a compact Hausdorff space.

Q. (Urysohn Metrization Theorem [23]) Let X be a normal topological space, let $\mathcal{B}$ be a base for the topology on X, and for each pair (U, V) of sets in $\mathcal{B}$ such that $U^- \subset V$ let $f_{U,V}$ be a Urysohn function on X separating U^- and $X \backslash V$ (Th. 9.32). Show that the collection $\{f_{U,V}\}$ of such functions (one for each pair (U, V) in $\mathcal{B}$ such that $U^- \subset V$) is hyperseparating on X. Complete the proof of the following theorem: *A regular topological space satisfying the second axiom of countability is metrizable.*

R. (Compact metric spaces revisited; cf. Problem 8T)

(i) For a compact Hausdorff space X the following conditions are equivalent:

(1) X is metrizable,
(2) X satisfies the second axiom of countability,
(3) X is homeomorphic to a subspace of a Tikhonov cube $\prod_{\gamma \in \Gamma} Y_\gamma$ with countable index set Γ.

(ii) For a compact metric space X the following conditions are equivalent:

(1) X is totally disconnected,
(2) For any positive number ε there is a (finite) partition of X into closed-open subsets of diameter less than ε,
(3) X is homeomorphic to a subset of the Cantor set.

(iii) A compact metric space is homeomorphic to the Cantor set if and only if it is totally disconnected and perfect (Prob. 6O).

S. A notion that plays a fairly prominent role throughout the theory of metric spaces, and one that can play the same role in the theory of topological spaces in the presence of the first axiom of countability, is that of a subsequence of a given sequence. Since nets replace sequences in the general theory, it is natural to ask if there is a corresponding notion of *subnet*, and there is. If $\{x_\lambda\}_{\lambda\in\Lambda}$ is a net in a set X, if Γ is another directed set, and if N is an arbitrary mapping of Γ into Λ, then $\{x_{N(\gamma)}\}_{\gamma\in\Gamma}$ is also a net in X. If N has the property that for every index λ_0 in Λ there exists an index γ_0 in Γ such that $N(\gamma) \geq \lambda_0$ for all $\gamma \geq \gamma_0$, then the net $\{x_{N(\gamma)}\}_{\gamma\in\Gamma}$ is called a *subnet* of $\{x_\lambda\}_{\lambda\in\Lambda}$. Thus a *subsequence* of a given sequence $\{x_n\}$ is a subnet of that sequence in which the second directed set Γ is required to be the same as the directed set indexing the given sequence, viz., $\mathbb{N}$ or $\mathbb{N}_0$. Give an example of a subnet of a sequence that is not a subsequence.

(i) Verify that if a net $\{x_\lambda\}$ converges to a limit a_0 in a topological space X, then every subnet of $\{x_\lambda\}$ also converges to a_0.

(ii) A point a in a topological space X is a *cluster point* of a net $\{x_\lambda\}$ in X if for an arbitrary neighborhood V of a and an arbitrary index λ_0 there exists an index λ such that $\lambda \geq \lambda_0$ and $x_\lambda \in V$. Show that a point a in X is a cluster point of a net $\{x_\lambda\}$ in X if and only if some subnet of $\{x_\lambda\}$ converges to a (cf. Proposition 6.5). (Hint: Turn the product $\Lambda \times \mathcal{V}_a$ into a directed set and define Γ to be the subset of $\Lambda \times \mathcal{V}_a$ consisting of those pairs (λ, V) such that $x_\lambda \in V$.)

(iii) Conclude that a topological space X is compact if and only if every net in X possesses a convergent subnet.

T. A topological space X is said to be *locally compact* if every point of X has a compact neighborhood. Examples of locally compact spaces are $\mathbb{R}^n, \mathbb{C}^n$, all compact Hausdorff spaces, and all open subspaces of compact Hausdorff spaces.

(i) Show that if x is a point of a locally compact Hausdorff space X, then the compact neighborhoods of x form a neighborhood base at x, and conclude that X is completely regular. (Hint: If L is a compact neighborhood of x and V an arbitrary neighborhood of x, then L is closed (Prop. 9.43) and $W = V^\circ \cap L^\circ$ is an open neighborhood of x such that W^- is contained in L and is therefore compact and normal (Prop. 9.44).)

(ii) On the other hand, a locally compact Hausdorff space need not be normal. (Hint: Recall Example X.)

(iii) If K is a compact subset of a locally compact Hausdorff space X and U is an open subset of X containing K, then there exists a continuous mapping f of X into [0,1] such that $\chi_K \leq f \leq \chi_{X\setminus U}$. (Hint: Construct an open set V such that $K \subset V \subset V^- \subset U$ and such that V^- is compact, and apply Urysohn's lemma.)

(iv) The idea of a nowhere dense set clearly makes sense in an arbitrary topological space X, and so therefore do the notions of sets of first and second category in X. Show that if U is a nonempty open subset of a locally compact Hausdorff space X, then U is of second category in X. (Hint: Follow the proof of Theorem 8.20.)

U. By a *compactification* of a topological space X is meant a pair $(\widehat{X}, \alpha)$, where $\widehat{X}$ is a compact Hausdorff space and α is a homeomorphism of X onto a dense subspace of $\widehat{X}$. (According to Problem P, a topological space admits a compactification if and only if it is completely regular.) If there exists a compactification $(\widehat{X}, \alpha)$ of a topological space X such that $\widehat{X} \backslash \alpha(X)$ is compact, then X is necessarily a locally compact Hausdorff space. Show conversely that if X is a locally compact Hausdorff space, then there exists a compactification $(\widehat{X}, \alpha)$ of X with the property that $\widehat{X} \backslash \alpha(X)$ is a *singleton*, and that this compactification is unique up to a uniquely determined homeomorphism. (That is, if $(\widehat{X}, \alpha)$ and $(\widehat{X}', \alpha')$ are two such compactifications of the same space X, then there exists a unique homeomorphism Φ of $\widehat{X}$ onto $\widehat{X}'$ such that commutativity holds in the diagram

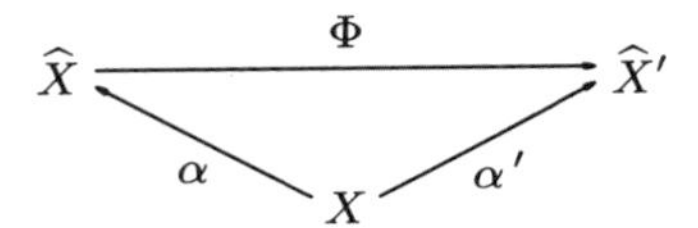

i.e., such that $\Phi \circ \alpha = \alpha'$.) The essentially unique pair $(\widehat{X}, \alpha)$ is called the *one-point compactification* of X. The single point $\widehat{X} \backslash \alpha(X)$ is known as the *point at infinity*, and is frequently denoted by ∞.

V. If $(\widehat{X}, \alpha)$ is a compactification of a topological space X, and if ϕ is a mapping of X into a space Y, then a mapping $\widehat{\phi} : \widehat{X} \to Y$ is a *lifting* of ϕ to $\widehat{X}$ if $\widehat{\phi} \circ \alpha = \phi$, i.e., if commutativity holds in the diagram

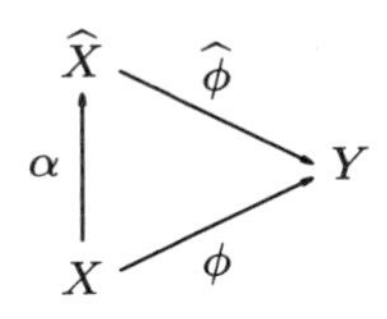

Show that a continuous lifting to $\widehat{X}$ of a continuous mapping of X into a Hausdorff space Y is unique if it exists.

(i) Verify that if X is a completely regular space, then among the compactifications of X there is one $(\check{X}, \beta)$ with the property that *every continuous mapping of X into $[0,1]$ can be lifted to $\check{X}$*. Show further that, in fact, every continuous mapping of X into an arbitrary compact Hausdorff space Y can be lifted to $\check{X}$. (Hint: Take for β the topological embedding of X in the Tikhonov cube of Problem P, where $\mathcal{F}$ is the family of *all* continuous mappings of X into $[0,1]$. Deal first with the case in which Y is itself a Tikhonov cube.)

(ii) Show that the compactification $(\check{X},\beta)$ of (i) is unique in the sense that if $(\check{X}',\beta')$ is another compactification of X with the property that every continuous mapping of X into $[0,1]$ can be lifted to $\check{X}'$, then there exists a unique homeomorphism Φ of $\check{X}$ onto $\check{X}'$ such that commutativity holds in the diagram

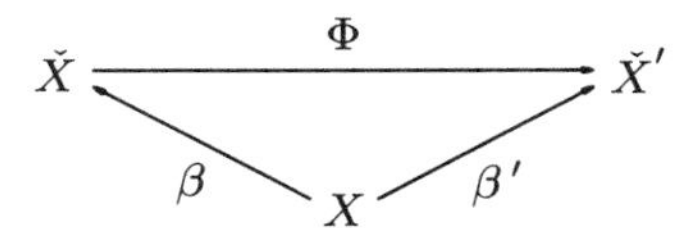

(The essentially unique pair $(\check{X},\beta)$ is called the *Stone–Čech compactification* of X.)

In dealing with these essentially unique compactifications of a topological space X it is a matter of indifference whether, as was done above, we think of the compactification as a separate compact Hausdorff space $\widehat{X}$ along with an embedding α of X in $\widehat{X}$, or we think of α as having been used to replace the subspace $\alpha(X)$ by X itself, so that X becomes literally a subspace of $\widehat{X}$, and α becomes the inclusion mapping of X into $\widehat{X}$. We have here an exact parallel of the situation encountered in completing a metric space; cf. Proposition 8.1.

W. Suppose that X is a topological space and $\sim$ is an equivalence relation on X. Let π denote, as usual, the projection of X onto the quotient space $X/\sim$ (see Chapter 1), and define a subset U of $X/\sim$ to be *open* if and only if $\pi^{-1}(U)$ is open in X. The collection of open sets so obtained forms a topology on $X/\sim$ (called the *quotient topology*), and the projection π of X onto the space $X/\sim$ equipped with the quotient topology is continuous.

(i) Let ϕ be a mapping of X into a topological space Y that respects the equivalence relation $\sim$, that is, such that $x \sim y$ implies $\phi(x) = \phi(y)$. Then ϕ can be factored through $X/\sim$, so that $\phi = \widetilde{\phi}\circ\pi$, where $\widetilde{\phi}$ maps $X/\sim$ into Y. (Such a mapping $\widetilde{\phi}$ is a *lifting* of ϕ to $X/\sim$.) Show that if $X/\sim$ is equipped with the quotient topology, then ϕ is continuous when and only when $\widetilde{\phi}$ is.

(ii) Let σ be a pseudometric on a set X, and let $(X/\sim,\rho)$ be the metric space associated with (X,σ), so that $x \sim y$ means $\sigma(x,y) = 0$ (see Chapter 6, in particular Proposition 6.20, for definitions and terminology). Verify that if we declare a subset U of X to be open if and only if for each point x_0 of U there is a positive real number ε such that $\{y \in X : \sigma(y,x_0) < \varepsilon\} \subset U$, then the collection $\mathcal{T}$ of all such open sets is a topology on X. (The topology $\mathcal{T}$ is called the *pseudometric topology induced by* σ.) Verify also that the metric topology induced on $X/\sim$ by ρ coincides with the quotient topology on $X/\sim$ obtained by factoring out the equivalence relation $\sim$ on the topological space $(X, \mathcal{T})$.

X. Let X be a nonempty Hausdorff space possessing the following properties:

(S1) X has property (COB) (that is, possesses a topological base consisting of closed-open sets; see Problem O),

(S2) X satisfies the second axiom of countability,

(S3) No nonempty open subset of X is compact (equivalently, every compact subset of X is nowhere dense in X).

Examples of such spaces are the space $\mathbb{Q}$ of rational numbers and the complementary space $\mathbb{R}\backslash\mathbb{Q}$ of irrational numbers, each in its relative topology.

(i) Verify that every closed-open subspace of X also possesses properties (S1), (S2) and (S3). Prove that X possesses a countable topological base consisting of closed-open sets. (Hint: Review Problem 6S.) Prove too that X can be partitioned into a countably infinite collection $\mathcal{P}$ of nonempty closed-open subsets. (Hint: There exists an infinite sequence $\{E_n\}_{n=1}^{\infty}$ of closed-open sets in X that covers X and is such that no finite sequence $\{E_1, \ldots, E_N\}$ covers X.)

(ii) Prove that X is metrizable. (Hint: Use Problem Q.) Show also that if ρ is an arbitrary metric that induces the topology on X, and if ε is a given positive number, then the partition $\mathcal{P}$ in (i) can be taken to consist of sets of diameter less than ε.

Y. A nonempty Hausdorff space X is called an *S-space* if it possesses properties (S1), (S2) and (S3) of Problem X and is an absolute G_δ, i.e., is metrized by a metric with respect to which it is complete (Ex. 8O).

(i) Every nonempty closed-open subspace of an S-space is an S-space. The product of a nonempty countable collection of S-spaces is an S-space. The space $\mathbb{R}\backslash\mathbb{Q}$ of irrational numbers is an S-space, while the space $\mathbb{Q}$ of rational numbers is not.

(ii) Prove that the space S of all infinite sequences $\{k_n\}_{n=0}^{\infty}$ of positive integers is an S-space in the topology of pointwise convergence. (See Problem 6H.)

(iii) Prove that if X is an S-space and Y is an arbitrary nonempty, complete, separable metric space, then there exists a continuous mapping of X onto Y. Prove also, in the same vein, that any two S-spaces are homeomorphic. (Hint: Review the methods used in Problems R and 8T.)

Bibliography

1. Baire, R., Sur les fonctions de variables réeles, Ann. di Mat., *3* (1899), 1–122.
2. Bourbaki, N., *Éléments de mathématique: Livre I, Théorie des ensembles*, Act. Sci. et Ind., No. 846, Paris, 1939.
3. Bourbaki, N. and J. Dieudonné, Note de tératopologie. II., Rev. Sci., *77* (1939), 180–181.
4. Burali-Forti, C., Una questione sui numeri transfiniti, Rendiconti Palermo, *11* (1897), 154–164.
5. Burali-Forti, C., Sulle classi ben ordinate, Rendiconti Palermo, *11* (1897), 260.
6. Cantor, Georg, Beiträge zur Begründung der transfiniten Mengenlehre. I., Math. Ann., *46* (1895), 481–512.
7. Galilei, Galileo, *Two new sciences* (transl. Crew and de Salvio), New York, 1914.
8. Gödel, Kurt, *The consistency of the continuum hypothesis*, Annals of Math. Studies, No. 3, Princeton, 1940.
9. Hahn, Hans, Über halbstetige und unstetige Funktionen, Sitzungsber. Akad. Wiss. Ber. (Wien), *103* (1917), 91–110.
10. Halmos, Paul R., *Finite-dimensional vector spaces* (second edition), Princeton, 1958.
11. Halmos, Paul R., *Naive set theory*, New York, 1974.
12. Hardy, G. H., *A course in pure mathematics*, Cambridge, 1938.
13. Jacobson, Nathan, *Lectures in abstract algebra*, Princeton, 1951.
14. Johnson, R. E., *First course in abstract algebra*, New York, 1953.

15. König, J., Zum Kontinumproblem, Math. Ann., *60* (1905), 177–180.
16. Landau, Edmund, *Grundlagen der Analysis*, Leipzig, 1930.
17. Lavrentiev, M., Contribution à la théorie des ensembles homéomorphes, Fund. Math., *6* (1924), 149–160.
18. Lebesgue, H., *Leçons sur l'intégration*, Paris, 1928.
19. Milne, Stephen C., Peano curves and smoothness of functions, Adv. in Math., *35* (1980), 129–157.
20. Moore, E. H., On certain crinkly curves, Trans. Amer. Math. Soc., *1* (1900), 72–90.
21. Peano, G., Sur une courbe, qui remplit toute une aire plane, Math. Ann., *36* (1890), 157–160.
22. Tihonov, A., Über die topologische Erweiterung von Räumen, Math. Ann., *102* (1930), 544–561.
23. Urysohn, Paul, Zum Metrisationsproblem, Math. Ann., *94* (1925), 309–315.

Index

A

absolute convergence 188
absolute G_δ 191, 196f, 276
absolute value 27, 39, 44f (*see also* modulus)
accumulation point 107ff, 128, 226f
addition 6, 25, 46 (*see also* linear operations)
  coordinatewise 50
  of cardinal numbers 71ff, 78
  of complex numbers 37, 43, 47, 137
  of ordered sets 81
  of ordinal numbers 82, 90f
  of polynomials 49
  of real numbers 25f, 47, 137
  pointwise 56
  termwise 50
adherent point 107ff, 115, 177, 180, 226f, 233ff, 262
  of a collection of sets 158, 253
  of a filter base 158, 171, 253, 260, 271
affine
  combination 63f
  mapping 146, 182
  variety 63f
aleph 66, 87f, 142
  naught 66
  one 87
algebra 57f, 61
  associative 57
  complex 57f, 60f
  elementary 63
  linear 57, 60
  real 57ff
  unital 57f, 61
algebra homomorphism 61
algebra isomorphism 61
algebraic
  number 77
  system of generators 48
  topology 208
alternating series 156
arithmetic progression 164
Ascoli's theorem 221f
associative
  algebra 57
  binary operations 6, 18, 46
  law 10, 25, 72, 74f, 92
  product 32
axiom of choice 10, 19, 69, 74, 87,

axiom of choice (*cont.*)
93ff
axiom of completeness 28
axiom of countability 119
first 119, 235ff, 257f, 262, 270, 273
second 119f, 130, 199, 204, 235f, 256, 271f, 276
axiom of products 10, 19, 78, 93, 257

B

Baire, René 153, 191f
Baire
categories 193, 274
category theorem 195f
class 194f, 197, 217
metric 126ff, 222
ball 109, 121, 151, 157f, 167, 171f, 175, 192, 199
closed 98, 108, 176, 208, 234
open 98f, 101, 105, 107f, 115, 117ff, 132, 152, 180, 192, 196, 232, 234, 239, 243
unit 208
Banach-Knaster-Tarski lemma 13
Banach space 175f, 182, 188, 202, 206f, 217
complex 188, 207
real 207
base
of a filter 158, 171, 179f, 233ff, 260, 262, 267
of an expansion 34f
of open sets (topological base) 118ff, 130f, 219, 229ff, 236, 254, 256, 260, 266, 269, 271f, 276
basis 49, 52, 55, 59
Hamel 49f, 55, 59, 61, 76, 79
natural 50
Bendixson kernel 132, 217
big square 260f
bilinear functional 58, 62f, 123
positive (semi) definite 63, 123
symmetric 58, 62f, 123
binary
expansion (*see* expansion)
notation (*see* notation)
operation 6, 46f
binomial theorem 41
bound 12 (*see also* lower, upper)
boundary 130, 193, 208, 226f
point 226
bounded 11 (*see also* mapping, sequence, subset, etc.)
above 11ff, 16, 19, 21, 28f, 36, 67f, 70, 74, 86, 88, 92f, 143
below 11ff, 28, 67f, 140
pointwise 217, 222
uniformly 218
bounded linear transformation 138, 217f
Bourbaki, Nicolas 94
Burali-Forti paradox 85

C

calculus 124, 159
advanced 160
elementary 53, 102, 206
Cantor, Georg 65f, 70, 72, 85
Cantor-Bernstein theorem 68f
Cantor-Lebesgue function 148f, 161, 168, 185
Cantor set 111ff, 119, 129, 132, 144, 148f, 161, 166f, 185, 187, 210f, 220f, 249, 272f
generalized 113, 129, 144, 183ff
planar 113f, 211
Cantor's paradox 75, 85
Cantor's theorem 70, 203f
cardinal number 65ff, 80ff, 101, 119, 128, 130f, 142, 236
countable 80
finite 65
infinite 67, 71f, 74, 77f, 86ff
of an ordinal number 86
uncountable 71, 87
carrier 96, 225, 267
Cartesian product (*see* product)
category argument 194
Cauchy
criterion 174, 179, 187f, 213
filter (base) 179f, 189, 213
net 213
sequence (*see* sequence)

Cauchy-Schwarz inequality 123f, 139
cell 109, 216
  closed 108f
  open 115, 119
cellular partition 199, 215f
center 98f, 105ff, 118, 188, 196, 232
central θth part 113, 183f, 209
  cross-shaped 114, 209
central third 111f, 186, 209
chained (collection of sets) 8, 115, 248f, 269
characteristic function 22, 71, 139, 156, 169, 242
choice function 10
circle 110, 188
class 75
closed (collection of sets) 225, 231, 233, 265 (*see also* ball, cell, subset, etc.)
closed limit 110, 128, 133, 219
  inferior (superior) 109f, 132
closure 108ff, 117, 226f, 231f, 235, 249, 264f
  relative 117, 193, 232, 247
cluster point
  metric 133
  of a net 171, 273
  of a sequence 105ff, 110, 132f, 174, 201, 203
coarser (finer)
  filter 170, 233f, 237, 253, 260, 265
  partition 22, 42, 184
  topology 228, 233, 237, 252, 268
codomain 5f, 190, 262, 264
coefficient 188
cofinal 19, 22, 29, 76f, 271
coherent 21, 144f, 161
commutative
  binary operation 6, 46
  law 10, 25, 75, 92
commutativity 274f
compact (*see* metric space, subset)
compactification 274f
  one-point 274
  Stone-Čech 275
compactness 250, 252, 270
comparable topologies 252, 266
complement
  of a linear manifold 59
  of a set 15, 115ff, 129, 140, 193, 196ff, 207, 226, 230, 242, 247
complemented 217
complete (*see* lattice, metric space)
complete normality 269f
completion 177f, 191, 212, 275
complex (*see also* algebra, linear space, etc.)
  conjugate 44, 52, 123
  field 37, 47, 57, 62
  polynomial 49f, 53, 146, 258
complex number 37ff
  conjugate 44, 123
  polar representation 45
  pure imaginary 37, 44
  real 37, 44
  standard form 37, 44
complex number system 4, 37, 44
complexification 52f, 55
complex-valued function 44, 51ff, 58, 60, 126, 137, 145f, 160, 217, 258
  bounded 126
  continuous 145f, 258
component 249f, 271
composition
  of linear transformations 57
  of mappings 17f, 57, 101, 145f, 154
  of relations 17
condensation point 132
congruence modulo a linear manifold 51
conjugate
  complex 44, 52, 123
  Hölder 124
  linear functional 58
connectedness 248 (*see also* topological space)
constituent 269f
  interval 116, 143, 249, 269
contiguous
  interval 116, 143, 149, 161, 167, 185
  ray 116
continuity 135ff, 236ff
  Lipschitz-Hölder 167f, 190, 215f, 222
  relative 139, 144, 238ff
  uniform 147ff, 167, 189, 205, 218,

uniform continuity (*cont.*)
223
continuous mapping 135ff (*see also* real-valued function, etc.)
continuously differentiable 53, 172f
continuum problem 71, 88
generalized 88
convergence 174, 234
absolute 188
coordinatewise 102, 256, 258
of a filter (base) (*see* filter (base))
of a net (*see* net)
of a sequence (*see* sequence)
of a series 102, 188, 247
pointwise 155f, 167, 194, 202, 218, 220, 222, 260
termwise 126f
uniform 103, 146ff, 185, 187f, 202, 210, 215f, 220, 222, 247
convex
body 207f
combination 64
hull 64
set 64
coordinate 9, 49f, 56
function 160
coset 51, 64
countable (*see* cardinal number, set, etc.)
countably compact (*see* topological space)
countably determined 76f
countably infinite (*see* set)
covering 7, 76, 229ff, 250, 266, 268
closed 116, 144f, 194, 268, 276
countable 194, 204, 271, 276
finite 198f, 204, 250f, 253, 268, 270f, 276
locally finite 145
open 116, 144f, 204, 240, 250ff, 268, 271, 276
cube
closed 234
open 119, 234
rational 119
Tikhonov 272, 275
curve 149, 167, 186, 208ff, 215f
Peano (space-filling) 150, 161, 168, 185, 187, 208, 210, 215, 217
planar 149, 215

D

Darboux sum 43
Dedekind cut 178
definition
inductive (by mathematical induction) 31, 54, 67f, 89f, 106, 108, 111, 113f, 182, 185, 187, 196, 200, 210, 215f, 245f
transfinite (by transfinite induction) 31, 89f, 131
degree 50
De Moivre's theorem 45
denary expansion (*see* expansion)
derivation order 131f
derivative 53, 206
of a set 131
partial 160, 172
derived set 108f, 128, 131, 226
diagonal 4ff, 186, 209, 211, 215f, 261
process 201, 218, 222
sequence 200f
diameter 98f, 127, 198ff, 272, 276
difference
of ideal numbers 36
of integers 32
of rational numbers 33
of real numbers 26
of real-valued functions 40
of sets 3, 197
symmetric 3, 130
difference quotient 206
differentiability 160
differentiable
function 53f, 172
mapping 173
differential 160
digit 34f
dimension 49f, 52, 59
Hamel 76, 79
Dini, Ulisse 220
direct sum 50
directed set (*see* set)
disc 188, 203
disconnection 247, 250
discontinuity 135ff, 236f

discontinuity (*cont.*)
  relative 139
discontinuous mapping 135ff (*see also* real-valued function, etc.)
discrete
  metric 98, 101, 127, 167, 211, 226
  metric space 165, 211, 217
  set 128, 164
  topology 226, 228, 234, 236
disjoint 7, 21, 74, 78, 91, 116, 130
disjointification 23, 73
distance 96, 98f, 108
distributive law 10, 26, 75, 92
division 26, 145
  algorithm 52
  by zero 26
  of ordinal numbers 93
divisor of zero 10
domain of definition 5f, 20f, 78, 89, 93, 159, 165, 194
doubleton 3, 111
duality 11f, 14f, 109f, 129f, 132, 153, 162f, 166, 220, 226, 230f, 237, 241f, 264f
Durchschnitt 129
dyadic fraction 185, 244f

E

edge 119, 199, 216, 234
element 3
embedding 258, 274f
empty (*see* set, mapping, etc.)
entry 213
entrywise linear operations 60
envelope 163f, 198, 265
equality 4f
equicontinuity 221f
equiconvergence 177
equivalence
  class 7, 19f, 122, 177, 212, 249
  relation 7f, 19f, 22, 104, 116, 122, 177, 212, 249, 275
equivalent metrics 104f, 117, 126f, 133, 136, 142, 145, 179, 183, 191, 196, 213, 224f
Euclidean space 97, 108, 146, 150, 160, 172, 199, 204, 207, 216, 234
  (*see also* metric, norm)
evaluation 10, 62, 222
expansion
  binary 35, 148f, 161, 218
  denary 35
  p-ary 35
  ternary 35, 112, 148f, 217
exponent 167f, 190, 216
exponentiation 76
extended real number 36f, 43f, 103f, 106, 150ff, 169, 173, 241f
  system (*see* number system)
extended real-valued function 150ff, 162ff, 173, 241f, 264
  semicontinuous 150ff, 163ff, 168ff, 173, 195, 219f, 241f, 264f
extension
  by continuity 161, 178, 185, 189f, 214, 263
  of a function 156, 159, 161, 167, 178, 263
  of a relation 20
  piecewise linear 146, 186f, 209, 215f
extension ordering 20, 69, 83, 93

F

F_σ 129, 173, 197f, 217, 220, 242
factor 147, 256f, 259
factorization (through a quotient space) 55, 275
fermé 129
field 26, 33, 37, 46
  of complex numbers 37, 47, 57, 62
  of rational numbers 33
  of real numbers 25f, 47, 49, 57
  ordered 28, 34
  scalar 47, 50, 54ff, 59, 61f
field extension 37
filter 170f, 228, 232ff, 253, 265, 267, 270
  coarser (finer) 170, 233f, 237, 253, 260, 265, 270
  convergent 171, 234, 253
  neighborhood 232ff, 253, 262, 264ff, 270
  tail 171, 213

filter base 158f, 171f, 233ff, 260, 262, 267, 270
  Cauchy 179f, 189, 213
  convergent 158f, 171f, 179f, 189f, 234, 262
  countable 270
  tail 171, 213
finite
  cardinal number (*see* cardinal number)
  covering 144, 198f, 204, 250ff, 268, 270
  dimensional space (*see* linear space)
  intersection property 171, 253, 267, 270
  ordinal number (*see* ordinal number)
  real number 36
  sequence (*see* sequence)
  set (*see* set)
finite-valued function 153ff, 162, 195
first axiom of countability 119, 235ff, 257f, 262, 270, 273
fixed point 5, 13, 94, 182, 214
fraction
  dyadic 185, 244f
  p-adic 35, 109
  triadic 112
fractional part 33
full algebraic dual 51, 57, 79
full space of linear transformations 56, 60f
function 5ff (*see also* mapping, real-valued function, etc.)
  choice (selection) 10
  coordinate 160
  lower 163
  of Baire class one (two, zero) 217
  rational 146, 258
  symmetric 99f, 121, 123
  upper 163
functional
  bilinear 58, 62f, 123
  conjugate linear 58
  linear 51, 54, 60, 79
  sesquilinear 58, 62f, 123
function lattice 40, 60

G

G_δ 129, 173, 189ff, 213, 217, 220, 242
  absolute 191, 196f, 276
Galilei, Galileo 72
Gebiet 129
generator (*see* system of generators)
graph 4, 20
greatest element 11, 13, 19, 67, 70, 80ff, 89, 94, 143, 170, 206, 230f
group 18
  symmetric 18

H

Hahn interpolation theorem 155, 157, 245
half-open (*see* interval)
Hamel
  basis 49f, 55, 59, 61, 76, 79
  dimension 76, 79
Hausdorff metric 122, 133, 222f
Hausdorff space 243, 251f, 257, 259, 262f, 266ff, 276 (*see also* topological space)
  compact 252, 260f, 271ff
  locally compact 273f
Hausdorff-Lennes condition 247ff, 269f
Heine-Borel theorem 204ff
Hermitian symmetric 58, 63
Hilbert parallelotope 203
Hilbert-Schmidt norm 139, 183
Hölder conjugate 124
Hölder inequality 124f
homeomorphic copy 259
homeomorphism 142ff, 154, 165f, 179, 190f, 207ff, 221, 239
  local 165f
homomorphism 61
  unital 61
hypernormal 94
hyperseparating 272

I

ideal 61f
  left (right) 60
  of sets 194
  two-sided 60
ideal numbers 36, 219
identity 57
  mapping 5, 9, 17, 42, 44, 101, 142, 184, 190f, 240, 252, 254, 262
  relation 7
image
  inverse 4, 191
  of a point 5
  of a set 4
imaginary part 37, 44
inclusion
  mapping 5, 66, 178, 254, 275
  ordering 11f, 16, 20, 23, 58, 68, 82f, 170, 228
independent (sets of vectors) 59
indexed
  intersection 10, 18
  product 9f, 19, 22
  sum 7, 48
  union 10, 18
indexed family 8ff, 13, 18, 43f, 48f, 56, 73, 75, 77f, 87, 90f, 93, 131, 169, 212, 229, 238, 254ff
  countable 73, 256ff, 272
  nested 140, 169
indexing 8f, 85
  self- 9
index set (*see* set)
induced
  mapping 19, 220, 223, 239
  metric (*see* metric)
  topology (*see* topology)
induction
  mathematical 29ff, 39, 67f, 72, 89, 92, 105, 112, 182, 185, 187, 196, 200, 210, 245ff, 261, 268
  transfinite 31, 131, 250
inductive
  definition (*see* definition)
  partially ordered set 23
  set (*see* set)
inequality
  Cauchy-Schwarz 123f, 139
  Hölder 124f
  Minkowski 124f
  triangle 40, 45, 96, 99f, 104, 109, 115, 121ff, 147, 174, 180
infimum 11ff, 21, 28, 140, 152, 169, 230, 242, 264
infinite
  cardinal number (*see* cardinal number)
  dimensional linear space (*see* linear space)
  series 102, 187f
  set (*see* set)
initial
  number 87, 90ff, 131
  segment 82ff, 88f, 94, 143f, 230f, 241
  value problem 214
inner product 123
integers 3f, 32f
  nonnegative 3, 66, 72, 80f, 90
  positive 3f, 40f, 62, 66, 72, 87, 127, 164, 199, 211, 276
integral part 33
interior 129, 143, 185, 193f, 197, 207, 226
  point 129, 192, 226
intersection 3ff
  indexed 10, 18
interval 27, 33f, 42, 53, 108, 112ff, 119, 131, 140ff, 149, 161, 164, 167ff, 183ff, 207ff, 214, 216, 246, 249f, 266, 269
  closed 27, 42, 66, 69, 108, 111ff, 156, 173, 205f, 214, 240, 267
  constituent 116, 143, 249, 269
  contiguous 116, 143, 149, 161, 167, 185
  half-open 27, 29, 33, 69
  nondegenerate 66, 69, 111, 113, 194, 240
  open 27, 30, 40f, 53, 66, 69, 99, 103f, 112, 115f, 119, 131, 143, 149, 153, 155ff, 164ff, 179, 185, 206, 230f, 249, 268
  parameter 149, 185, 208, 210, 214, 216
  unit 27, 41, 66, 69, 111, 148f, 153, 156, 166, 187, 208, 215, 267

invariant metric 123
inverse 57
  image 4, 191
  inclusion ordering 12, 15, 171, 235
  mapping 17, 29, 54, 101, 142, 165, 239
  of a linear transformation 54
  ordering 12, 90
  relation 17
inversely induced
  mapping 19
  topology 254ff, 272
invertible 57f, 60
irrational number 139, 164f, 193, 236, 276
isolated point 128, 165, 193, 211, 250
isometry 101, 137, 142, 177ff, 190, 212
isomorphic 23
  algebras 61f
  linear spaces 56
isomorphism 23f, 213
  algebra 61
  linear space 56, 60f
  of metric spaces 101
  spatially implemented 61

K

kernel 54
  Bendixson 132, 217
König's theorem 78
Kronecker delta 50, 125

L

lattice 12f, 21, 27
  boundedly complete 12f, 20, 28
  complete 12f, 20, 22, 58, 228
  of functions (function lattice) 40, 60
  of partitions 42f
least element 11, 13, 16, 29ff, 67, 70, 80ff, 88, 92, 94, 143, 170, 206, 230
Lebesgue, Henri 217
level set 7f, 142, 240
lexicographical ordering 92f
lifting (of a mapping) 274f
limit 150
  at a point 157, 189, 261ff
  closed 110, 128, 133
  coordinatewise 102, 258
  from above (below) 103
  from the left (right) 103, 206
  metric 122, 133
  of a filter 171
  of a filter base 158f
  of a net 103, 234, 260
  of a sequence 101
  pointwise 152ff, 167, 194f, 217, 222
  termwise 102, 127f
limit inferior 14f, 106f, 162, 264
  closed 109f, 132
limit ordinal (number) 90ff, 131, 250
limit superior 14f, 106f, 162, 264
  closed 109f, 132
Lindelöf space 271
line 63f, 194, 207
linear
  algebra (*see* linear algebra)
  combination 48, 168
  dependence 124
  functional (*see* linear functional)
  independence 48
  isomorphism (*see* linear space)
  manifold 48
  operations (*see* linear operations)
  parametrization 146, 185f, 208ff, 214
  space (*see* linear space)
  submanifold 48f, 58, 63f
  transformation (*see* linear transformation)
linear algebra 57
  associative 57
  complex (real) 57f
  unital 57f
linear functional 51, 54, 79
  positive 60
  self-conjugate 60
linearly independent set 49, 52, 59, 79
linearly ordered set 15, 143 (*see also* simply ordered set)

linear operations 50
  coordinatewise 50
  entrywise 60
  pointwise 51, 207
  termwise 50
linear space 46ff, 79, 125f (*see also* vector space)
  complex 47, 51ff, 125, 127
  finite dimensional 49ff, 55f, 60f, 76
  infinite dimensional 49f, 76, 79
  of functions 51ff
  of linear transformations 56, 60f
  of sequences 125, 127, 176, 202f
  real 47, 50ff
  self-conjugate 52f, 60
linear space isomorphism 56, 60f, 182
  spatially implemented 61
linear transformation 54, 59, 64, 137f, 160, 182, 213
  bounded 138, 217f
line segment 64, 146, 149, 167, 172, 185f, 194, 207ff, 214, 216
Lipschitz
  condition 136f, 147, 167
  constant 136f, 147f, 182, 189f, 223
Lipschitz-Hölder (L.-H.)
  condition 167
  constant 167f, 216
  continuity 167f, 190, 215f, 222
Lipschitzian (mapping) 136ff, 147f, 166, 189, 223
local
  homeomorphism 165f
  property 144
locally
  bounded 165, 172
  closed 165
  compact (*see* Hausdorff space, topological space)
  Lipschitz-Hölder continuous 167
  Lipschitzian 145, 148f, 168, 172f
  open 165
lower
  bound 11ff (*see also* infimum)
  envelope 163f, 198, 265
  function 163
  limit (*see* limit inferior)
lower semicontinuity 150ff, 163f, 168f, 195, 220, 241f, 265
lowest terms 164

M

map 5 (*see also* function, mapping, etc.)
mapping 5ff (*see also* function)
  affine 146, 182
  codomain of 5f, 190, 262, 264
  domain of (definition of) 5f, 20f, 78, 89, 93, 159, 165, 194
  empty 21, 89, 101
  identity 5, 9, 17, 42, 44, 101, 142, 184, 190f, 240, 252, 254, 262
  inclusion 5, 66, 178, 254, 275
  induced 19, 220, 223, 239
  inverse 17, 29, 54, 101, 142, 165, 239
  inversely induced 19
  one-to-one 17f, 21, 41, 54, 65ff, 86, 94, 101, 116, 133, 141f, 165, 179, 190f, 207f, 239, 252, 254, 258, 266
  onto 5, 17f, 68f, 72ff, 191, 211, 215, 221, 276
  partial 219f
  piecewise linear 146, 184ff, 209f, 215f
  range of 5, 20f, 78, 83f, 98, 149
mappings between metric spaces
  closed 142, 165, 205ff, 223
  continuous 135ff
  continuously differentiable 172f
  contractive 136f, 166
  differentiable 173
  discontinuous 135
  homeomorphism 142ff, 146, 154, 165f, 179, 190f, 207ff, 211, 221
  isometric 101, 137, 142, 177ff, 190, 212
  isomorphism 101
  Lipschitz-Hölder continuous 167f, 190, 215f, 222
  Lipschitzian 136ff, 147f, 166, 182, 189f, 214, 223
  locally bounded 165, 172

mappings between metric spaces (*cont.*)
  locally closed 165
  locally homeomorphic 165f
  locally Lipschitz-Hölder continuous 167
  locally Lipschitzian 145, 148f, 168, 172f
  locally open 165
  open 142, 165, 207
  point of continuity of 135, 166, 173, 197
  point of discontinuity of 135f, 166, 173, 197f, 217, 220
  relatively continuous 139, 141f
  strongly contractive 182f, 214
mappings between ordered sets
  monotone decreasing 12, 131
  monotone increasing 12f, 29, 41f, 68, 110, 169, 227
  order anti-isomorphism 22, 240
  order isomorphism 12, 15f, 20f, 24, 29f, 67f, 82ff, 93, 143, 185, 239
  strictly monotone 12, 22, 29, 41f, 130
mappings between topological spaces
  closed 239, 252
  continuous 236ff, 252ff, 263, 266, 274ff
  homeomorphism 239f, 252, 257f, 272ff
  open 239, 252, 256f
  point of continuity of 236ff, 263
  point of discontinuity of 236ff
  relatively (dis)continuous 238, 240
mappings into metric spaces
  bounded 98ff, 147, 165, 175f, 205, 212
  boundedly equivalent 212
  pointwise bounded 217, 222
  uniformly bounded 218
mappings into ordered sets
  bounded 12
  bounded above 12, 100, 103
  bounded below 12, 104
  infimum of 12
  supremum of 12, 98, 138
mappings into topological spaces 254ff, 275
mathematical induction 29ff, 39, 67f, 72, 89, 92, 105, 112, 182, 185, 187, 196, 200, 210, 245f, 248, 261, 268
matrix 55, 60, 138, 160, 183, 213f
maximal element 11, 16, 22, 59, 70, 74, 78, 86, 92f, 170, 253, 269f
maximum
  element (*see* greatest element)
  value 206, 220, 252
maximum principle 16, 23, 95
mean value theorem 172, 206
member 3
mesh 42, 205
method of successive approximations 182f
metric 96ff, 225f, 238, 276
  associated 122, 133, 177
  Baire 126ff, 222
  defined by a norm 123, 176
  discrete 98, 101, 127, 167, 211
  Euclidean 97f, 101, 139, 183
  Hausdorff 122, 133, 222f
  induced 254
  of pointwise convergence 127
  of uniform convergence 103, 126, 128, 147, 175f, 183, 188, 206f, 212, 214, 222
  product 120, 127, 136, 218ff, 258
  relative 98, 127, 191, 213, 232, 238
  usual 97ff, 102, 124, 137, 139, 234
  zero 101
metric limit 122, 133
metric space 96ff, 224f, 227, 232, 234f, 238, 240, 242f, 258, 275
  associated 122, 275
  bounded 104
  compact 202ff, 220ff, 271ff
  complete 175ff, 276
  complete separable 276
  countable 217
  countably compact 271
  discrete 165, 211, 217
  of bounded mappings 101f, 125, 128, 175f, 183, 212
  perfect 197f, 217, 273
  separable 108f, 120f, 128, 131f, 142, 197ff, 217, 219, 276
  totally bounded 199, 202, 221

metric space (*cont.*)
  totally disconnected 272
metric topology (*see* topology)
metrizability 244, 269 (*see also* topological space)
minimal element 11
minimum
  element (*see* least element)
  value 206, 252
Minkowski inequality 124f
modulus 45
monotone (*see* function, mapping, etc.)
multiplication 6, 25, 39
  of cardinal numbers 71, 76
  of complex numbers 37f, 145
  of extended real numbers 36
  of ordinal numbers 92f
  pointwise 58

N

natural
  basis 50
  numbers 29ff
  projection 8, 51, 55, 61, 275
  topology 236
negative
  of a real number 26, 38
  of a vector 46
  part 27, 40
  real number 26, 39
neighborhood 232ff, 251, 253, 255ff, 260, 262ff, 270, 273
  closed 234, 263, 267
  compact 273
  open 273
  punctured 160
neighborhood base 233ff, 250, 257, 267, 273
neighborhood filter 232ff, 253, 262, 264ff, 270
nested
  family 83, 86, 89, 140, 169
  sequence 114, 153, 180, 196, 200, 203, 211, 236, 245
  set 16, 21, 158
net 13f, 43f, 103ff, 135, 158, 171, 213, 234ff, 241f, 255ff, 267, 273
  along a filter base 171
  bounded 103, 106f
  Cauchy 213
  convergent 103f, 106f, 135, 158, 171, 213, 234ff, 255f, 258ff, 267, 273
  monotone 106
  of extended real numbers 106
  of finite sums 43f
neutral element 6, 18, 26, 37, 46
norm 123ff, 138f, 183, 188, 202
  Euclidean 124
  Hilbert-Schmidt 139, 183
  of a linear transformation 138
  sup 126, 188, 245
  usual 124
normal
  element 94
  space (*see* topological space)
normed space 101f, 123, 137f, 145f, 167, 175, 182, 187f, 217
notation
  binary 69
  place holder 34
  ternary 69, 112f, 149
n-tuple 9, 50, 56, 101
nullity 60
null space 54, 60 (*see also* kernel)
number (*see* cardinal number, real number, etc.)
number class 87f
  first 87
number system 4
  complex 4, 37, 44
  extended real 36f, 150, 152, 169, 173, 231, 240ff
  rational 4, 28, 33, 73, 109, 116, 139, 178, 193, 217
  real 4, 15, 25, 28, 33, 36, 38, 47, 49, 57, 97, 102, 178

O

one-point compactification 274
one-sided derivative 53
one-to-one correspondence 5, 19,

one-to-one correspondence (*cont.*) 101, 142f, 185, 211, 215f (*see also* mapping)
open (*see* ball, cell, set, etc.)
open mapping theorem 207
order
 anti-isomorphism 22, 240
 isomorphic 12, 67f, 80ff, 90, 143
 isomorphism 12, 15f, 20f, 24, 29f, 67f, 82ff, 93, 143, 185, 239
 topology (*see* topology)
ordered
 field 28, 34
 set 10ff (*see also* partially ordered set, etc.)
ordering
 inclusion 11f, 16, 20, 23, 58, 68, 82f, 170, 228
 inverse inclusion 12, 15, 171, 235
 lexicographical 92f
 partial 10ff, 20, 22, 68, 84, 86 (*see also* partially ordered set)
 simple 22, 69, 84 (*see also* simply ordered set)
ordinal number 81ff, 130ff, 250, 260f, 266, 270f
 countable 131f
 finite 81
 limit 90ff, 131, 250
 transfinite 89
 uncountable 87
ordinal number segment 84f, 89, 92, 130f, 250, 260f, 270
origin 46, 50, 207f
oscillation 173
 at a point 173, 189, 195, 197
 over a partition 173, 184, 205

P

p-adic fraction 35, 109
paradox 72, 75
 Burali-Forti 85
 Cantor's 75, 85
parameter 186
parameter interval 149, 185, 208, 210, 214, 216
part
 fractional 33
 imaginary 37, 44
 integral 33
 negative 27, 40
 positive 27, 40
 real 37, 44
partial derivative 160, 172
partial mapping 219f
partial ordering 10ff, 20, 22, 68, 84, 86
 associated 20
partially ordered set 10ff, 19ff, 59, 69f, 74, 80, 83, 86, 92ff, 170
 inductive 23
 weakly 20
partial sum 91, 102, 155, 188
partition 7f, 22, 33, 39, 59, 74f, 79, 91, 116, 199f, 247, 249, 272, 276
 cellular 199, 215f
 coarser (finer) 22, 42, 184
 determined 22
 of an interval 42, 111, 113, 146, 173, 183ff, 199, 205, 208ff, 215f
 of unity 268
Peano, Giuseppe 149, 216
Peano curve 149f
Peano postulates 28, 41
perfect
 metric space 197f, 217, 273
 set (*see* subset of a metric space)
permutation 16, 18, 126
ϕ-tower 94
Picard, Émile 214
piecewise linear
 extension 146, 186f, 209, 215f
 mapping 146, 184ff, 209f, 215f
plane 194
point 3ff
 adherent 107ff, 115, 177, 180, 226f, 233ff, 262
 at infinity 274
 cluster 105ff, 110, 132f, 171, 174, 201ff, 273
 condensation 132
 fixed 5, 13, 94, 182, 214
 interior 129, 192, 226
 isolated 128, 165, 193, 211, 250
 of accumulation 107ff, 128, 226f
 of a partition 42, 184, 186

point (*cont.*)
  of continuity 135ff, 166, 173, 197, 236, 238, 263f
  of discontinuity 135ff, 166, 173, 197f, 217, 220, 236, 238
  of semicontinuity 264, 268
pointwise
  bounded 217, 222
  Cauchy 222
  convergent 155f, 167, 194, 202, 218, 220, 222, 260
  linear operations 51, 207
polar
  angle 45
  distance 160
  representation 45
polarization identity 63
polynomial 49ff, 61
  complex 49f, 53, 146, 258
  rational 77
  real 49ff, 58
positive
  linear functional 60
  part 27, 40
  (semi) definite bilinear functional 63, 123
positive integers 3f, 40f, 62, 66, 72, 87, 127, 164, 199, 211, 276 (*see also* natural numbers)
  relatively prime 164
power
  class 4, 70f, 75, 170, 173, 220f, 225, 228
  of a set 66
  of the continuum 66, 69, 128
  series 188
predecessor 80, 90
predicate 3, 14, 92, 218
principle
  maximum 16, 23, 95
  of inductive definition 31
  of mathematical induction 29, 88, 92
  of transfinite definition 89
  of transfinite induction 88
product 6, 9, 57, 60f
  associative 32
  Cartesian 3, 5, 7, 9, 52, 71, 73, 92, 114, 120, 126, 128, 255, 273
  indexed 9f, 19, 22
  inner 123
  of cardinal numbers 71ff
  of complex numbers 37f, 43f, 47
  of extended real numbers 36, 43
  of integers 32
  of intervals 199
  of linear transformations 57
  of matrices 61
  of metric spaces 120, 126f, 137, 147, 218, 220
  of natural numbers 30
  of ordered sets 92f
  of ordinal numbers 92f
  of polynomials 49
  of rational numbers 33
  of real numbers 25f, 37, 39, 42, 47
  of topological spaces 255ff, 272, 276
  of vectors by scalars 46ff
  row-by-column 61
product topology 255ff
projection 9, 137, 147, 218, 255f
  natural 8, 51, 55, 61, 275
property (COB) 271f, 276
pseudodistance 121
pseudometric 121f, 133f, 177, 275
  space 121f
  topology 276
punctured neighborhood 160
pure imaginary 37, 44

Q

quadratic form 62f
quotient
  algebra 60, 62
  space 7, 51f, 60, 62, 122, 177, 275
  topology 275f

R

radius 98f, 105ff, 115, 118, 128, 140, 144, 160ff, 171, 192, 196, 207, 217, 232
  of convergence 188
range 5, 20f, 78, 83f, 98, 149

range (*cont.*)
of a linear transformation 54
rank 60
ratio 113f, 183, 185, 209f
rational
field 33
function 146, 258
number 33ff, 41f, 116, 119, 164f
number system (*see* number system)
polynomial 77
ray 27ff, 41, 104, 194, 249, 269
closed 27f, 108, 138, 148
left 27, 231
open 27, 115f, 231, 249
right 27f, 138, 231
ray topology 230f, 240f, 253
real (*see also* algebra, linear space, etc.)
line 97, 179
part 37, 44
polynomial 49ff, 58
real number
algebraic 77
irrational 118, 164f, 189, 193, 236
negative 26, 39
positive 26, 28f, 39, 41f
rational 33ff, 41f, 116, 119, 164f
transcendental 77
real number field 25f, 47, 49, 57
real number system (*see* number system)
real-valued function 40, 42, 51ff, 60, 99f, 121, 123, 137, 148, 161ff, 168f, 172f, 183ff, 195, 197f, 212, 217, 264
bounded 99, 126, 153f, 157, 179, 183ff, 214, 244, 273ff
continuous 140ff, 152ff, 161, 167ff, 183ff, 206f, 214, 220, 242, 244, 246, 252, 268f, 273ff
continuously differentiable 53, 172
extended (*see* extended real-valued function)
finite 153ff, 162, 195
monotone decreasing 99f, 104, 125, 162, 168f, 173
monotone increasing 99f, 104, 148, 162, 169, 173, 185
strictly increasing 29, 41f, 104
uniformly continuous 167
Urysohn 245, 268, 272
reciprocal 26, 29, 38, 164
rectangle 113, 186f, 209f, 214f
refinement
of a filter 170, 233f
of a partition 22, 42, 184
of a topology 228, 233, 237, 252, 254, 268
reflexive relation 7, 10, 20, 68
regular space (*see* topological space)
regularity 264
relation 4ff
equivalence 7f, 19ff, 104, 116, 122, 177, 212, 249, 275
identity 7
inverse 17
reflexive 7, 10, 20, 68
symmetric 7, 17
transitive 7, 10, 20f, 68, 268
relative
closure 117, 193, 232, 247
metric 98, 127, 191, 213, 232, 238
topology (*see* topology)
relatively
closed 117, 232, 247
continuous 139, 144, 238ff
open 117, 119, 129, 231f, 251
remainder 90
restriction 20, 186, 197, 209, 211
r-net 198ff, 222
Rolle's theorem 206
root 45, 77
nth 41, 45
square 41

S

scalar 46, 123, 188
scalar field 47, 50, 54ff, 59, 61f
scalar-valued function 51
bounded 245f
continuous 217
of Baire class one (two, zero) 217
second axiom of countability 119f, 130, 199, 204, 235f, 256, 271f, 276
selection function 10

self-conjugate
  linear functional 53, 60
  linear space 52f, 60
self-indexing 9
semicontinuity 150ff, 163ff, 168f, 220, 241f (*see also* extended real-valued function)
  lower 195, 219f, 265
  upper 169f, 173, 264
separable
  metric space 108f, 121, 128, 130ff, 142, 197ff
  topological space 236
separating (collection of mappings) 9, 258
separation 242f, 245, 251, 267ff
sequence 9, 14, 31, 67, 91, 126
  diagonal 200f
  empty 113
  finite 31f, 42, 112, 120, 126, 133, 149
  infinite 9, 14, 68f, 79, 125, 127, 132, 134, 202, 218
  monotone 77, 261
  of coefficients 188
  of curves 210, 216
  of digits 35, 221
  of predicates 218
  of ratios 113f, 185, 210
  of sequences 200
  of terms 102
  of vectors 102, 125, 208
  strictly monotone 71
sequence of complex numbers 125, 127
  bounded 125, 176
  Cauchy 175
  convergent 175
sequence of functions 185, 187, 194, 217
  bounded 195
  boundedly equivalent 212
  convergent 155f, 185, 215ff, 220, 222, 242
  equicontinuous 221f
  monotone 148, 152ff, 166, 195, 202, 220
  pointwise convergent 155f, 167, 194, 202, 242
  uniformly convergent 103, 147f, 175, 185, 187, 202, 210, 215f, 220
sequence of points in a metric space 101ff, 176, 181f, 205, 218
  bounded 174f, 195
  Cauchy 174ff, 196, 200ff, 212, 218, 221
  convergent 101ff, 117, 120, 126ff, 135f, 139, 151, 158, 165, 174ff, 182, 190, 196, 201f, 205, 207, 212, 234, 258
  equiconvergent 128, 178, 205, 211
  of bounded variation 181, 185
  uniformly scattered 200, 218
sequence of points in a topological space 235, 238, 241, 262, 273
  convergent 234, 236f, 258
sequence of positive integers 127
  strictly increasing 14, 132f, 220
sequence of real numbers 102, 106f, 127, 218f
  bounded 107, 188, 207
  Cauchy 177f
  convergent 102, 177, 207, 218
  monotone 104, 131, 201
sequence of sets 14f, 23, 73, 109ff, 114, 122, 126, 129, 132f, 180, 191, 194, 196f, 203f, 218f, 222, 276
  monotone decreasing 14f, 128, 180, 195, 197, 203f, 220, 271
  monotone increasing 14f, 128, 197, 204
  nested 114, 153, 180, 196, 200, 203, 211, 236, 245
series 91, 102, 154ff, 181, 187f, 247
  absolutely convergent 188
  alternating 156
  telescoping 155
  uniformly convergent 154, 247
sesquilinear functional 58, 62f
  Hermitian symmetric 58, 63
  positive (semi)definite 63, 123
set 3ff (*see also* subset)
  compact convex 207
  convex 64
  countable 67, 73, 77, 116, 120, 128, 132, 193f, 197, 204, 217, 222, 235, 242
  countably infinite 66f, 73f, 77, 79,

countably infinite (*cont.*)
128, 204, 276
directed 13, 15, 23, 43, 76, 103, 171, 173, 235, 255, 273
empty 3, 98, 101, 115, 169f, 204, 217, 220, 228, 231
finite 65, 70f, 76, 80f, 90, 108, 228
index 8f, 91, 93, 253, 260
inductive 28f, 92
infinite 66, 76, 81
level 7f, 142, 240
linearly independent 49, 52, 59, 79
linearly ordered 15, 143 (*see also* simply ordered set)
nondenumerable 67 (*see also* uncountable set)
ordered 10ff (*see also* partially ordered set, etc.)
totally ordered 15 (*see also* simply ordered set)
uncountable 67, 132, 217, 235, 257
well-ordered (*see* well-ordered set)
σ-ideal 194
simple arc 208, 210f
simple ordering 22, 69, 84
simply ordered set 15f, 21, 23, 26, 59, 67f, 70, 74, 78, 80, 86, 90, 92ff, 103, 143, 230f, 239f, 266, 269f
dense 67f
densely ordered 67
singleton 3, 5, 20, 70, 82, 99, 101, 109, 142, 170, 180, 193, 211, 228, 240, 242f, 257, 274
open 250
slice 257
somme 129
space
Banach (*see* Banach space)
Euclidean 97, 108, 146, 150, 160, 172, 199, 204, 207, 216, 234
Hausdorff (*see* Hausdorff space)
(ℓ_p) 125, 128, 176, 202f
Lindelöf 271
linear (*see* linear space, vector space)
metric (*see* metric space)
normal (*see* topological space)
normed 101f, 123, 137f, 145f, 167, 175, 182, 187f, 217
of bounded continuous mappings 183, 206, 222
of bounded mappings 100ff, 126, 128, 175f, 183, 206, 212
of bounded sequences 125
of bounded subsets 121f, 133, 222f
of continuous mappings 183, 206, 222
of irrational numbers 276
of rational numbers 276
of sequences of positive integers 127, 276
pseudometric 121f
quotient 7f, 20, 51, 55, 60, 62, 122, 275
S- 276
T_i- 228, 243f, 266ff
topological (*see* topological space)
unitary 97
vector (*see* linear space, vector space)
square root 41
S-space 276
standard form (of a complex number) 37, 44
Stone-Čech compactification 275
strongly contractive (mapping) 182f, 214
subalgebra 60f
subbase 230, 266
subcovering 204, 250f, 271
subcurve 187, 210
subfield 34, 37
subinterval 42, 111ff, 184ff, 199, 209ff, 215f
sublattice 13, 20, 40
submanifold (*see* linear)
subnet 273
subrectangle 215
subsequence 14, 105f, 110, 117, 126, 132f, 174f, 182, 200ff, 218f, 222, 273
subset 3ff
subset of a metric space
bounded 98f, 104, 128, 132f, 143f, 167, 199, 201ff, 217, 222f
closed 108ff, 117, 128ff, 140ff, 147, 156, 164, 166, 176, 179ff, 189,

closed (*cont.*)
193f, 197, 201ff, 214, 217ff, 243
compact 201ff, 220, 222
dense 108f, 116, 120, 128, 140f, 161, 169, 177f, 194ff, 212, 217, 222
dense in itself 128f, 197
derived 108f, 128, 131, 226
discrete 128, 164
F_σ 129, 173, 197f, 217, 220
fermé 129
G_δ 129, 173, 189ff, 196f, 213, 217, 220, 276
nowhere dense 191ff, 196f
of first category 192ff, 220
of second category 192ff, 217, 220
open 115ff, 129f, 140ff, 149ff, 165ff, 172, 192ff, 204, 207, 213f, 217, 220, 225, 243
perfect 129, 132, 143, 221
relatively closed 117
relatively open 117, 119, 121, 141, 163, 193, 204
totally bounded 199ff, 222

subset of an ordered set
bounded 11, 143f, 149, 167, 206
bounded above 11, 13, 16, 19, 21, 28f, 36, 67f, 74, 86, 88, 92f, 143
bounded below 11, 13, 28, 67f, 140
convex 269f

subset of a topological space
closed 226ff
closed-open 247f, 271f, 276
compact 250ff, 273f, 276
connected 247ff, 271
dense 226f, 235f, 240, 266, 274
derived 226f
disconnected 247
nowhere dense 274, 276
of first (second) category 274
open 224ff
relatively closed 232, 247, 261, 269
relatively open 231f, 251

subspace of a metric space 98, 117, 121, 127, 129, 139, 178f, 193, 204, 232

subspace of a topological space 231f, 238, 247, 250ff, 261, 266ff, 275
closed 268
closed-open 276
compact 250
connected 248
dense 274
open 269, 273

subsquare 215
subtraction 137
successor 41, 80f, 90, 92
sum 6, 198
direct 50
indexed 7, 48
of an infinite series 102, 247
of cardinal numbers 71, 73
of extended real numbers 43
of integers 32
of linear transformations 56
of natural numbers 30
of ordered sets 81
of ordinal numbers 82, 91
of real numbers 25, 39, 42
vector 48

superset 171
sup norm 126, 188, 245
supremum 11ff, 21, 28, 70, 78, 83, 89ff, 98, 138, 143f, 152, 154, 230, 242, 245, 254, 264
symmetric
bilinear functional 58, 62f, 123
function 99f, 121, 123
group 18
relation 7, 17

system of generators
of a filter 170
of a topology 228, 266

T

tail 14f, 108, 174, 179, 196, 201ff, 261
filter (base) 171, 213
of a net 106f, 171

telescoping series 155
term 9, 102
terminal segment 230
termwise 50, 102, 126f
ternary notation (*see* notation)
theorem of Lavrentiev 190
theorem of Weierstrass 206

thread (a Cantor set) 211
Tietze extension theorem 156, 245f
Tikhonov, A. N. 268
Tikhonov cube 272, 275
Tikhonov plank 261
Tikhonov's theorem 260f
topological
  base 118ff, 130f, 219, 229ff, 236, 254, 256, 260, 266, 269, 271f, 276
  concept 224, 236
  property 117, 135, 142, 179, 196, 224
  subbase 230, 266
topological space 225ff
  compact 250ff, 259f, 271, 273
  compact Hausdorff 251f, 260f, 272ff
  completely normal 261, 269f
  completely regular 267f, 272ff
  connected 247f
  countably compact 271
  disconnected 247
  Hausdorff 243, 251f, 257, 259ff, 266ff, 271ff
  Lindelöf 271
  locally compact 273f
  metrizable 225, 235, 240, 243f, 269ff, 276
  nonmetrizable 235, 270f
  normal 243ff, 251, 261, 267ff, 271ff
  of extended real numbers 231, 240
  regular 263f, 267f, 271f
  separable 236
  totally disconnected 250, 271ff
topologization 225
topology 224ff
  algebraic 208
  coarser (finer) 228, 233, 237, 252, 254, 268
  coarsest 228, 254
  discrete 226, 228, 234, 236
  general 224, 242, 250, 258f
  generated 228ff, 266
  indiscrete 226ff, 230, 240, 254, 262
  inversely induced 254ff, 272
  metric 225, 231ff, 258, 267, 276
  metrizable 225f, 235, 240, 254, 258, 276
  natural 236
  of pointwise convergence 260, 276
  order 230f, 239, 250, 260f, 266, 269ff
  product 255ff
  pseudometric 276
  quotient 275f
  ray 230f, 240f, 253
  relative 231f, 236, 238, 240, 254, 259, 261, 266, 271, 276
  T_1- 228, 266
  usual 231, 234, 268
totally bounded 199ff, 221f
totally ordered set 15 (*see also* simply ordered set)
trace 159, 233, 262
transcendental number 77
transfinite
  definition 89f
  induction 31, 131, 250
  ordinal number 89
transformation 5, 218 (*see also* mapping, function)
  linear (*see* linear transformation)
transitive relation 7, 10, 20f, 68, 268
translate 64, 208
translation 101
triadic fraction 112
triangle inequality 40, 45, 96ff
  for a norm 123ff
  in a metric space 96ff, 147, 174, 180
  in a pseudometric space 121ff
trichotomy law 21, 26
trisection 214ff
trivial
  linear space 51
  submanifold 48
T_i-space
  $i = 1$ 228, 243f, 266ff, 271f
  $i > 1$ 267ff
type I (ordinal number) 90f, 131, 250

U

ultrafilter 253, 259f, 270
uncountable cardinal number 71, 87
uncountable ordinal number 87, 260,

uncountable ordinal number (*cont.*) 271
  first (least) 87
uncountable set (*see* set)
uniform boundedness theorem 217
uniform convergence (*see* convergence)
uniformly bounded 218
uniformly convergent
  sequence (*see* sequence of functions)
  series 154, 247
uniformly scattered 200, 218
union 3ff
  indexed 10, 18
unit 57
  ball 208
  interval (*see* interval)
  sphere 207
  square 114, 149, 210f, 215
unital
  algebra 57f, 61
  homomorphism 61
unitary space 97
upper
  bound 11ff, 28f, 86 (*see also* supremum)
  envelope 163f, 198, 265
  function 163
  limit (*see* limit superior)
upper semicontinuity 150ff, 163f, 168ff, 173, 220, 241f, 264
Urysohn function 245, 268, 272
Urysohn metrization theorem 272
Urysohn's lemma 244ff, 274
usual
  metric (*see* metric)
  norm 124
  topology 231, 234, 268

V

value 5
vanishing (of a function) 160
  "big oh" ("little oh") 160
vector 46, 102, 203, 217f
vector space 46ff, 54, 57ff, 123 (*see also* linear space)
  complex 47, 49f, 52ff, 62f, 79
  finite dimensional 49
  infinite dimensional 53f
  left 47
  real 47, 49f, 52, 54, 57, 63f, 79
vector sum 48

W

weight 130f
well-ordered set 16, 22, 31, 80ff, 90ff
well-ordering 86

Z

Zermelo's well-ordering theorem 86f, 93f
zero metric 101
Zorn's lemma 16, 23, 70, 74, 78, 86f, 93f, 270

Graduate Texts in Mathematics

continued from page ii

65 WELLS. Differential Analysis on Complex Manifolds. 2nd ed.
66 WATERHOUSE. Introduction to Affine Group Schemes.
67 SERRE. Local Fields.
68 WEIDMANN. Linear Operators in Hilbert Spaces.
69 LANG. Cyclotomic Fields II.
70 MASSEY. Singular Homology Theory.
71 FARKAS/KRA. Riemann Surfaces. 2nd ed.
72 STILLWELL. Classical Topology and Combinatorial Group Theory. 2nd ed.
73 HUNGERFORD. Algebra.
74 DAVENPORT. Multiplicative Number Theory. 2nd ed.
75 HOCHSCHILD. Basic Theory of Algebraic Groups and Lie Algebras.
76 IITAKA. Algebraic Geometry.
77 HECKE. Lectures on the Theory of Algebraic Numbers.
78 BURRIS/SANKAPPANAVAR. A Course in Universal Algebra.
79 WALTERS. An Introduction to Ergodic Theory.
80 ROBINSON. A Course in the Theory of Groups.
81 FORSTER. Lectures on Riemann Surfaces.
82 BOTT/TU. Differential Forms in Algebraic Topology.
83 WASHINGTON. Introduction to Cyclotomic Fields.
84 IRELAND/ROSEN. A Classical Introduction to Modern Number Theory. 2nd ed.
85 EDWARDS. Fourier Series. Vol. II. 2nd ed.
86 VAN LINT. Introduction to Coding Theory. 2nd ed.
87 BROWN. Cohomology of Groups.
88 PIERCE. Associative Algebras.
89 LANG. Introduction to Algebraic and Abelian Functions. 2nd ed.
90 BRØNDSTED. An Introduction to Convex Polytopes.
91 BEARDON. On the Geometry of Discrete Groups.
92 DIESTEL. Sequences and Series in Banach Spaces.
93 DUBROVIN/FOMENKO/NOVIKOV. Modern Geometry—Methods and Applications. Part I. 2nd ed.
94 WARNER. Foundations of Differentiable Manifolds and Lie Groups.
95 SHIRYAYEV. Probability.
96 CONWAY. A Course in Functional Analysis. 2nd ed.
97 KOBLITZ. Introduction to Elliptic Curves and Modular Forms. 2nd ed.
98 BRÖCKER/TOM DIECK. Representations of Compact Lie Groups.
99 GROVE/BENSON. Finite Reflection Groups. 2nd ed.
100 BERG/CHRISTENSEN/RESSEL. Harmonic Analysis on Semigroups: Theory of Positive Definite and Related Functions.
101 EDWARDS. Galois Theory.
102 VARADARAJAN. Lie Groups, Lie Algebras and Their Representations.
103 LANG. Complex Analysis. 3rd ed.
104 DUBROVIN/FOMENKO/NOVIKOV. Modern Geometry—Methods and Applications. Part II.
105 LANG. $SL_2(\mathbf{R})$.
106 SILVERMAN. The Arithmetic of Elliptic Curves.
107 OLVER. Applications of Lie Groups to Differential Equations. 2nd ed.
108 RANGE. Holomorphic Functions and Integral Representations in Several Complex Variables.
109 LEHTO. Univalent Functions and Teichmüller Spaces.
110 LANG. Algebraic Number Theory.
111 HUSEMÖLLER. Elliptic Curves.
112 LANG. Elliptic Functions.
113 KARATZAS/SHREVE. Brownian Motion and Stochastic Calculus. 2nd ed.
114 KOBLITZ. A Course in Number Theory and Cryptography. 2nd ed.
115 BERGER/GOSTIAUX. Differential Geometry: Manifolds, Curves, and Surfaces.
116 KELLEY/SRINIVASAN. Measure and Integral. Vol. I.
117 SERRE. Algebraic Groups and Class Fields.
118 PEDERSEN. Analysis Now.
119 ROTMAN. An Introduction to Algebraic Topology.
120 ZIEMER. Weakly Differentiable Functions: Sobolev Spaces and Functions of Bounded Variation.

121 LANG. Cyclotomic Fields I and II. Combined 2nd ed.
122 REMMERT. Theory of Complex Functions. *Readings in Mathematics*
123 EBBINGHAUS/HERMES et al. Numbers. *Readings in Mathematics*
124 DUBROVIN/FOMENKO/NOVIKOV. Modern Geometry—Methods and Applications. Part III.
125 BERENSTEIN/GAY. Complex Variables: An Introduction.
126 BOREL. Linear Algebraic Groups.
127 MASSEY. A Basic Course in Algebraic Topology.
128 RAUCH. Partial Differential Equations.
129 FULTON/HARRIS. Representation Theory: A First Course. *Readings in Mathematics*
130 DODSON/POSTON. Tensor Geometry.
131 LAM. A First Course in Noncommutative Rings.
132 BEARDON. Iteration of Rational Functions.
133 HARRIS. Algebraic Geometry: A First Course.
134 ROMAN. Coding and Information Theory.
135 ROMAN. Advanced Linear Algebra.
136 ADKINS/WEINTRAUB. Algebra: An Approach via Module Theory.
137 AXLER/BOURDON/RAMEY. Harmonic Function Theory.
138 COHEN. A Course in Computational Algebraic Number Theory.
139 BREDON. Topology and Geometry.
140 AUBIN. Optima and Equilibria. An Introduction to Nonlinear Analysis.
141 BECKER/WEISPFENNING/KREDEL. Gröbner Bases. A Computational Approach to Commutative Algebra.
142 LANG. Real and Functional Analysis. 3rd ed.
143 DOOB. Measure Theory.
144 DENNIS/FARB. Noncommutative Algebra.
145 VICK. Homology Theory. An Introduction to Algebraic Topology. 2nd ed.
146 BRIDGES. Computability: A Mathematical Sketchbook.
147 ROSENBERG. Algebraic *K*-Theory and Its Applications.
148 ROTMAN. An Introduction to the Theory of Groups. 4th ed.
149 RATCLIFFE. Foundations of Hyperbolic Manifolds.
150 EISENBUD. Commutative Algebra with a View Toward Algebraic Geometry.
151 SILVERMAN. Advanced Topics in the Arithmetic of Elliptic Curves.
152 ZIEGLER. Lectures on Polytopes.
153 FULTON. Algebraic Topology: A First Course.
154 BROWN/PEARCY. An Introduction to Analysis.
155 KASSEL. Quantum Groups.
156 KECHRIS. Classical Descriptive Set Theory.